Nonlinear Control Systems with Recent Advances and Applications

Nonlinear Control Systems with Recent Advances and Applications

Editors

Ahmad Taher Azar
Amjad J. Humaidi
Ibraheem Kasim Ibraheem
Giuseppe Fusco
Quanmin Zhu

Basel • Beijing • Wuhan • Barcelona • Belgrade • Novi Sad • Cluj • Manchester

Editors

Ahmad Taher Azar
College of Computer and
Information Science (CCIS)
Prince Sultan University
Riyadh
Saudi Arabia

Amjad J. Humaidi
Control and Systems
Engineering Department
University of Technology
Baghdad
Iraq

Ibraheem Kasim Ibraheem
Department of ICT and
Natural Sciences
Norwegian University of
Science and Technology
Alesund
Norway

Giuseppe Fusco
Department of Electrical and
Information Engineering
University of Cassino and
Southern Lazio
Cassino
Italy

Quanmin Zhu
School of Engineering
University of the West of
England
Bristol
United Kingdom

Editorial Office
MDPI
St. Alban-Anlage 66
4052 Basel, Switzerland

This is a reprint of articles from the Special Issue published online in the open access journal *Entropy* (ISSN 1099-4300) (available at: www.mdpi.com/journal/entropy/special_issues/Nonlinear_Control_Systems).

For citation purposes, cite each article independently as indicated on the article page online and as indicated below:

Lastname, A.A.; Lastname, B.B. Article Title. *Journal Name* **Year**, *Volume Number*, Page Range.

ISBN 978-3-7258-0150-3 (Hbk)
ISBN 978-3-7258-0149-7 (PDF)
doi.org/10.3390/books978-3-7258-0149-7

© 2024 by the authors. Articles in this book are Open Access and distributed under the Creative Commons Attribution (CC BY) license. The book as a whole is distributed by MDPI under the terms and conditions of the Creative Commons Attribution-NonCommercial-NoDerivs (CC BY-NC-ND) license.

Contents

About the Editors vii

Preface ix

Jinghui Pan, Kaixiang Peng and Weicun Zhang
From Nonlinear Dominant System to Linear Dominant System: Virtual Equivalent System Approach for Multiple Variable Self-Tuning Control System Analysis
Reprinted from: *Entropy* **2023**, *25*, 173, doi:10.3390/e25010173 1

Saim Ahmed, Ahmad Taher Azar and Mohamed Tounsi
Design of Adaptive Fractional-Order Fixed-Time Sliding Mode Control for Robotic Manipulators
Reprinted from: *Entropy* **2022**, *24*, 1838, doi:10.3390/e24121838 17

Yang Li, Quanmin Zhu, Jianhua Zhang and Zhaopeng Deng
Adaptive Fixed-Time Neural Networks Control for Pure-Feedback Non-Affine Nonlinear Systems with State Constraints
Reprinted from: *Entropy* **2022**, *24*, 737, doi:10.3390/e24050737 39

Shengya Meng, Shihong Li, Heng Chi, Fanwei Meng and Aiping Pang
H_∞ Observer Based on Descriptor Systems Applied to Estimate the State of Charge
Reprinted from: *Entropy* **2022**, *24*, 420, doi:10.3390/e24030420 53

Ibrahim A. Hameed, Luay Hashem Abbud, Jaafar Ahmed Abdulsaheb, Ahmad Taher Azar, Mohanad Mezher, Anwar Ja'afar Mohamad Jawad, et al.
A New Nonlinear Dynamic Speed Controller for a Differential Drive Mobile Robot
Reprinted from: *Entropy* **2023**, *25*, 514, doi:10.3390/e25030514 65

Zhiwei Fan, Kai Jia, Lei Zhang, Fengshan Zou, Zhenjun Du, Mingmin Liu, et al.
A Cartesian-Based Trajectory Optimization with Jerk Constraints for a Robot
Reprinted from: *Entropy* **2023**, *25*, 610, doi:10.3390/e25040610 95

Hong Shen, Qin Wang and Yang Yi
Event-Triggered Tracking Control for Adaptive Anti-Disturbance Problem in Systems with Multiple Constraints and Unknown Disturbances
Reprinted from: *Entropy* **2023**, *25*, 43, doi:10.3390/e25010043 116

Ilham Toumi, Billel Meghni, Oussama Hachana, Ahmad Taher Azar, Amira Boulmaiz, Amjad J. Humaidi, et al.
Robust Variable-Step Perturb-and-Observe Sliding Mode Controller for Grid-Connected Wind-Energy-Conversion Systems
Reprinted from: *Entropy* **2022**, *24*, 731, doi:10.3390/e24050731 135

Guanyu Lai, Weizhen Liu, Weijun Yang, Huihui Zhong, Yutao He and Yun Zhang
An Incremental Broad-Learning-System-Based Approach for Tremor Attenuation for Robot Tele-Operation
Reprinted from: *Entropy* **2023**, *25*, 999, doi:10.3390/e25070999 166

Qiang Chen, Yong Zhao and Lixia Yan
Adaptive Orbital Rendezvous Control of Multiple Satellites Based on Pulsar Positioning Strategy
Reprinted from: *Entropy* **2022**, *24*, 575, doi:10.3390/e24050575 184

About the Editors

Ahmad Taher Azar

Prof. Ahmad Azar is a full professor at Prince Sultan University, Riyadh, Kingdom of Saudi Arabia. He is also a full professor at the Faculty of Computers and Artificial Intelligence, Benha University, Egypt. He is the leader of the Automated Systems and Soft Computing Lab (ASSCL), Prince Sultan University, Saudi Arabia. Prof. Azar specializes in artificial intelligence (AI), machine learning, control theory and applications, robotics, computational intelligence, reinforcement learning, and dynamic system modeling. He has published/co-published over 500 research papers, book chapters, and conference proceedings in prestigious peer-reviewed journals. He is the editor of several books published by Springer, Elsevier, and IGI-Global on control systems, robotics, dynamic system modeling, artificial and computational intelligence, machine learning, fuzzy logic systems, and chaos modeling. Prof. Ahmad Azar serves as a reviewer for several international journals. He is a member of many international and peer-reviewed conference and international program committees. Prof. Azar is actively involved in the academic community as a reviewer for international journals and a committee member for international conferences. He has received several awards and honors, including the Egyptian State Encouragement Award in Engineering Sciences and the Abdul Hameed Shoman Arab Researchers Award in Machine Learning and Big Data Analytics. He has also been recognized as one of the top computer scientists in Saudi Arabia and as one of the top 2% of scientists in the world in artificial intelligence by Stanford University. Additionally, Prof. Azar holds various leadership positions in professional organizations, such as the Vice Chair of the International Federation of Automatic Control (IFAC) Technical Committee on Control Design. He is actively involved in IEEE committees and task forces related to emerging technologies, robotics, fuzzy logic, and computational collective intelligence.

Amjad J. Humaidi

Amjad Jaleel Humaidi received his B.Sc. and M.Sc. degrees in control engineering from Al-Rasheed College of Engineering and Science, University of Technology, Baghdad, Iraq, in 1992 and 1997, respectively. He received his Ph.D. degree from the University of Technology in 2006 with a specialization in control and automation. He is presently a staff member in the control and systems engineering departments. His fields of interest include adaptive control, nonlinear control, nonlinear observers, intelligent control, optimization, identification, and real-time image processing.

Ibraheem Kasim Ibraheem

Ibrahim A. Hameed (Senior Member, IEEE) received his Ph.D. in AI from Korea University, South Korea, in 2010, and a Ph.D. degree in field robotics from Aarhus University, Denmark, in 2012. He is currently a professor and the Deputy Head of Research and Innovation, Department of ICT and Natural Sciences, Faculty of Information Technology and Electrical Engineering, Norwegian University of Science and Technology (NTNU), AAlesund, Norway. His research interests include control systems, robotics, AI, and machine learning.

Giuseppe Fusco

In 1988, Giuseppe Fusco received the Laurea in Electrical Engineering at the University of Naples Federico II. In 1990, he won a grant entitled "Ottimizzazione delle reti elettriche di trazione urbana e suburbana con metodi non lineari" from the Ansaldo Trasporti. In 1990, he started working as a scientific assistant at the Laboratory of Industrial Engineering at the University of Cassino. In 1995, he served as an assistant professor, and since 2003, he has been an associate professor in control engineering. He has been a member of several Italian and European projects, such as ARCAS, IGREENGrid, AEROARMS, WiMUST, DeXRoV, ROBUST, C4E, Robilaut, and Canopies. He is a patentee concerning an innovative relay that detects the unintentional islanding operation of a part of a distribution grid. Since 2009, he has been the European regional editor for the "International Journal of Modelling, Identification and Control". He is a co-author of about 110 international journal and conference papers. He is the co-author of the Springer book titled "Adaptive Voltage Control in Power Systems: Modeling Design and Applications". His main research interest is in the control of electric power systems and smart grids.

Quanmin Zhu

Quanmin Zhu is a professor of control systems at the School of Engineering, University of the West of England, Bristol, UK. He obtained his MSc from Harbin Institute of Technology, China, in 1983 and his PhD from the Faculty of Engineering, University of Warwick, UK, in 1989, and worked as a postdoctoral research associate at the Department of Automatic Control and Systems Engineering, University of Sheffield, UK, between 1989 and 2004. His main research interest is in nonlinear system modeling, identification, and control. He has published over 300 papers on these topics, edited various books with Springer, Elsevier, and other publishers, and provided consultancy to various industries. Currently, Professor Zhu is serving as the editor of Elsevier's book series Emerging Methodologies and Applications in Modeling, Identification, and Control.

Preface

Nonlinear control, characterized by robust stabilization and adaptive tracking, has emerged as an indispensable field in the realm of control systems. The complexities inherent in nonlinearly controlled systems, coupled with unknown uncertainties and time-varying disturbances, necessitate sophisticated methodologies to ensure effective control. The last few decades have witnessed remarkable strides in the evolution of design techniques tailored for nonlinear systems. This progress, underpinned by diverse mathematical tools, has paved the way for applications spanning energy, health care, robotics, biology, and big data research.

The landscape of nonlinear control systems is rife with intriguing challenges, promising an intellectually stimulating future. As we navigate the intricacies of diverse control tasks, especially those emanating from the integration of cutting-edge technologies in communication and computation, this field is poised for unprecedented growth. The nexus of nonlinear control with emerging technologies not only presents challenges but also opens avenues for innovative solutions that transcend traditional boundaries.

This reprint endeavors to encapsulate the state-of-the-art developments in nonlinear control, both in theoretical underpinnings and practical applications. While the existing literature boasts a wealth of valuable results, the synthesis of control strategies for a broader class of nonlinear systems, as well as their application across diverse domains, remains an ongoing challenge. Our collective ambition is to contribute to the resolution of these challenges by presenting a collection of articles that showcase novel approaches to nonlinear control. Through this comprehensive compilation, we aim to provide insights into the latest advancements, bridging the gap between theory and application.

The overarching goal of this reprint is to serve as a beacon for researchers, practitioners, and enthusiasts in the field of control systems. By delving into the intricacies of nonlinear control, we hope to inspire fresh perspectives and foster collaborative efforts that will shape the future trajectory of this dynamic and evolving discipline.

Ahmad Taher Azar, Amjad J. Humaidi, Ibraheem Kasim Ibraheem, Giuseppe Fusco, and Quanmin Zhu

Editors

Article

From Nonlinear Dominant System to Linear Dominant System: Virtual Equivalent System Approach for Multiple Variable Self-Tuning Control System Analysis

Jinghui Pan, Kaixiang Peng and Weicun Zhang *

School of Automation and Electrical Engineering, University of Science and Technology Beijing, Beijing 100083, China
* Correspondence: weicunzhang@263.net

Abstract: The stability and convergence analysis of a multivariable stochastic self-tuning system (STC) is very difficult because of its highly nonlinear structure. In this paper, based on the virtual equivalent system method, the structural nonlinear or nonlinear dominated multivariable self-tuning system is transformed into a structural linear or linear dominated system, thus simplifying the stability and convergence analysis of multivariable STC systems. For the control process of a multivariable stochastic STC system, parameter estimation is required, and there may be three cases of parameter estimation convergence, convergence to the actual value and divergence. For these three cases, this paper provides four theorems and two corollaries. Given the theorems and corollaries, it can be directly concluded that the convergence of parameter estimation is a sufficient condition for the stability and convergence of stochastic STC systems but not a necessary condition, and the four theorems and two corollaries proposed in this paper are independent of specific controller design strategies and specific parameter estimation algorithms. The virtual equivalent system theory proposed in this paper does not need specific control strategies, parameters and estimation algorithms but only needs the nature of the system itself, which can judge the stability and convergence of the self-tuning system and relax the dependence of the system stability convergence criterion on the system structure information. The virtual equivalent system method proposed in this paper is proved to be effective when the parameter estimation may have convergence, convergence to the actual value and divergence.

Keywords: virtual equivalent system; stochastic multivariable STC; stability; convergence

1. Introduction

The stability and convergence analysis of stochastic self-tuning control systems is more difficult than that of deterministic self-tuning control systems. It is difficult to analyze and understand in theory, which makes it very difficult for engineers and technicians to analyze the stability and convergence of such systems in practice.

References [1–4] studied the stability and convergence of the self-tuning control system consisting of the minimum variance control strategy and the stochastic gradient parameter estimation algorithm. References [5,6] provided the stability and convergence results of the self-tuning control system consisting of the minimum variance control strategy and the least squares parameter estimation algorithm. References [7,8] presented the stability and convergence results of the self-tuning control system consisting of the pole-placement control strategy and the weighted least squares parameter estimation algorithm. As a summary of the stability and convergence of STC, the results regarding the minimum variance control strategy do not require parameter estimation convergence; however, the results regarding other control strategies, such as pole placement, require parameter estimation convergence [9,10].

The above results are all for the minimum phase object, and the common feature is that the convergence of parameter estimation is not required. If the controlled object is of a non-minimum phase, the self-tuning algorithm of the minimum variance type cannot be used for control, but pole assignment and other control strategies need to be used. The corresponding stability and convergence analysis are more difficult than the adaptive control system of the minimum variance type. The existing results are basically obtained under the premise of parameter estimation convergence (can converge to the real value or non-real value); there are also some results that do not require the convergence of parameter estimation but can only guarantee the stability and robustness of the system, and cannot guarantee the convergence of the system.

For the controlled object with non-minimum phase certainty, [11,12] proved the stability and convergence of pole placement self-tuning control by introducing additional excitation signals. The stability and convergence of pole assignment self-tuning control can also be obtained by modifying the estimation model without introducing an additional excitation signal [13]. The literature [1,14] analyzes the stability and convergence of the pole assignment algorithm with a "key technology lemma". For non-minimum phase random controlled objects, refs. [15,16] adopted the method of adding an "attenuated excitation signal" to ensure the stability and convergence of random pole assignment self-tuning control, while [7,8] provided that it does not need an external excitation signal, but uses a self-convergent weighted least squares parameter estimation algorithm to ensure the stability and convergence of random pole assignment self-tuning control. The literature [17–20] analyzed the stability and convergence of the adaptive decoupling control system, and the literature [21] analyzed the stability and convergence of generalized minimum variance self-tuning control for minimum phase objects and some non-minimum phase objects based on the Lyapunov function. The literature [22] proposed a theory of virtual equivalent systems, but it mainly focused on single-variable systems and did not study multivariable systems.

The disadvantage of the minimum variance self-tuning control method is that it is not suitable for non-minimum phase objects. The main reason is that the unstable pole of the regulator cannot be exactly canceled with the zero point of the object, resulting in the instability of the system. In addition, even if the generalized minimum variance self-tuning controller is used, in order to ensure the closed-loop stability of the system, the control weight factor is usually determined through trial and error. This constraint also introduces significant inconvenience to specific applications. The computation amount of the random gradient algorithm is much less than that of the least squares algorithm, but its convergence speed is very slow. Moreover, under the conditions of strong, persistent excitation, the parameter estimation error of the system using the stochastic gradient algorithm converges to zero uniformly, but under other conditions, it is very difficult to prove the convergence of the stochastic gradient algorithm. The least squares estimation method is simple in its algorithm and is easy to implement. It does not need to know the statistical information of the measurement error, but its accuracy is difficult to improve. Its limitations are reflected in two aspects: first, it can only estimate the deterministic constant value vector but cannot estimate the time process of the random vector; second, it can only ensure the minimum mean square error of the measurement, but it does not ensure the best estimation error of the estimator, and the accuracy is not high. The stochastic self-tuning system using the pole placement method requires high accuracy of the model, has the problem of modeling error, and requires the convergence of parameter estimation.

In view of the characteristics and shortcomings of the above control algorithms, this paper, based on the theory of virtual equivalent systems, weakens the conditions required by the stability and convergence criteria of the stochastic self-tuning system, mainly eliminating the direct dependence on the order information of the controlled object, reducing the requirements for parameter estimation errors, and eliminating the dependence of the pole placement self-tuning control strategy on the convergence of parameter estimation, The difficulty of analysis is transferred from the system structure to the compensation signal, thus reducing the difficulty of the original problem.

New self-tuning control schemes are still emerging, such as the robust multi-model adaptive control system, fuzzy parameter self-tuning PID method, intelligent AC contactor self-tuning control technology, self-tuning control method of simulation turntable based on the accurate identification of the model parameters, sliding model adaptive control method [23–27], and the traditional approach cannot prove its stability and convergence due to the lack of a general theory of adaptive control. In general, it is expected that the stability and convergence analysis methods of stochastic STC systems are independent of specific control strategies and parameter estimation methods. Some scholars have made efforts in this field to develop a general theory [28–31], but the results are not satisfactory.

There are many related achievements that are difficult to enumerate one by one. In recent years, many adaptive control schemes related to stability and convergence have achieved good results in practical applications [32], but there are no theoretical analysis results.

New adaptive control schemes are emerging, and it is difficult to analyze the stability and convergence of each adaptive control system one by one. For this reason, people have been expecting to find a unified stability and convergence analysis method [23,33–36]. However, despite some sporadic results [32,34,37,38], the expected unified analysis method and theory still need to be explored. The concept of the virtual equivalent system and its corresponding analysis methods are generated under such a background [25,39–41]. Weicun Zhang, one of the authors of this paper, proposed the concept of the virtual equivalent system and then analyzed the stability and convergence of various self-tuning control systems in a unified framework, converting the nonlinear system into an equivalent linear system with an infinitesimal nonlinear compensation signal.

In this paper, we will consider three cases of parameter estimation: (1) The parameter estimation converges to the real value; (2) the parameter estimation converges to the non-real value; (3) and the parameter estimation may not converge. The second and third cases do not require the structural information of the plant. Considering the particularity of the minimum variance self-tuning control, a criterion with an intuitive explanation is given for this kind of STC system.

It is worth pointing out that for a general multiple-input multiple-output stochastic self-tuning control system, not only the minimum variance control system but also the convergence of parameter estimation is not a necessary condition for the stability and convergence of the STC system.

Through the theoretical and experimental research in this paper, it is concluded that for the self-tuning control system of nonlinear controlled objects (deterministic or stochastic, minimum phase or non-minimum phase), to ensure its stability and convergence, only the boundedness of parameter estimation, slow time variation and the output approximation effect of the estimation model (i.e., the parameter estimation error is relatively infinitesimal) are required, and the control strategy meets the stability and tracking requirements according to the principle of deterministic equivalence.

2. Virtual Equivalent System of Stochastic Self-Tuning Control System

For the convenience of description, we first consider the following multivariable stochastic system ΣP with known structural information but unknown parameters (for the discussion of general stochastic systems containing colored noise, see Section 4).

$$\Sigma P : A(q^{-1})y(k) = q^{-d}B(q^{-1})u(k) + \omega(k) \tag{1}$$

where, $y(k)$, $u(k)$ and $\omega(k)$, are the output signals, the input signal and the noise signal with the appropriate dimension of the plant to be controlled, respectively.

Assuming that
$$y(k) = 0, u(k) = 0, \omega(k) = 0, \forall k < 0,$$

$$\lim_{n \to \infty} \frac{1}{n} \sum_{i=1}^{n} \|\omega(i)\|^2 = R < \infty, \ a.s. \tag{2}$$

$$A(q^{-1}) = I + A_1 q^{-1} + \ldots + A_n q^{-n}, n \geq 1$$
$$B(q^{-1}) = B_0 + B_1 q^{-1} + \ldots + B_m q^{-m}, m \geq 1 \tag{3}$$

Introducing symbols

$$\theta^T = [-A_1, \ldots - A_n, B_0, ..B_m],$$

$$\phi^T(k-d) = [y(k-1), \ldots y(k-n), \ldots, u(k-d), ..u(k-d-m)],$$

Then, we have

$$y(k) = \phi^T(k-d)\theta + \omega(k) \quad (4)$$

The estimated model is denoted as $\Sigma P_m(k)$, and its parameter matrix is

$$\Sigma P_m(k) : \hat{\theta}^T(k) = [-\hat{A}_1, \ldots, -\hat{A}_n, \hat{B}_0, .., \hat{B}_m]$$

The performance of the parameter estimation can be expressed by the (posterior) model output error

$$e(k) = y(k) - \phi^T(k-d)\hat{\theta}(k) - \omega(k) = \phi^T(k-d)\theta_0 - \phi^T(k-d)\hat{\theta}(k). \quad (5)$$

The self-tuning controller is denoted by $\Sigma C(k)$, and can also be treated as a matrix $\theta_c(k)$. The controller can be obtained by various design methods, such as pole placement. Different control strategies actually represent different mapping, i.e.,

$$f : \hat{\theta}(k) \to \theta_c(k), \text{ or } \theta_c(k) = f(\hat{\theta}(k)),$$

Additionally, the control law is generally denoted by

$$u(k) = \phi_c^T(k)\theta_c(k),$$

where

$$\phi_c^T(k) = [y_r(k), y_r(k-1) \ldots y(k), y(k-1) \ldots, u(k-1), \ldots].$$

where $y_r(k)$ is a known bounded reference signal.

The above-described self-tuning control system is shown in Figure 1, which is abbreviated as $(\Sigma C(k), \Sigma P)$.

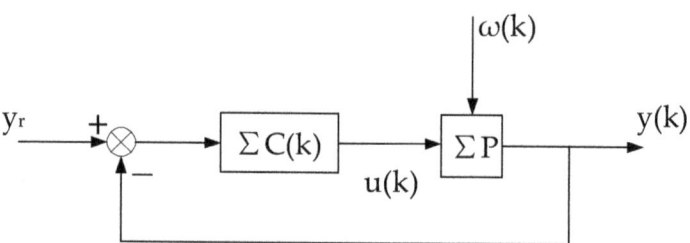

Figure 1. Stochastic self-tuning control system.

Accordingly, the real plant corresponds to an 'ideal' controller, i.e.,

$$\begin{cases} \theta_c = f(\theta) \\ u_0(k) = \phi_c^T(k)\theta_c \end{cases}.$$

This constant control system is abbreviated as $(\Sigma C, \Sigma P)$, as shown in Figure 2. On the basis of $(\Sigma C, \Sigma P)$, we can artificially construct a system that is equivalent in the input–output sense to the self-tuning control system. It consists of the constant control system of Figure 2 and a compensational signal $\Delta u(k)$, abbreviated as $(\Sigma C, \Sigma P, \Delta u(k))$, as shown in Figure 3.

$$\Delta u(k) = u(k) - u_0(k) = \phi_c^T(k)\theta_c(k) - \phi_c^T(k)\theta_c. \quad (6)$$

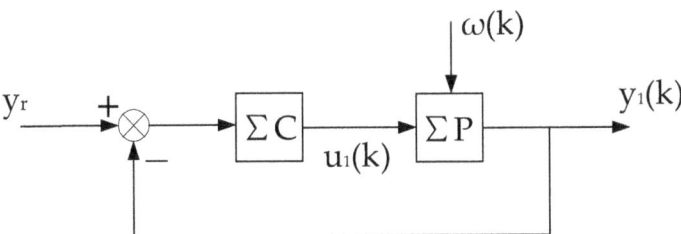

Figure 2. Constant control system corresponding to self-tuning control system.

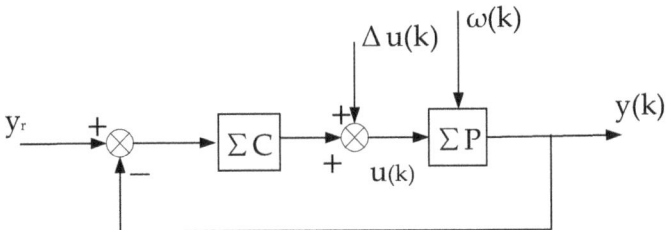

Figure 3. Virtual equivalent system of stochastic self-tuning control system.

Since θ_c is unknown, $\Delta u(k)$ cannot be calculated exactly, but it can be estimated (see the analysis below for details).

The above equivalent system is recorded as the "virtual equivalent system" of the self-tuning control system. The reason why it is recorded as "virtual" is that it exists but is unknown. One of the merits of the "virtual equivalent system" is that it quantitatively reflects the difference between a self-tuning control system and the corresponding constant control system. The definition of the convergence of the self-tuning control system is based on the constant system shown in Figure 2.

The stability of the self-tuning control system is defined by

$$\lim_{n\to\infty} \sup \frac{1}{n} \sum_{i=1}^{n} (\|y(i)\|^2 + \|u(i)\|^2) < \infty.$$

The convergence of the self-tuning control system is defined by

$$\lim_{n\to\infty} \frac{1}{n} \sum_{i=1}^{n} \|(y(i) - y_r(i))\|^2 = \lim_{n\to\infty} \frac{1}{n} \sum_{i=1}^{n} \|(y_1(i) - y_r(i))\|^2.$$

3. Main Results

3.1. Parameter Estimation Converges to the True Value

Considering a self-tuning control system based on an arbitrary control strategy and an arbitrary parameter estimation algorithm, the following results are obtained.

Theorem 1. *For the self-tuning control system of (1), if the following conditions are met.*
The parameter estimation converges to a true value.
The control strategy satisfies the stability requirements of the object with known parameters, then the close- loop system composed by $(\Sigma P, \Sigma C)$ is stable.
The mapping $f(\cdot)$ is continuous at $\hat{\theta}(k) = \theta$.
Then, the self-tuning control system is stable and convergent.

Proof. The stability of the system is proved by the method of contradiction, and then the convergence is proved.

Since the virtual equivalent system shown in Figure 3 is a linear constant structure (its time-varying nonlinear features are transferred to $\Delta u(k)$), it can be decomposed into two subsystems, one is the constant control system shown in Figure 2, and the other is the system shown in Figure 4.

$$y(k) = y_1(k) + y_2(k), \tag{7}$$

$$u(k) = u_1(k) + u_2(k). \tag{8}$$

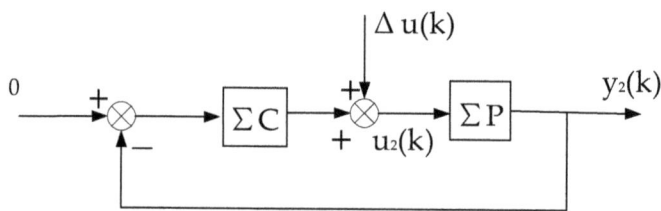

Figure 4. Decomposition subsystem of the virtual equivalent system 2.

By the superposition principle and considering the situation of the two subsystems separately, the system shown in Figure 2 is a conventional random control system. The second condition in Theorem 1 ensures that it is closed-loop stable, so there is

$$\limsup_{n \to \infty} \frac{1}{n} \sum_{i=1}^{n} (\|y_1(i)\|^2 + \|u_1(i)\|^2) < \infty, \tag{9}$$

$$\lim_{n \to \infty} \frac{1}{n} \sum_{i=1}^{n} \|(y_1(i) - y_r(i))\|^2 < \infty. \tag{10}$$

For the system shown in Figure 4, there is no influence of noise due to the closed-loop system being stable, so we have [42]

$$\sum_{k=1}^{n} \|y_2(k)\|^2 \leq M_1 \sum_{k=1}^{n} \|\Delta u(k)\|^2 + M_2, \ 0 < M_1 < \infty, \ 0 \leq M_2 < \infty, \tag{11}$$

$$\sum_{k=1}^{n} \|u_2(k)\|^2 \leq M_3 \sum_{k=1}^{n} \|\Delta u(k)\|^2 + M_4, \ 0 < M_3 < \infty, \ 0 \leq M_4 < \infty, \tag{12}$$

That is,

$$\sum_{k=1}^{n} \|y_2(k)\|^2 = O\left(\sum_{k=1}^{n} \|\Delta u(k)\|^2\right) + M_2, \sum_{k=1}^{n} \|u_2(k)\|^2 = O\left(\sum_{k=1}^{n} \|\Delta u(k)\|^2\right) + M_4.$$

By theorem condition (1) and condition (3) in Theorem 1, we have $\theta_c(k) \to \theta_c$, $\|\Delta u(k)\| = o(\|\phi_c(k)\|)$.

By the composition of $\phi_c(k)$, we know that $\|\phi_c(k)\| = O(\|\phi(k-d)\|) + M$, M is a bounded constant.

Furthermore, by the convergence of parameter estimation, we have

$$\frac{1}{n} \sum_{k=1}^{n} \|\Delta u(k)\|^2 = o\left(\frac{1}{n} \sum_{k=1}^{n} \left(\frac{1}{n} \sum_{k=1}^{n} \|\phi(k-d)\|^2\right)\right).$$

Thus,

$$\frac{1}{n} \sum_{k=1}^{n} \|y_2(k)\|^2 = o\left(\frac{1}{n} \sum_{k=1}^{n} \|\phi(k-d)\|^2\right) \tag{13}$$

$$\frac{1}{n}\sum_{k=1}^{n}||u_2(k)||^2 = o\left(\frac{1}{n}\sum_{k=1}^{n}||\phi(k-d)||^2\right) \tag{14}$$

Then, to prove the following formula

$$\limsup_{n\to\infty}\frac{1}{n}\sum_{i=1}^{n}(||y(i)||^2 + ||u(i)||^2) < \infty.$$

It suffices to prove that

$$\limsup_{n\to\infty}\frac{1}{n}\sum_{i=1}^{n}||\phi(k-d)||^2 < \infty.$$

We can construct $\phi_1(k-d)$ (corresponding to the system of Figure 2) and $\phi_2(k-d)$ (corresponding to the system of Figure 4), so that

$$\phi(k-d) = \phi_1(k-d) + \phi_2(k-d).$$

It can be seen from Equations (13) and (14) that

$$\frac{1}{n}\sum_{i=1}^{n}||\phi_2(k-d)||^2 = o\left(\frac{1}{n}\sum_{k=1}^{n}||\phi(k-d)||^2\right)$$

By the triangle inequalities, we have

$$\begin{aligned}\frac{1}{n}\sum_{i=1}^{n}||\phi(k-d)||^2 &= \frac{1}{n}\sum_{i=1}^{n}||\phi_1(k-d)+\phi_2(k-d)||^2 \\ &\leq \frac{1}{n}\sum_{i=1}^{n}||\phi_1(k-d)||^2 + \frac{1}{n}\sum_{i=1}^{n}||\phi_2(k-d)||^2 \\ &= \frac{1}{n}\sum_{i=1}^{n}||\phi_1(k-d)||^2 + o\left(\frac{1}{n}\sum_{k=1}^{n}||\phi(k-d)||^2\right).\end{aligned} \tag{15}$$

Furthermore, considering

$$\limsup_{n\to\infty}\frac{1}{n}\sum_{i=1}^{n}\left(||y_1(i)||^2 + ||u_1(i)||^2\right) < \infty$$

Thus, we obtain

$$\limsup_{n\to\infty}\frac{1}{n}\sum_{i=1}^{n}||\phi_1(k-d)||^2 < \infty.$$

Taking (15) into consideration, we obtain

$$\limsup_{n\to\infty}\frac{1}{n}\sum_{i=1}^{n}||\phi(k-d)||^2 < \infty.$$

Thus, combining the above formula with (13) and (14), it follows that

$$\frac{1}{n}\sum_{k=1}^{n}||y_2(k)||^2 = o(1), \tag{16}$$

$$\frac{1}{n}\sum_{k=1}^{n}||u_2(k)||^2 = o(1). \tag{17}$$

Next, we prove

$$\lim_{n\to\infty}\frac{1}{n}\sum_{i=1}^{n}||y(i)-y_r(i)||_2 = \lim_{n\to\infty}\frac{1}{n}\sum_{i=1}^{n}||y_1(i)-y_r(i)||^2.$$

Combining Cauchy's Inequality with (10) and (16), we have

$$0 \le \left\{\frac{1}{n}\sum_{i=1}^{n}\|y_1(i) - y_r(i)\| \cdot \|y_2(i)\|\right\}^2 \le \left\{\frac{1}{n}\sum_{i=1}^{n}\|y_1(i) - y_r(i)\|^2\right\} \cdot \left\{\frac{1}{n}\sum_{i=1}^{n}\|y_2(i)\|^2\right\} \to 0.$$

By the Squeeze Theorem, we obtain

$$\lim_{n\to\infty}\left\{\frac{1}{n}\sum_{i=1}^{n}\|y_1(i) - y_r(i)\| \cdot \|y_2(i)\|\right\}^2 = 0.$$

It follows that

$$\lim_{n\to\infty}\frac{1}{n}\sum_{i=1}^{n}\|y_1(i) - y_r(i)\| \cdot \|y_2(i)\| = 0. \tag{18}$$

Finally, let us consider

$$\lim_{n\to\infty}\frac{1}{n}\sum_{i=1}^{n}\|y(i) - y_r(i)\|^2 = \lim_{n\to\infty}\frac{1}{n}\sum_{i=1}^{n}\|(y_1(i) - y_r(i)) + y_2(i)\|^2.$$

According to the norm triangle inequality and (16) and (18), we have

$$\frac{1}{n}\sum_{i=1}^{n}\|(y_1(i) - y_r(i)) + y_2(i)\|^2 \le \frac{1}{n}\sum_{i=1}^{n}\|y_1(i) - y_r(i)\|^2 + \frac{1}{n}\sum_{i=1}^{n}\|y_2(i)\|^2 + \frac{2}{n}\sum_{i=1}^{n}\|y_1(i) - y_r(i)\| \cdot \|y_2(i)\|$$
$$\to \frac{1}{n}\sum_{i=1}^{n}\|y_1(i) - y_r(i)\|^2$$

Similarly,

$$\frac{1}{n}\sum_{i=1}^{n}\|(y_1(i) - y_r(i)) + y_2(i)\|^2 \ge \frac{1}{n}\sum_{i=1}^{n}\|y_1(i) - y_r(i)\|^2 + \frac{1}{n}\sum_{i=1}^{n}\|y_2(i)\|^2 - \frac{2}{n}\sum_{i=1}^{n}\|y_1(i) - y_r(i)\| \cdot \|y_2(i)\|$$
$$\to \frac{1}{n}\sum_{i=1}^{n}\|y_1(i) - y_r(i)\|^2$$

According to the Squeeze Theorem, we have

$$\lim_{n\to\infty}\frac{1}{n}\sum_{i=1}^{n}\|(y_1(i) - y_r(i)) + y_2(i)\|^2 = \lim_{n\to\infty}\frac{1}{n}\sum_{i=1}^{n}\|y_1(i) - y_r(i)\|^2,$$

i.e.,

$$\lim_{n\to\infty}\frac{1}{n}\sum_{i=1}^{n}\|y(i) - y_r(i)\|^2 = \lim_{n\to\infty}\frac{1}{n}\sum_{i=1}^{n}\|y_1(i) - y_r(i)\|^2.$$

That completes the proof of Theorem 1. □

3.2. Parameter Estimation Converges to Non-True Value

Considering the fact that the structure information of the controlled object is unknown, the order of the estimated model can be lower than the order of the real controlled object, which is often the case in practical engineering applications.

Theorem 2. *For the self-tuning control system of the plant (1), if*

(1) *The parameter estimate converges to θ_0, the estimated model $\Sigma P_m(k)$ is consistently controllable,*

$$\sum_{k=1}^{n}\left\|y(k) - \phi^T(k-d)\hat{\theta}(k) - \omega(k)\right\|^2 = o(1 + \sum_{k=1}^{n}\|\phi(k-d)\|^2).$$

(2) *The control strategy satisfies the stability requirements for the known objects of the parameters;*

(3) The mapping $f(\cdot)$ is continuous at $\hat{\theta}(k) = \theta_0$. Then, the self-tuning control system is stable and convergent.

Proof. The stability of the system is proved by the method of contradiction, and then the convergence is proved.

In order to prove Theorem 2, we need to build another virtual equivalent system, as shown in Figure 5.

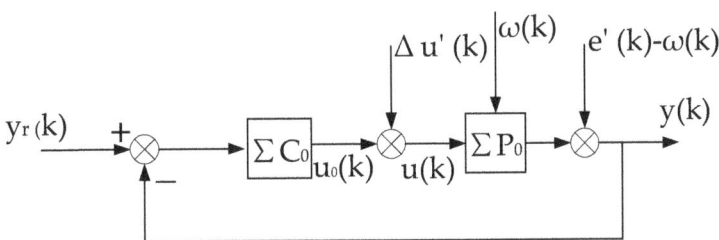

Figure 5. Virtual equivalent system when parameter estimates converge to non-true values.

ΣP_0 represents the model corresponding to the convergence value θ_0 of the parameter estimation, and ΣC_0 represents the controller corresponding to ΣP_0.

The virtual equivalent system shown in Figure 5 is different from the virtual equivalent system shown in Figure 3. First, ΣC_0 and ΣP_0 are different from ΣC and ΣP, respectively. Second, the definitions of $e'(k)$ and $\Delta u'(k)$ are different from (5) and (6), respectively. In Figure 5

$e'(k) = y(k) - \phi^T(k-d)\theta_0 = y(k) - \phi^T(k-d)\hat{\theta}(k) + \phi^T(k-d)\hat{\theta}(k) - \phi^T(k-d)\theta_0$, is

$$e'(k) = e(k) + \phi^T(k-d)\hat{\theta}(k) - \phi^T(k-d)\theta_0, \tag{19}$$

$$\Delta u'(k) = u(k) - u_0(k) = \phi_c^T(k)\theta_c(k) - \phi_c^T(k)\theta_{c0}. \tag{20}$$

It is known from condition (1) in Theorem 2.

$$\sum_{k=1}^{n} \|e(k) - \omega(k)\|^2 = o(1 + \sum_{k=1}^{n} \|\phi(k-d)\|^2). \tag{21}$$

Then, we have

$$e'(k) - \omega(k) = e(k) - \omega(k) + \phi^T(k-d)\hat{\theta}(k) - \phi^T(k-d)\theta_0.$$

It is also known by condition (1) that $\hat{\theta}(k) \to \theta_0$, so we have

$$\sum_{k=1}^{n} \|e'(k) - \omega(k)\|^2 = o(1 + \sum_{k=1}^{n} \|\phi(k-d)\|^2). \tag{22}$$

Further, from condition (3), we have

$$\|\Delta u'(k)\| = o(\beta + \|\phi(k-d)\|). \tag{23}$$

Thus, $\Delta u(k)$ has the same properties as in the proof of Theorem 1, i.e.,

$$\frac{1}{n}\sum_{k=1}^{n} \|\Delta u(k)\|^2 = o(\frac{1}{n}\sum_{k=1}^{n} \|\phi(k-d)\|^2).$$

Decomposing the system shown in Figure 5 into three subsystems (as shown in Figures 6–8, respectively), it is known from condition (2) that the subsystem, as shown in

Figure 6, is stable; and the rest of the proof process is similar to that of Theorem 1 (details are omitted to save space). □

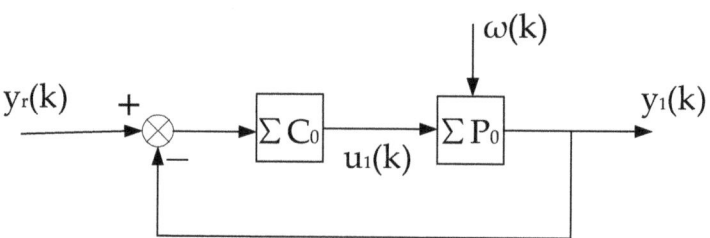

Figure 6. Decomposition system of virtual equivalent system 1.

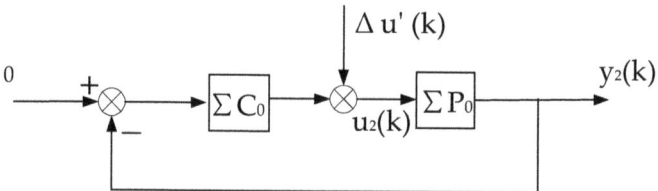

Figure 7. Decomposition system of virtual equivalent system 2.

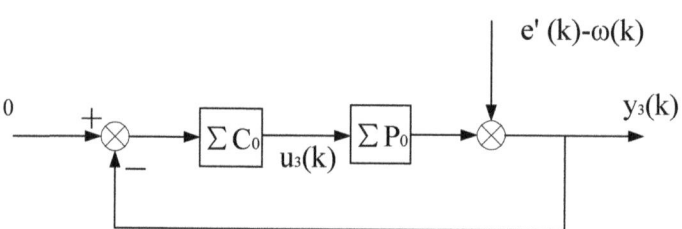

Figure 8. Decomposition system of virtual equivalent system 3.

3.3. Parameter Estimation May Not Converge

This section demonstrates two theorems for the STC, consisting of the minimum variance control strategy and the arbitrary control strategy. As described in Section 3.2, the structural information of the estimated model could be inconsistent with the real control plant.

First, let us consider the minimum variance control strategy and explain why this particular type of self-tuning control system does not require the convergence of parameter estimation.

Theorem 3. *For the minimum variance type self-tuning control system of plant (1), any feasible parameter estimation algorithm can be used if the following conditions are met.*

(1) $B(q^{-1})$ *is Hurwitz stable polynomial, and* $|B_0| \neq 0$.
(2) *Control strategy* $u(k)$ *exists.*
(3) *Parameter estimation satisfies.* $\sum_{k=1}^{n} \|e(k) - \omega(k)\|^2 = o(\alpha + \sum_{k=1}^{n} \|\phi(k-d)\|^2)$, α *is a non-zero constant. The self-tuning control system is then stable and convergent.*

Proof. Using the virtual equivalent system shown in Figure 3, under the condition that condition (2) and condition (3) are satisfied, it can be proved that the minimum variance self-tuning control has the following special properties [22].

$$\Delta u(k) = B_0^{-1}[\phi^T(k-1)(\theta_0 - \hat{\theta}(k))]. \tag{24}$$

Further,
$$\Delta u(k) = B_0^{-1}[e(k) - \omega(k)].$$
Therefore, $\Delta u(k)$ has the following property
$$\frac{1}{n}\sum_{k=1}^{n}\|\Delta u(k)\|^2 = o(\frac{1}{n}\sum_{k=1}^{n}\|\phi(k-d)\|^2).$$

Decompose the virtual equivalent system of the minimum variance self-tuning control system into two subsystems, as shown in Figures 3 and 4. The rest of the proof process is similar to that of Theorem 1, and the details are omitted.

In fact, the key to the proof process of the three theorems is to prove the property of $\Delta u(k)$ or $\Delta u'(k)$ in the virtual equivalent system. In Theorems 1 and 2, considering the arbitrary (linear) control strategy, the mapping relationship between the estimated parameters and the controller parameters is complicated. Therefore, parameter estimation convergence is required to ensure the properties of $\Delta u(k)$. In the minimum variance control strategy, the controller parameters can be directly represented by the estimated parameters. Additionally, we have $\Delta u(k) = B_0^{-1}[e(k) - \omega(k)]$; therefore, parameter estimation convergence is not required. Only
$\sum_{k=1}^{n}\|e(k) - \omega(k)\|^2 = o(\alpha + \sum_{k=1}^{n}\|\phi(k-d)\|^2)$ is needed to ensure the stability and convergence of the minimum variance controller. □

Corollary 1. *Considering a minimum-variance type self-correcting control system using any feasible parameter estimation algorithm, if*
(1) $B(q^{-1})$ *is a Hurwitz stable polynomial, and* $|B_0| \neq 0$.
(2) *Control law* $u(k)$ *exists.*
(3) *The parameter estimation error is bounded; that is,* $\frac{1}{n}\sum_{k=1}^{n}\|e(k) - \omega(k)\|^2 \leq M' < \infty$, *then the self-tuning control system is stable.*

The stability and convergence of a self-tuning control system consisting of an arbitrary control strategy when parameter estimation may not converge are considered below.

Theorem 4. *Self-tuning control system for the controlled plant (1), if*
(1) $\|\hat{\theta}(k)\| \leq M < \infty$, $\|\hat{\theta}(k) - \hat{\theta}(k-1)\| \to 0, l$ *is a finite value.*
(2) $\sum_{k=1}^{n}\|e(k) - \omega(k)\|^2 = o(\alpha + \sum_{k=1}^{n}\|\phi(k-d)\|^2)$, α *is a non-zero constant.*
(3) *The control strategy satisfies the stability requirements for the known parameters and tracks the reference signal*
(4) *The controller parameter is a continuous function of the parameter estimates; that is,* $\theta_c(k)$ *is a continuous function of* $\hat{\theta}(k)$.
If the above conditions are met, the self-tuning control system is stable and convergent.

Remark 1. *To ensure that the parameter estimates are bounded, a projection approach can be used, see references [43–46].*

Proof. Considering another virtual equivalent system, as shown in Figure 9, where $P_m(k)$ and $C(k)$ are corresponding to $\Sigma P_m(k)$ and $\Sigma C(k)$, respectively.

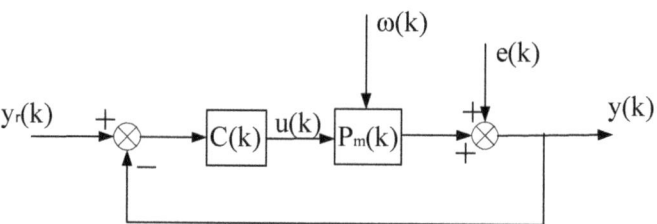

Figure 9. Virtual equivalent system I when parameter estimation may not converge.

The system shown in Figure 9 is further converted into the virtual equivalent system shown in Figure 10. From conditions (1) and (4) of Theorem 4, the interval between t_k and t_{k-1} can be chosen to be sufficiently large, such that to maintain the property of the required $\Delta u'(k)$. Therefore, the system as shown in Figure 10 is a "slow switching" system.

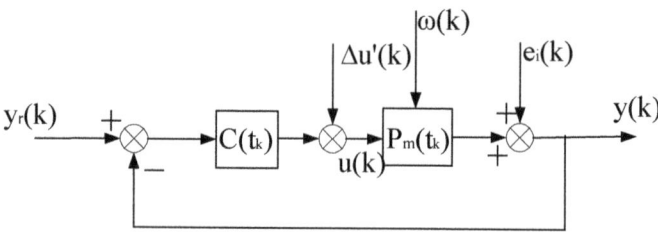

Figure 10. Virtual Equivalent System II when parameter Estimation may not convergence.

Next, the virtual equivalent system shown in Figure 10 is decomposed into three subsystems, as shown in Figures 11–13, respectively. Figure 11 is a stochastic system. Figures 12 and 13 are deterministic systems. By conditions (1) and (2), we have

$$\sum_{k=1}^{n} \|e_i(k) - \omega(k)\|^2 = o(\alpha + \sum_{k=1}^{n} \|\phi(k-d)\|^2).$$

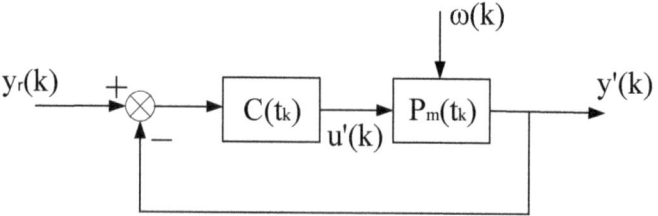

Figure 11. Decomposed subsystem 1.

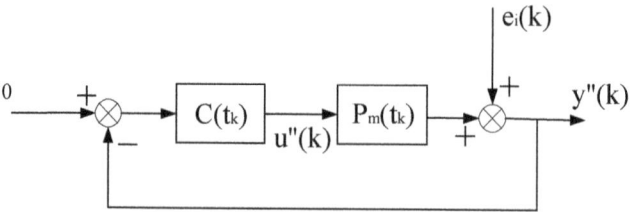

Figure 12. Decomposed subsystem 2.

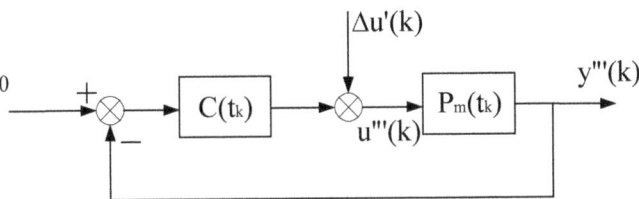

Figure 13. Decomposed subsystem 3.

Based on the results of the "slow switching" stochastic system [43–46] and conditions (1) and (3), it is known that the system shown in Figure 11 is stable and tracking. The rest of the proof process is similar to that of Theorem 1.

Further, considering the low-order modeling situation, we have the following result.

Corollary 2. *Consider a self-tuning control system consisting of any feasible parameter estimation algorithm and control strategy if the following conditions hold.*

(1) $\|\hat{\theta}(k)\| \leq M < \infty$, $\|\hat{\theta}(k) - \hat{\theta}(k-1)\| \to 0$, l is a finite value.

(2) $\frac{1}{n} \sum\limits_{k=1}^{n} \|e(k) - w(k)\|^2 \leq M' < \infty$.

(3) *The control strategy satisfies the stability requirements for the known objects of the parameters and tracks the reference.*

(4) *The controller parameter is a continuous function of the parameter estimate; that is, $\theta_c(k)$ is a continuous function of $\hat{\theta}(k)$.*

Then, the self-tuning control system is stable and convergent. □

4. Extended Results

The above results can be extended to the colored noise situation. The difficulty is that the noise must be estimated together with the parameter estimation.

Considering a general multivariate stochastic system ΣP.

$$A(q^{-1})y(k) = q^{-d}B(q^{-1})u(k) + C(q^{-1})w(k), \tag{25}$$

where $y(k)$, $u(k)$, $w(k)$ has the same meanings as in (1).

$$A(q^{-1}) = I + A_1 q^{-1} + \ldots + A_n q^{-n}, n \geq 1$$
$$B(q^{-1}) = B_0 + B_1 q^{-1} + \ldots + B_m q^{-m}, m \geq 1$$
$$C(q^{-1}) = I + C_1 q^{-1} + \ldots + C_r q^{-r}, r \geq 1$$

Introducing symbols

$$\theta^T = [-A_1, \ldots, -A_n, B_0, \ldots, B_m, C_1, \ldots, C_r],$$
$$\phi_0^T(k-d) = [y(k-1), \ldots y(k-n), \ldots, u(k-d), \ldots u(k-d-m), w(k-1), \ldots, w(k-r)].$$

Then, we have

$$y(k) = \phi_0^T(k-d)\theta + w(k).$$

The elements of the parameter matrix have changed to.

$$\hat{\theta}^T(k) = [-\hat{A}_1, \ldots, -\hat{A}_n, \hat{B}_0, \ldots, \hat{B}_m, \hat{C}_1, \ldots, \hat{C}_r].$$

Since $\phi_0^T(k-d)$ contains unknown noise terms, it is necessary to estimate the noise terms while estimating the parameters so that the regression matrix (vector) of the parameter estimation is as follows:

$$\phi^T(k-d) = [y(k-1),\ldots y(k-n),\ldots,u(k-d),..u(k-d-m),\hat{w}(k-1),\ldots,\hat{w}(k-r)],$$

where

$$\hat{w}(k) = y(k) - \phi^T(k-d)\hat{\theta}(k).$$

The other symbols are the same as before. The self-tuning controller is denoted as $\Sigma C(k)$, and can also be regarded as a matrix $\theta_c(k)$, which can be obtained by various control design methods. The self-tuning control system is abbreviated as $(\Sigma C(k), \Sigma P)$, and the corresponding non-adaptive control system is abbreviated as $(\Sigma C, \Sigma P)$. The virtual equivalent system of the self-tuning control system is abbreviated as $(\Sigma C, \Sigma P, \Delta u(k))$, which can still be illustrated by Figure 3.

If the calculation of the control law does not require the estimation of noise, the calculation of $u(k)$, $\Delta u(k)$, $\phi_c(k)$ will not cause noise estimation problems.

If the calculation of the control law requires the estimation of noise, there will be the following problem. The regression matrices (vector) used to calculate $u(k)$ and calculate $u_0(k)$ are different; that is,

$$u(k) = \phi_c^T(k)\theta_c(k), \quad u_0(k) = \phi_{c0}^T(k)\theta_c.$$

where

$$\phi_c^T(k) = [y_r(k), y_r(k-1)\ldots y(k), y(k-1)\ldots, u(k-1),\ldots,\hat{w}(k-1),\ldots],$$
$$\phi_{c0}^T(k) = [y_r(k), y_r(k-1)\ldots y(k), y(k-1)\ldots, u(k-1),\ldots,w(k-1),\ldots].$$

However, due to condition (2) in Theorem 3.

$$\sum_{k=1}^{n}\|e(k) - w(k)\|^2 = o(\alpha + \sum_{k=1}^{n}\|\phi(k-d)\|^2).$$

That is equivalent to (by definition, $\hat{w}(k)$ is $e(k)$)

$$\sum_{k=1}^{n}\|\hat{w}(k) - w(k)\|^2 = o(\alpha + \sum_{k=1}^{n}\|\varphi(k-d)\|^2).$$

Thus, the difference between $\phi_c(k)$ and $\phi_{c0}(k)$ can be merged into $\Delta u(k)$ without affecting the property of $\Delta u(k)$. Therefore, the above results (with white noise) still hold true for the general stochastic system (25).

Remark 2. *For simulation verification, see reference [47].*

5. Conclusions

Based on the equivalent system concept, a unified analysis of multivariable stochastic self-tuning control (STC) systems is presented. In this paper, by the virtual equivalent system, the difficulty of analyzing the stability and convergence of the stochastic self-tuning control system is transferred from the system structure to the compensational signal, which reduces the difficulty of the original problem, making the stability and convergence analysis of the stochastic self-tuning control system more intuitive and easier to understand. We investigated three situations, i.e., parameter estimation converges to the true value, parameter estimation converges to a non-true value, and parameter estimation may not converge.

Author Contributions: Conceptualization, J.P.; methodology, J.P., W.Z. and K.P.; software, W.Z.; validation, J.P.; formal analysis, K.P.; investigation, J.P. and K.P.; resources, W.Z. and K.P.; data curation, J.P. and W.Z.; writing—original draft preparation, J.P.; writing—review and editing, J.P. and K.P.; visualization, K.P. and W.Z.; supervision, W.Z.; project administration, W.Z.; funding acquisition, W.Z. and K.P. All authors have read and agreed to the published version of the manuscript.

Funding: This research was funded by the National Key R&D Program of China, grant number 2019YFB2005804.

Institutional Review Board Statement: Not applicable.

Data Availability Statement: Data are contained within the article. The data presented in this study can be requested from the authors.

Conflicts of Interest: The authors declare no conflict of interest.

References

1. Goodwin, G.C.; Sin, K.S. *Adaptive Filtering, Prediction and Control*; Prentice Hall: Englewood, IL, USA, 1984.
2. Goodwin, G.C.; Ramadge, P.J.; Caines, P.E. Discrete time adaptive control. *SIAM J. Control Optim.* **1981**, *19*, 829–853. [CrossRef]
3. Bidikli, B.; Bayrak, A. A self-tuning robust full-state feedback control design for the magnetic levitation system. *Control Eng. Pract.* **2018**, *78*, 175–185. [CrossRef]
4. Zou, Z.; Zhao, D.; Liu, X.; Guo, Y.; Guan, C.; Feng, W.; Guo, N. Pole-placement self-tuning control of nonlinear Hammerstein system and its application to pH process control. *Chin. J. Chem. Eng.* **2015**, *23*, 1364–1368. [CrossRef]
5. Guo, L.; Chen, H.F. Stability and Optimality of Self-tuning Regulator. *Sci. China (Ser. A)* **1991**, *9*, 905–913.
6. Guo, L.; Chen, H.F. The Astrom-Wittenmark self-tuning regulator revisited and ELS-based adaptive trackers. *IEEE Trans. Autom. Control* **1991**, *36*, 802–812.
7. Guo, L. Self-convergence of weighted least-squares with applications to stochastic adaptive control. *IEEE Trans. Autom. Control* **1996**, *41*, 79–89.
8. Nassiri-Toussi, K.; Ren, W. Indirect adaptive pole-placement control of MIMO stochastic systems: Self-tuning results. *IEEE Trans. Autom. Control* **1997**, *42*, 38–52. [CrossRef]
9. Yagmur, O. Clonal selection algorithm based control for two-wheeled self-balancing mobile robot. *Simul. Model. Pract. Theory* **2022**, *118*, 102552.
10. Dario, R. Pole-zero assignment by the receptance method: Multi-input active vibration control. *Mech. Syst. Signal Process.* **2022**, *172*, 108976.
11. Anderson, B.D.O.; Johnstone, R.M.G. Global adaptive pole positioning. *IEEE Trans. Autom. Control* **1985**, *30*, 11–22. [CrossRef]
12. Elliott, H.; Cristi, R.; Das, M. Global stability of adaptive pole placement algorithms. *IEEE Trans. Autom. Control* **1985**, *30*, 348–356. [CrossRef]
13. Lozano, R.; Zhao, X.H. Adaptive pole placement without excitation probing signals. *IEEE Trans. Autom. Control* **1994**, *39*, 47–58. [CrossRef]
14. Chan, C.Y.; Sirisena, H.R. Convergence of adaptive pole-zero placement controller for stable non-minimum phase systems. *Int. J. Control* **1989**, *50*, 743–754. [CrossRef]
15. Lai, T.L.; Wei, C.Z. Extended least squares and their applications to adaptive control and prediction in linear systems. *IEEE Trans. Autom. Control* **1986**, *31*, 898–906.
16. Chen, H.F.; Guo, L. Asymptotically optimal adaptive control with consistent parameter estimates. *SIAM J. Control Optim.* **1987**, *25*, 558–575. [CrossRef]
17. Wittenmark, B.; Middleton, R.H.; Goodwin, G.C. Adaptive decoupling of multivariable systems. *Int. J. Control* **1987**, *46*, 1993–2009. [CrossRef]
18. Chai, T.Y. The global convergence analysis of a multivariable decoupling self-tuning controller. *Acta Autom. Sin.* **1989**, *15*, 432–436.
19. Chai, T.Y. Direct adaptive decoupling control for general stochastic multivariable systems. *Int. J. Control* **1990**, *51*, 885–909. [CrossRef]
20. Chai, T.Y.; Wang, G. Globally convergent multivariable adaptive decoupling controller and its application to a binary distillation column. *Int. J. Control* **1992**, *55*, 415–429. [CrossRef]
21. Patete, A.; Furuta, K.; Tomizuka, M. Stability of self-tuning control based on Lyapunov function. *Int. J. Adapt. Control. Signal Process.* **2008**, *22*, 795–810. [CrossRef]
22. Zhang, W. On the stability and convergence of self-tuning control–virtual equivalent system approach. *Int. J. Control* **2010**, *83*, 879–896. [CrossRef]
23. Fekri, S.; Athans, M.; Pascoal, A. Issues, progress and new results in robust adaptive control. *Int. J. Adapt. Control. Signal Process.* **2006**, *20*, 519–579. [CrossRef]
24. Wang, Y.; Li, P.; Tang, J. MPPT control of photovoltaic power generation system based on fuzzy parameter self-tuning PID method. *Electr. Power Autom. Equip.* **2008**, *28*, 4.
25. Tang, L.; Xu, Z. The Control Technology of Self-correction for Intelligent AC Contactors. *Proc. CSEE* **2015**, *35*, 1516–1523.

26. Chen, S.; Wu, J.; Yang, B.; Ma, J. A Self-Tuning Control Method for Simulation Turntable Based on Precise Identification of Model Parameters. CN Patent 201710271289.4, 18 February 2020.
27. Shao, K.; Zheng, J.; Wang, H.; Wang, X.; Lu, R.; Man, Z. Tracking Control of a Linear Motor Positioner Based on Barrier Function Adaptive Sliding Mode. *IEEE Trans. Ind. Inform.* **2021**, *17*, 7479–7488. [CrossRef]
28. Nassiri-Toussi, K.; Ren, W. A unified analysis of stochastic adaptive control: Potential self-tuning. In Proceedings of the American Control Conference, Seattle, DC, USA, 21–23 June 1995.
29. Nassiri-Toussi, K.; Ren, W. A unified analysis of stochastic adaptive control: Asymptotic self-tuning. In Proceedings of the 34th IEEE Conference on Decision and Control, New Orleans, LA, USA, 13–15 December 1995; p. 3.
30. Morse, A.S. Towards a unified theory of parameter adaptive control: Tunability. *IEEE Trans. Autom. Control* **1990**, *35*, 1002–1012. [CrossRef]
31. Morse, A.S. Towards a unified theory of parameter adaptive control-part II: Certainty equivalence and implicit tuning. *IEEE Trans. Autom. Control* **1992**, *37*, 15–29. [CrossRef]
32. Katayama, T.; McKelvey, T.; Sano, A.; Cassandras, C.G.; Campi, M.C. Trends in systems and signals. *Annu. Rev. Control* **2006**, *30*, 5–17. [CrossRef]
33. Li, Q.Q. Adaptive control. *Comput. Autom. Meas. Control* **1999**, *7*, 56–60.
34. Li, Q.Q. *Adaptive Control System Theory, Design and Application*; Science Press: Beijing, China, 1990.
35. Aström, K.J.; Wittenmark, B. *Adaptive Control*; Addison-Wesley: Upper Saddle River, NJ, USA, 1995.
36. Ioannou, P.A.; Sun, J. *Robust Adaptive Control*; Prentice-Hall: Englewood Cliffs, NJ, USA, 1996.
37. Kumar, P.R. Convergence of adaptive control schemes using least-squares parameter estimates. *IEEE Trans. Autom. Control* **1990**, *35*, 416–424. [CrossRef]
38. van Schuppen, J.H. Tuning of Gaussian stochastic control systems. *IEEE Trans. Autom. Control* **1994**, *39*, 2178–2190. [CrossRef]
39. Zhang, W.C. The convergence of parameter estimates is not necessary for a general self-tuning control system-stochasticplant. In Proceedings of the 48th IEEE Conference on Decision and Control, Shanghai, China, 15–18 December 2009.
40. Zhang, W.C.; Li, X.L.; Choi, J.Y. A unified analysis of switching multiple model adaptive control—Virtual equivalent system approach. In Proceedings of the 17th IFAC World Congress, Seoul, Republic of Korea, 6–11 July 2008; Volume 41, pp. 14403–14408.
41. Zhang, W.C. Virtual equivalent system theory for self-tuning control. *J. Harbin Inst. Technol.* **2014**, *46*, 107–112.
42. Feng, C.; Shi, W. *Adaptive Control*; Publishing House of Electronics Industry: Beijing, China, 1986.
43. Chatterjee, D.; Liberzon, D. Stability analysis of deterministic and stochastic switched systems via a comparison principle and multiple Lyapunov functions. *SIAM J. Control. Optim.* **2006**, *45*, 174–206. [CrossRef]
44. Chatterjee, D.; Liberzon, D. On stability of stochastic switched systems. In Proceedings of the 43rd Conference on Decision and Control, Nassau, Bahamas, 14–17 December 2004; Volume 4, pp. 4125–4127.
45. Prandini, M. Switching control of stochastic linear systems: Stability and performance results. In Proceedings of the 6th Congress of SIMAI, Chia Laguna, Cagliari, Italy, 27–31 May 2002.
46. Prandini, M.; Campi, M.C. Logic-based switching for the stabilization of stochastic systems in presence of unmodeled dynamics. In Proceedings of the 40th IEEE Conference on Decision and Control, Orlando, FL, USA, 4–7 December 2001.
47. Zhang, W.C.; Wei, W. Virtual equivalent system theory for adaptive control and simulation verification. *Sci. Sin. Inf.* **2018**, *48*, 947–962. [CrossRef]

Disclaimer/Publisher's Note: The statements, opinions and data contained in all publications are solely those of the individual author(s) and contributor(s) and not of MDPI and/or the editor(s). MDPI and/or the editor(s) disclaim responsibility for any injury to people or property resulting from any ideas, methods, instructions or products referred to in the content.

Article

Design of Adaptive Fractional-Order Fixed-Time Sliding Mode Control for Robotic Manipulators

Saim Ahmed [1,2], Ahmad Taher Azar [1,2,3,*] and Mohamed Tounsi [1,2]

1. College of Computer and Information Sciences, Prince Sultan University, Riyadh 11586, Saudi Arabia
2. Automated Systems and Soft Computing Lab (ASSCL), Prince Sultan University, Riyadh 11586, Saudi Arabia
3. Faculty of Computers and Artificial Intelligence, Benha University, Benha 13518, Egypt
* Correspondence: aazar@psu.edu.sa

Abstract: In this investigation, the adaptive fractional-order non-singular fixed-time terminal sliding mode (AFoFxNTSM) control for the uncertain dynamics of robotic manipulators with external disturbances is introduced. The idea of fractional-order non-singular fixed-time terminal sliding mode (FoFxNTSM) control is presented as the initial step. This approach, which combines the benefits of a fractional-order parameter with the advantages of NTSM, gives rapid fixed-time convergence, non-singularity, and chatter-free control inputs. After that, an adaptive control strategy is merged with the FoFxNTSM, and the resulting model is given the label AFoFxNTSM. This is done in order to account for the unknown dynamics of the system, which are caused by uncertainties and bounded external disturbances. The Lyapunov analysis reveals how stable the closed-loop system is over a fixed time. The pertinent simulation results are offered here for the purposes of evaluating and illustrating the performance of the suggested scheme applied on a PUMA 560 robot.

Keywords: robotic manipulators; adaptive fixed-time control; fractional-order sliding mode control; unknown dynamics

---

Citation: Ahmed, S.; Azar, A.T.; Tounsi, M. Design of Adaptive Fractional-Order Fixed-Time Sliding Mode Control for Robotic Manipulators. *Entropy* **2022**, *24*, 1838. https://doi.org/10.3390/e24121838

Academic Editor: Udo Von Toussaint

Received: 16 November 2022
Accepted: 7 December 2022
Published: 16 December 2022

Publisher's Note: MDPI stays neutral with regard to jurisdictional claims in published maps and institutional affiliations.

Copyright: © 2022 by the authors. Licensee MDPI, Basel, Switzerland. This article is an open access article distributed under the terms and conditions of the Creative Commons Attribution (CC BY) license (https://creativecommons.org/licenses/by/4.0/).

1. Introduction

The latest advancements in the domain of control systems are having a significant impact on the field of mechatronics and robotic system design and development. The topic of controlling a robotic manipulator is investigated in the field of control theory. Specifically, it is a highly non-linear system that also possesses a high degree of mechanical instability. Due to this, the system in question needs to be able to maintain a high level of stability, while still having the capacity to monitor accurately its course in the face of external disturbance and uncertainty [1]. Despite the fact that a large variety of viable solutions have been proposed for uncertain robotic systems that are subject to external disturbances, it is impossible to avoid the uncertain parameters when operating under real-world conditions. Due to this, it is difficult for a system to be precisely regulated if the controller is impacted in any way by the disturbance. As a direct result of this, there is a growing interest in the creation of robust control systems, which have been the subject of substantial research and are currently being deployed in a wide variety of industries [2]. Moreover, a robust adaptive control mechanism is built to compensate for the unknown uncertainties and disturbances so that the system continues to function effectively. The advantage of the approach behind robust adaptive control is that the control system itself needs to be robust in order to guarantee the attainment of the necessary level of both performance and stability.

Sliding mode control, commonly known as SMC, is a type of control strategy that is both non-linear and robust [3]. It can effectively deal with non-linear systems that are uncertain, have confined disturbances, and have a low sensitivity to changes in the system's parameters. Terminal SMC (TSMC) was introduced in [4] with the objective of achieving robust finite-time stability. TSMC offers accurate tracking and increased precision.

However, delayed convergence and singularity are problematic. As a result, SMC approaches were created as solutions to these issues in order to achieve rapid convergence with fast terminal SMC (FTSMC) and eliminate singularities with non-singular terminal SMC (NTSMC) [5,6]. Moreover, the initial values of the non-linear system have a significant impact on the amount of time required for the finite-time system to converge, and this amount of time always increases as the initial values of the non-linear system increase. Fixed-time stability is, therefore, an option that can be utilized to precisely compute the time of convergence irrespective of the initial conditions [7,8].

The theory of fractional-order (Fo) calculus, which has been around for the past three centuries and deals with derivatives and integrals of non-integer order [9–13], was recently rediscovered by scientists and engineers and is being utilized in various domains such as material sciences [14], bioengineering [15], finance [16], and electronic circuits [17,18], including the field of control theory [19–24]. The numerous control techniques such as proportional–integral–derivative (PID) control, the SMC method, and various fuzzy and neural network schemes have all implemented their respective control techniques using a Fo controller [25–29]. Dadras [30] is credited with being the first author to present the ideas of Fo in combination with finite-time TSMC. Moreover, the adaptive scheme with fractional-order non-singular fast TSMC (FOTSMC) was introduced with the intention of controlling the robotic manipulator. This was done so as to address the issue of dealing with unknown dynamics [31]. Recently, several Fo fixed-time SMC schemes have been developed for applications such as micro-gyroscopes [32], chaotic systems [33], unmanned surface vessesl [34], nonholonomic mobile robots [35], and multimachine power systems [36].

Control engineering applications are increasingly gravitating toward the use of adaptive control, which is a well-known control technology that is gaining popularity [37,38]. It demonstrates an unusual capacity for adaptation in the face of system uncertainty and external disturbances, and it helps improve the tracking performance of closed-loop systems [39,40]. A robust adaptive strategy based on a class of high-order SMC was devised for a fractional chaotic system in the presence of non-linearity [41]. Several adaptive finite-time FoSMC techniques have been suggested for use with the robotic manipulator, which also takes into account the presence of uncertainties and disturbances. In the study in [25], a robust adaptive finite-time FoFTSM was built for the robotic system. Within this model, unknown dynamics were estimated by employing an adaptive controller. It was suggested to estimate the unknowable dynamics of the non-linear robot using an output feedback adaptive super-twisting finite-time FoSMC [31]. Moreover, a fixed-time disturbance observer-based adaptive FoNFTSM has been designed for indeterminate manipulators under unknown disturbances [42].

It is fascinating to note that each of the aforementioned papers concentrated their attention largely on the adaptive scheme for the estimate of the upper bounds of uncertain dynamics by applying finite-time FoNTSM control. It is generally agreed that the most significant benefit of using fixed-time non-singular TSMC (FxNTSM) control is that it eliminates the risk of singularity, possesses high robustness in the face of both internal and external disturbances, and ensures that convergence time is independent of the initial values. This research has shown that very few works provide adaptive FxNTSM control, and that no research whatsoever has been conducted on adaptive FoFxTSMC. Within the scope of this study, the fixed-time convergence of robotic manipulator systems that are vulnerable to external disturbances is explored. Specifically, the research focuses on the effects of the unknown dynamics of the systems. Considering all of this, the adaptive fractional-order fixed-time non-singular terminal SMC is designed, which is also known as AFoFxNTSM, for uncertain robotic manipulators that are influenced by external disturbances. The most important contributions given by this work are organized into the following points:

1. Based on the characteristics of fractional-order fixed-time non-singular terminal SMC, a sliding surface with good tracking performance, reduced control input chattering, and rapid convergence is designed.

2. The fractional-order control is applied in an attempt to improve the performance of the closed system.
3. It is proposed to use adaptive control with FoFxNTSM, so that the unknown dynamics are compensated for in order to produce the robust and sustainable performance for the PUMA 560 robotic manipulator.
4. The Lyapunov theory is utilized in order to carry out an investigation into the system's fixed-time stability.

The remaining parts of this work are organized as follows: The preliminaries are presented in Section 2. The modeling of the system, the control design, and its stability are explained in Section 3. The adaptive control approach and its stability are presented in Section 4. The numerical simulations to validate the performance of the proposed method are presented in Section 5. Section 6 is devoted to discussing the simulation findings. Section 7 delivers the conclusion of the paper.

2. Preliminaries

Definition 1. *For fractional calculus, the Riemann–Liouville (RL) definition is often employed [43]. Consequently, the Fo integral and derivative are given as follows. The following equation gives the RL fractional integral, as well as the derivative of the α_{th} – order function $f(t)$ in relation to t and a, provided by*

$$_a\mathcal{I}_t^\alpha f(t) = \frac{1}{\Gamma(\alpha)} \int_a^t \frac{f(\tau)}{(t-\tau)^{1-\alpha}} d\tau \tag{1}$$

$$_a\mathcal{D}_t^\alpha f(t) = \frac{d^\alpha f(t)}{dt^\alpha} = \frac{1}{\Gamma(1-\alpha)} \frac{d}{dt} \int_a^t \frac{f(\tau)}{(t-\tau)^\alpha} d\tau \tag{2}$$

where $n-1 < \alpha < n$, $m \in \mathbb{N}$ and $\Gamma(\cdot)$ is the Gamma function, described by Euler as

$$\Gamma(\alpha) = \int_0^\infty e^{-t} t^{\alpha-1} dt$$

whereas $\mathcal{D}$ and $\mathcal{I}$ represent, respectively, the fractional integral and the derivative of the function.

Lemma 1. *Consider the following non-linear system [44]*

$$\dot{x}(t) = f(t,x), \quad x(0) = x_0 \tag{3}$$

where $f(t,x)$ is a continuous non-linear function. For fixed-time stability with fast time convergence, the Lyapunov function $V(x)$ satisfies that

a. $V(x) = 0 \Leftrightarrow x = 0$
b. $\dot{V}(x) \leq -\varsigma_1 V^{\eta_1}(x) - \varsigma_2 V(x)^{\eta_2}$

where $\varsigma_1, \varsigma_2 > 0$, $0 < \eta_1 < 1$ and $\eta_2 > 1$. Then, the system is fixed-time stable and the convergence time can be computed as

$$T \leq \frac{1}{\varsigma_1(1-\eta_1)} + \frac{1}{\varsigma_2(\eta_2-1)} \tag{4}$$

Lemma 2. *With the fractional derivative such as $_a\mathcal{D}_t^{\alpha_1} f(t) = \frac{1}{\Gamma(1-\alpha_1)} \frac{d}{dt} \int_a^t \frac{f(\tau)}{(t-\tau)^{\alpha_1}} d\tau$ with $f(t) \in \mathbb{R}$, $0 \leq \alpha_1 < 1$, and its sign function, then, for the fractional derivative of the sign function [45], one obtains $_a\mathcal{D}_t^{\alpha_1} sign(f(t)) \begin{cases} > 0 & \text{if } f(t) > 0, t > 0 \\ < 0 & \text{if } f(t) < 0, t > 0 \end{cases}$.*

3. Fractional-Order Fixed-Time Non-Singular Terminal Sliding Control Design

This part begins with an introduction to the dynamics of the robot manipulator and continues with a study of the characteristics of a fractional-order non-singular fixed-time sliding surface and the development of a control design called FoFxNTSM. In addition to this, a study of the suggested FoFxNTSM's stability using the Lyapunov theorem is presented.

The following is a description of the dynamic equation of the $n-DOF$ robotic manipulator [46].

$$M(q)\ddot{q} + C(q,\dot{q})\dot{q} + G(q) = \tau(t) + \tau_f(t) + \tau_d(t) \tag{5}$$

where $q \in \mathbb{R}^n$ is the joints position, $\dot{q} \in \mathbb{R}^n$ is the joint velocity, and $\ddot{q} \in \mathbb{R}^n$ is the joint acceleration. $M(q) \in \mathbb{R}^{n \times n}$ represents the inertia matrix and satisfies that $m_1(M(q)) \leq \|M(q)\| \leq m_2(M(q))$, with m_1 and m_2 illustrating the positive min and the max eigenvalues of the matrix $M(q)$. $C(q,\dot{q}) \in \mathbb{R}^{n \times n}$ denotes the coriolis, centripetal, and friction forces matrix; $G(q) \in \mathbb{R}^n$ is the gravitational vector. $\tau_f \in \mathbb{R}^n$ is system's uncertainty, $\tau_d \in \mathbb{R}^n$ is a representation of the unknown external disturbance, $\tau(t) \in \mathbb{R}^n$ is the input torque at the joints.

The dynamic Equation (5) can be rewritten as

$$\ddot{q} = M^{-1}(q)\tau - M^{-1}(q)[C(q,\dot{q})\dot{q} + G(q)] + \Im(q,\dot{q},\ddot{q},\tau_d) \tag{6}$$

where $\Im(q,\dot{q},\ddot{q},\tau_d) = M^{-1}(q)\left[\tau_d(t) + \tau_f(t)\right]$ represents the uncertainties and external disturbances.

Using Equation (6), the trajectory tracking error can be expressed as

$$\ddot{\varepsilon} = M^{-1}(q)\tau + \partial(q,\dot{q}) + \Im(q,\dot{q},\ddot{q},\tau_d) \tag{7}$$

where $\partial(q,\dot{q}) = -M^{-1}(q)[C(q,\dot{q})\dot{q} + G(q)] - \ddot{q}_d$ denotes the known system dynamics. The tracking error is represented by the equation $\varepsilon = q - q_d$, where q represents the actual position vectors and q_d represents the desired position vectors.

Assumption 1. Conditional bounds on the uncertainty and external disturbance are expressed by (8), which is shown below:

$$\|\Im(q,\dot{q},\ddot{q},\tau_d)\| \leq \iota_1 + \iota_2\|q\| + \iota_3\|\dot{q}\|^2 \tag{8}$$

where ι_1, ι_2, and ι_3 are unknown constants of the uncertainties' and disturbances' upper bounds.

3.1. FoFxNTSM Surface

The aforementioned techniques served as inspiration for the development of the fractional-order non-singular terminal sliding mode control, which can be built to provide the robust and precise tracking performance of the $n-DOF$ robotic manipulators in a fixed time. Therefore, based on the features of fractional-order calculus, the proposed sliding surface is given as

$$s(t) = \dot{\varepsilon}(t) + \delta_1 \sqrt[1/\beta_1]{|\varepsilon|sign(\varepsilon)} + \delta_2 \sqrt[1/\beta_2]{|\varepsilon|sign(\varepsilon)} + \delta_3 \mathcal{D}^{\alpha-1}[|\varepsilon|sign(\varepsilon)] \tag{9}$$

where $s(t) \in \mathbb{R}^n$ is the sliding surface, and $\delta_1 \in \mathbb{R}^+$ and $\delta_2 \in \mathbb{R}^+$ are positive constants. To be more specific, β_1 and β_2 are the set of constants, such that $0 < \beta_1 < 1$, $1 < \beta_2$, and $0 < \alpha < 1$.

$$\dot{s}(t) = \ddot{\varepsilon}(t) + \beta_1\delta_1|\varepsilon|^{\beta_1-1}\dot{\varepsilon} + \beta_2\delta_2|\varepsilon|^{\beta_2-1}\dot{\varepsilon} + \delta_3\mathcal{D}^{\alpha}[|\varepsilon|sign(\varepsilon)] \tag{10}$$

$$\begin{aligned}\dot{s}(t) &= M^{-1}(q)\tau + \partial(q,\dot{q}) + \Im(q,\dot{q},\ddot{q},\tau_d) \\ &+ \delta_1 K(\varepsilon)\dot{\varepsilon} + \beta_2\delta_2|\varepsilon|^{\beta_2-1}\dot{\varepsilon} + \delta_3\mathcal{D}^{\alpha}[|\varepsilon|sign(\varepsilon)]\end{aligned} \tag{11}$$

where $K(\varepsilon) = \begin{cases} \beta_1|\varepsilon|^{\beta_1-1} & if\ \varepsilon \neq 0 \\ 0 & if\ \varepsilon = 0 \end{cases}$.

Now that the construction of the sliding manifold is complete, the robust performance against uncertainty and external disturbances is achieved using the proposed FoFxNTSM control design for $n-DOF$ robotic manipulators.

Throughout the course of the sliding mode, when $s(t) = 0$, the following dynamics can be derived from (9) as

$$\dot{\varepsilon}(t) = -\delta_1 \sqrt[1/\beta_1]{|\varepsilon|sign(\varepsilon)} - \delta_2 \sqrt[1/\beta_2]{|\varepsilon|sign(\varepsilon)} - \delta_3 \mathcal{D}^{\alpha-1}[|\varepsilon|sign(\varepsilon)] \quad (12)$$

The Lyapunov function is defined as follows

$$V_1(t) = 0.5\varepsilon(t)^T \varepsilon(t) \quad (13)$$

With (13), the $\dot{V}_1(t)$ can be computed as

$$\dot{V}_1(t) = \varepsilon(t)^T \dot{\varepsilon}(t) = \varepsilon(t)^T \left[-\delta_1 \sqrt[1/\beta_1]{|\varepsilon|sign(\varepsilon)} - \delta_2 \sqrt[1/\beta_2]{|\varepsilon|sign(\varepsilon)} - \delta_3 \mathcal{D}^{\alpha-1}[|\varepsilon|sign(\varepsilon)] \right] \quad (14)$$

By simplifying (14), one has

$$\dot{V}_1(t) = -\delta_1 \varepsilon(t)^T \sqrt[1/\beta_1]{|\varepsilon|sign(\varepsilon)} - \delta_2 \varepsilon(t)^T \sqrt[1/\beta_2]{|\varepsilon|sign(\varepsilon)} - \delta_3 |\varepsilon(t)|^T \mathcal{D}^{\alpha-1}\left[|\varepsilon|sign^2(\varepsilon)\right] \quad (15)$$

$$\dot{V}_1(t) \leq -\delta_1 \|\varepsilon\|^{\beta_1+1} - \delta_2 \|\varepsilon\|^{\beta_2+1} \quad (16)$$

$$\dot{V}_1(t) \leq -2^{\frac{\beta_1+1}{2}} \delta_1 V_1^{\frac{\beta_1+1}{2}} - 2^{\frac{\beta_2+1}{2}} \delta_2 V_1^{\frac{\beta_2+1}{2}} \quad (17)$$

In accordance with Lemma 1, the sliding surface (9) converges to zero in a fixed time, and the amount of time it takes to get there is bounded by

$$T_1 = \frac{1}{2^{\frac{\beta_1+1}{2}} \delta_1 \left(1-\frac{\beta_1+1}{2}\right)} + \frac{1}{2^{\frac{\beta_2+1}{2}} \delta_2 \left(\frac{\beta_2+1}{2}-1\right)}$$
$$= \frac{2}{2^{\frac{\beta_1+1}{2}} \delta_1 (1-\beta_1)} + \frac{2}{2^{\frac{\beta_2+1}{2}} \delta_2 (\beta_2-1)} \quad (18)$$

3.2. FoFxNTSM Control Design

For the purpose of controlling a robotic manipulator in the presence of known bounded uncertainties and external disturbances, the FoFxNTSM control law can be designed as follows

$$\tau(t) = \tau_{nm}(t) + \tau_{sw}(t) \quad (19)$$

where $\tau_{nm}(t)$ refers to the control input that is employed in the control of the known dynamics and $\tau_{sw}(t)$ refers to the control input that is utilized to deal with uncertain dynamics.

$$\tau_{nm} = -M(q) \left\{ \partial(q,\dot{q}) + \delta_1 K(\varepsilon)\dot{\varepsilon} + \beta_2 \delta_2 |\varepsilon|^{\beta_2-1}\dot{\varepsilon} + \delta_3 \mathcal{D}^{\alpha}[|\varepsilon|sign(\varepsilon)] \right\} \quad (20)$$

$$\tau_{sw} = -M(q) \left\{ \begin{array}{l} (\iota_1 + \iota_2\|q\| + \iota_3\|\dot{q}\|^2)sign(s) \\ +\delta_4 \sqrt[1/\varsigma_1]{|s|sign(s)} + \delta_5 \sqrt[1/\varsigma_2]{|s|sign(s)} + \delta_6 \mathcal{D}^{\alpha_1}sign(s) \end{array} \right\} \quad (21)$$

where $\delta_4 \in \mathbb{R}^+$, $\delta_5 \in \mathbb{R}^+$ and $\delta_6 \in \mathbb{R}^+$ are positive constants, and ς_1 and ς_2 are constants, such that $0 < \varsigma_1 < 1$, $1 < \varsigma_2$ and $0 \leq \alpha_1 < 1$, respectively.

3.3. Stability Analysis

The Lyapunov theorem is applied in this subsection to establish the closed-loop system stability.

Theorem 1. *Considering the described robotic manipulator (5), the suggested sliding manifold (9) and the designed FoFxNTSM controller (19) enable the intended angular position of the uncertain robotic manipulator to converge in a fixed amount of time with condition (8).*

Proof. The Lyapunov function is considered as follows

$$V_2(t) = 0.5 s(t)^T s(t) \quad (22)$$

where $\dot{V}_2(t)$ can be computed as

$$\dot{V}_2(t) = s(t)^T \dot{s}(t) \quad (23)$$

With $\dot{s}(t)$ from (10) substituted into Equation (23), one obtains

$$\dot{V}_2(t) = s(t)^T \left[\ddot{\varepsilon}(t) + \delta_1 K(\varepsilon)\dot{\varepsilon} + \beta_2 \delta_2 |\varepsilon|^{\beta_2-1}\dot{\varepsilon} + \delta_3 \mathcal{D}^\alpha [|\varepsilon| sign(\varepsilon)] \right] \quad (24)$$

By substituting $\ddot{\varepsilon}(t)$ from (7) in (24), one obtains

$$\dot{V}_2(t) = s(t)^T \left\{ \begin{array}{l} M^{-1}(q)\tau + \partial(q,\dot{q}) + \Im(q,\dot{q},\ddot{q},\tau_d) \\ + \delta_1 K(\varepsilon)\dot{\varepsilon} + \beta_2 \delta_2 |\varepsilon|^{\beta_2-1}\dot{\varepsilon} + \delta_3 \mathcal{D}^\alpha [|\varepsilon| sign(\varepsilon)] \end{array} \right\} \quad (25)$$

By substituting $\tau(t)$ from (19) in (25), one has

$$\dot{V}_2(t) = s(t)^T \left[\begin{array}{l} -\left\{ \begin{array}{l} (\iota_1 + \iota_2\|q\| + \iota_3\|\dot{q}\|^2)sign(s) + \partial(q,\dot{q}) \\ + \delta_1 K(\varepsilon)\dot{\varepsilon} + \beta_2 \delta_2 |\varepsilon|^{\beta_2-1}\dot{\varepsilon} + \delta_3 \mathcal{D}^\alpha [|\varepsilon| sign(\varepsilon)] \\ + \delta_4 \sqrt[1/\varsigma_1]{|s|} sign(s) + \delta_5 \sqrt[1/\varsigma_2]{|s|} sign(s) + \delta_6 \mathcal{D}^{\alpha_1} sign(s) \end{array} \right\} \\ + \partial(q,\dot{q}) + \Im(q,\dot{q},\ddot{q},\tau_d) + \delta_1 K(\varepsilon)\dot{\varepsilon} + \beta_2 \delta_2 |\varepsilon|^{\beta_2-1}\dot{\varepsilon} + \delta_3 \mathcal{D}^\alpha [|\varepsilon| sign(\varepsilon)] \end{array} \right] \quad (26)$$

The simplification of (26) yields

$$\dot{V}_2(t) = s(t)^T \left[\begin{array}{l} -\left\{ \begin{array}{l} (\iota_1 + \iota_2\|q\| + \iota_3\|\dot{q}\|^2)sign(s) + \delta_4 \sqrt[1/\varsigma_1]{|s|} sign(s) \\ + \delta_5 \sqrt[1/\varsigma_2]{|s|} sign(s) + \delta_6 \mathcal{D}^{\alpha_1} sign(s) \end{array} \right\} \\ + \Im(q,\dot{q},\ddot{q},\tau_d) \end{array} \right] \quad (27)$$

According to Assumption 1 and Lemma 2, one can easily obtain

$$\dot{V}_2(t) \leq -\delta_4 \|s\|^{\varsigma_1+1} - \delta_5 \|s\|^{\varsigma_2+1} \quad (28)$$

and (28) can be rewritten as

$$\dot{V}_2(t) \leq -2^{\frac{\varsigma_1+1}{2}} \delta_4 V_2(t)^{\frac{\varsigma_1+1}{2}} - 2^{\frac{\varsigma_2+1}{2}} \delta_5 V_2(t)^{\frac{\varsigma_2+1}{2}} \quad (29)$$

Therefore, the trajectory of the system reaches $s(t)$ in a fixed time. In accordance with Lemma 1, the time required for convergence can be expressed as

$$T_2 = \frac{1}{2^{\frac{\varsigma_1+1}{2}} \delta_4 \left(1 - \frac{\varsigma_1+1}{2}\right)} + \frac{1}{2^{\frac{\varsigma_2+1}{2}} \delta_5 \left(\frac{\varsigma_2+1}{2} - 1\right)} \quad (30)$$

Using relation $T_{s1} = T_1 + T_2$, the settling time T_{s1} can be formulated as

$$T_{s1} = \frac{2}{2^{\frac{\beta_1+1}{2}} \delta_1 (1 - \beta_1)} + \frac{2}{2^{\frac{\beta_2+1}{2}} \delta_2 (\beta_2 - 1)} + \frac{2}{2^{\frac{\varsigma_1+1}{2}} \delta_4 (1 - \varsigma_1)} + \frac{2}{2^{\frac{\varsigma_2+1}{2}} \delta_5 (\varsigma_2 - 1)} \quad (31)$$

As a result, it can be deduced from (31) that the suggested scheme is a fixed-time control scheme. □

4. Adaptive FoFxNTSM Control Design

The following describes how the control input utilizing an adaptive method is devised to account for the unknown dynamics and external disturbances.

$$\tau(t) = \tau_{ad}(t) \tag{32}$$

$$\tau_{ad}(t) = -M(q) \left\{ \begin{array}{l} (\hat{\imath}_1 + \hat{\imath}_2 \|q\| + \hat{\imath}_3 \|\dot{q}\|^2)sign(s) + \partial(q,\dot{q}) \\ +\delta_1 K(\varepsilon)\dot{\varepsilon} + \beta_2\delta_2 |\varepsilon|^{\beta_2-1}\dot{\varepsilon} + \delta_3 \mathcal{D}^\alpha[|\varepsilon|sign(\varepsilon)] \\ +\delta_4 \sqrt[1/c_1]{|s|}sign(s) + \delta_5 \sqrt[1/c_2]{|s|}sign(s) + \delta_6 \mathcal{D}^{\alpha_1}sign(s) \end{array} \right\} \tag{33}$$

where $\hat{\imath}_1$, $\hat{\imath}_2$, and $\hat{\imath}_3$ denote the estimation variable of ι_1, ι_2, and ι_3, respectively.

To compensate for unknown dynamics, the adaptive laws are proposed. In addition, the dead-zone method is applied to avoid the parameter drifting problem; thus, the adaptive laws are given as

$$\dot{\hat{\imath}}_i = \left\{ \begin{array}{ll} \gamma_i \|s\|\Delta\omega & if \ \|s\| > \omega \\ 0 & if \ \|s\| \leq \omega \end{array} \right. \& \quad i = 1,2,3 \tag{34}$$

where $\Delta\omega = \left[1, \|q\|, \|\dot{q}\|^2\right]$, $\omega > 0$ denotes the size of the dead zone, and γ_1, γ_2, and $\gamma_3 > 0$ are constants. The proposed model is given in Figure 1.

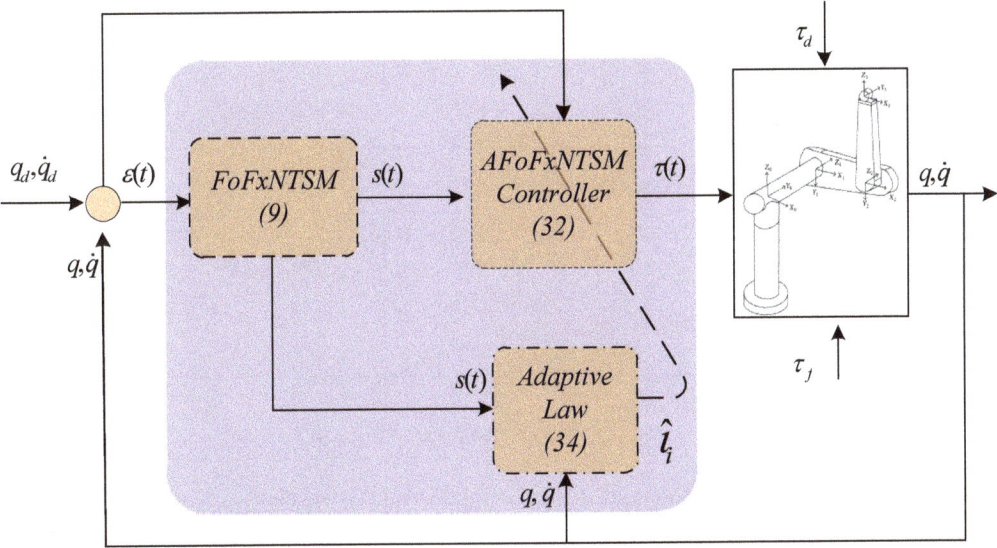

Figure 1. Control model of proposed scheme.

Compensating for the upper bounds of the unknown dynamics is dealt with the use of (34). Therefore, the AFoFxNTSM technique is what ultimately determines the tracking performance of the uncertain robot manipulators under disturbances.

Theorem 2. *Considering the given robotic manipulator (5) and its susceptibility to issues such as uncertainty and external disturbances, accordingly, the desired angular position of the robotic manipulator converges in a fixed time with the condition of Assumption 1, thanks to the suggested sliding surface (9), AFoFxNTSM control input (32), and adaptive laws (34).*

Proof. The following Lyapunov candidate is selected as

$$V_3(t) = 0.5 s(t)^T s(t) + \frac{0.5}{\gamma_1}\tilde{\imath}_1^2 + \frac{0.5}{\gamma_2}\tilde{\imath}_2^2 + \frac{0.5}{\gamma_3}\tilde{\imath}_3^2 \tag{35}$$

where $\tilde{\imath}_1 = \hat{\imath}_1 - \iota_1$, $\tilde{\imath}_2 = \hat{\imath}_2 - \iota_2$, $\tilde{\imath}_3 = \hat{\imath}_3 - \iota_3$ are estimation errors.

$\dot{V}_3(t)$ can be expressed as

$$\dot{V}_3(t) = s(t)^T \dot{s}(t) + \frac{1}{\gamma_1}\tilde{\imath}_1 \dot{\hat{\imath}}_1 + \frac{1}{\gamma_2}\tilde{\imath}_2 \dot{\hat{\imath}}_2 + \frac{1}{\gamma_3}\tilde{\imath}_3 \dot{\hat{\imath}}_3 \tag{36}$$

With the substitution of $\dot{s}(t)$ from (11) into (36), one can obtain

$$\dot{V}_3(t) = s(t)^T\left\{M^{-1}(q)\tau + \partial(q,\dot{q}) + \Im(q,\dot{q},\ddot{q},\tau_d) + \delta_1 K(\varepsilon)\dot{\varepsilon} + \beta_2\delta_2|\varepsilon|^{\beta_2-1}\dot{\varepsilon} + \delta_3 \mathcal{D}^\alpha[|\varepsilon|sign(\varepsilon)]\right\} \\ + \frac{1}{\gamma_1}\tilde{\imath}_1\dot{\hat{\imath}}_1 + \frac{1}{\gamma_2}\tilde{\imath}_2\dot{\hat{\imath}}_2 + \frac{1}{\gamma_3}\tilde{\imath}_3\dot{\hat{\imath}}_3 \tag{37}$$

With the substitution of $\tau(t)$ from (32) into (37), one can obtain

$$\dot{V}_3(t) = s(t)^T \left[-\left\{\begin{array}{l}(\hat{\imath}_1 + \hat{\imath}_2\|q\| + \hat{\imath}_3\|\dot{q}\|^2)sign(s) + \partial(q,\dot{q}) \\ + \delta_1 K(\varepsilon)\dot{\varepsilon} + \beta_2\delta_2|\varepsilon|^{\beta_2-1}\dot{\varepsilon} \\ + \delta_3\mathcal{D}^\alpha[|\varepsilon|sign(\varepsilon)] + \delta_4 \sqrt[1/\varsigma_1]{|s|}sign(s) \\ + \delta_5 \sqrt[1/\varsigma_2]{|s|}sign(s) + \delta_6 \mathcal{D}^{\alpha_1}sign(s) \\ + \partial(q,\dot{q}) + \Im(q,\dot{q},\ddot{q},\tau_d) + \delta_1 K(\varepsilon)\dot{\varepsilon} \\ + \beta_2\delta_2|\varepsilon|^{\beta_2-1}\dot{\varepsilon} + \delta_3\mathcal{D}^\alpha[|\varepsilon|sign(\varepsilon)]\end{array}\right\}\right] \\ + \frac{1}{\gamma_1}\tilde{\imath}_1\dot{\hat{\imath}}_1 + \frac{1}{\gamma_2}\tilde{\imath}_2\dot{\hat{\imath}}_2 + \frac{1}{\gamma_3}\tilde{\imath}_3\dot{\hat{\imath}}_3 \tag{38}$$

Simplifying (38) yields

$$\dot{V}_3(t) = s(t)^T \left[\begin{array}{l} -\left\{\begin{array}{l}(\hat{\imath}_1 + \hat{\imath}_2\|q\| + \hat{\imath}_3\|\dot{q}\|^2)sign(s) + \delta_4 \sqrt[1/\varsigma_1]{|s|}sign(s) \\ + \delta_5 \sqrt[1/\varsigma_2]{|s|}sign(s) + \delta_6 \mathcal{D}^{\alpha_1}sign(s)\end{array}\right\} \\ + \Im(q,\dot{q},\ddot{q},\tau_d) \end{array}\right] \\ + \frac{1}{\gamma_1}\tilde{\imath}_1\dot{\hat{\imath}}_1 + \frac{1}{\gamma_2}\tilde{\imath}_2\dot{\hat{\imath}}_2 + \frac{1}{\gamma_3}\tilde{\imath}_3\dot{\hat{\imath}}_3 \tag{39}$$

According to Lemma 2, (39) can be computed as

$$\dot{V}_3(t) \leq -\delta_4 \|s\|^{\varsigma_1+1} - \delta_5 \|s\|^{\varsigma_2+1} - \hat{\imath}_1\|s\| - \hat{\imath}_2\|q\|\|s\| - \hat{\imath}_3\|\dot{q}\|^2\|s\| + \|\Im(q,\dot{q},\ddot{q},\tau_d)\|\|s\| \\ + \frac{1}{\gamma_1}\tilde{\imath}_1\dot{\hat{\imath}}_1 + \frac{1}{\gamma_2}\tilde{\imath}_2\dot{\hat{\imath}}_2 + \frac{1}{\gamma_3}\tilde{\imath}_3\dot{\hat{\imath}}_3 \tag{40}$$

Using Assumption 1 and the substitution of (34) into (40), one can obtain

$$\dot{V}_3(t) \leq -\delta_4\|s\|^{\varsigma_1+1} - \delta_5\|s\|^{\varsigma_2+1} \tag{41}$$

As a result, the robotic manipulator that is utilized for the purpose of precise trajectory tracking is only capable of maintaining its fixed-time stability under specific circumstances. As a consequence of this, the proof of stability is investigated in great detail.

Following that, the fixed settling time is calculated, and Equation (41) can be expressed as [47]

$$\dot{V}_3(t) \leq -\delta_4\{2(V_3(t) - \Xi)\}^{\frac{\varsigma_1+1}{2}} - \delta_5\{2(V_3(t) - \Xi)\}^{\frac{\varsigma_2+1}{2}} \tag{42}$$

where $\Xi = \frac{0.5}{\gamma_1}\tilde{\imath}_1^2 + \frac{0.5}{\gamma_2}\tilde{\imath}_2^2 + \frac{0.5}{\gamma_3}\tilde{\imath}_3^2$

$$\dot{V}_3(t) \leq -\delta_{42} 2^{\frac{\varsigma_1+1}{2}} \{V_3(t) - \Xi\}^{\frac{\varsigma_1+1}{2}} - \delta_{52} 2^{\frac{\varsigma_2+1}{2}} \{V_3(t) - \Xi\}^{\frac{\varsigma_2+1}{2}} \tag{43}$$

$$\dot{V}_3(t) \leq -\delta_{42} 2^{\frac{\varsigma_1+1}{2}} \left\{1 - \frac{\Xi}{V_3(t)}\right\}^{\frac{\varsigma_1+1}{2}} V_3(t)^{\frac{\varsigma_1+1}{2}} - \delta_{52} 2^{\frac{\varsigma_2+1}{2}} \left\{1 - \frac{\Xi}{V_3(t)}\right\}^{\frac{\varsigma_2+1}{2}} V_3(t)^{\frac{\varsigma_2+1}{2}} \tag{44}$$

Calculating the fixed time using Lemma 1 yields the following

$$T_3 = \frac{1}{p_1\left(1 - \frac{\varsigma_1+1}{2}\right)} + \frac{1}{p_2\left(\frac{\varsigma_2+1}{2} - 1\right)} = \frac{2}{p_1(1 - \varsigma_1)} + \frac{2}{p_2(\varsigma_2 - 1)} \tag{45}$$

where $p_1 = \delta_{42} 2^{\frac{\varsigma_1+1}{2}} \left\{1 - \frac{\Xi}{V_3(t)}\right\}^{\frac{\varsigma_1+1}{2}}$, and $p_2 = \delta_{52} 2^{\frac{\varsigma_2+1}{2}} \left\{1 - \frac{\Xi}{V_3(t)}\right\}^{\frac{\varsigma_2+1}{2}}$. Calculating the settling time T_{s2} using the relation $T_{s2} = T_1 + T_3$ yields

$$T_{s2} = \frac{2}{p_1(1 - \varsigma_1)} + \frac{2}{p_2(\varsigma_2 - 1)} + \frac{2}{2^{\frac{\beta_1+1}{2}} \delta_1(1 - \beta_1)} + \frac{2}{2^{\frac{\beta_2+1}{2}} \delta_2(\beta_2 - 1)} \tag{46}$$

The resulting state trajectory tends to zero in a fixed amount of time. □

Remark 1. *When the proposed adaptive fractional-order fixed-time sliding mode control method is applied to the uncertain dynamics of the robotic system (5), which includes the fractional sliding surface (9), the proposed control input (32), and the adaptive laws (34), it is implied that the tracking error tends toward zero at a fixed time. The numerical simulation is provided in the following section.*

5. Simulation Results and Comparative Analyses

The PUMA 560 robotic manipulator is utilized to demonstrate the simulation performance in order to validate the AFoFxNTSM approach; its dynamics have been given in [48]. A 3 − DOF of the PUMA 560 manipulator is employed, and it operates in an environment containing external disturbances and uncertainties. In order to show the great performance of AFoFxNTSM, two different scenarios, one with known dynamics and one with unknown uncertainties and disturbances, are described, and MATLAB/Simulink is used to simulate the proposed method. To demonstrate further the efficacy of the suggested strategy, a comparison is made with adaptive fractional-order non-singular terminal sliding mode control (ATDENTSM) [49]. Therefore, the planned trajectories, external disturbance, and uncertainty levels are given as:

$q_d = [cos(t\pi/5) - 1,\ cos(t\pi/5 + \pi/2),\ cos(t\pi/5 + \pi/2) - 1]^T$
$\tau_f = [0.5\dot{q}_1 + sin(3q_1),\ 1.3\dot{q}_2 - 1.8sin(2q_2),\ -1.8\dot{q}_3 - 2sin(q_3)]^T$
$\tau_d = [20.5sin(\dot{q}_1),\ 21.1sin(\dot{q}_2),\ 10.15sin(\dot{q}_3)]^T$

To select the suitable Fo value, the position tracking errors at different values of α are demonstrated in Figure 2.

As seen in Figure 2, setting $\alpha = 0.9$ is a simple way to achieve the best results. On the other hand, at $\alpha = 0.1$ and $\alpha = 0.5$, the desired trajectories are not achieved in terms of tracking errors.

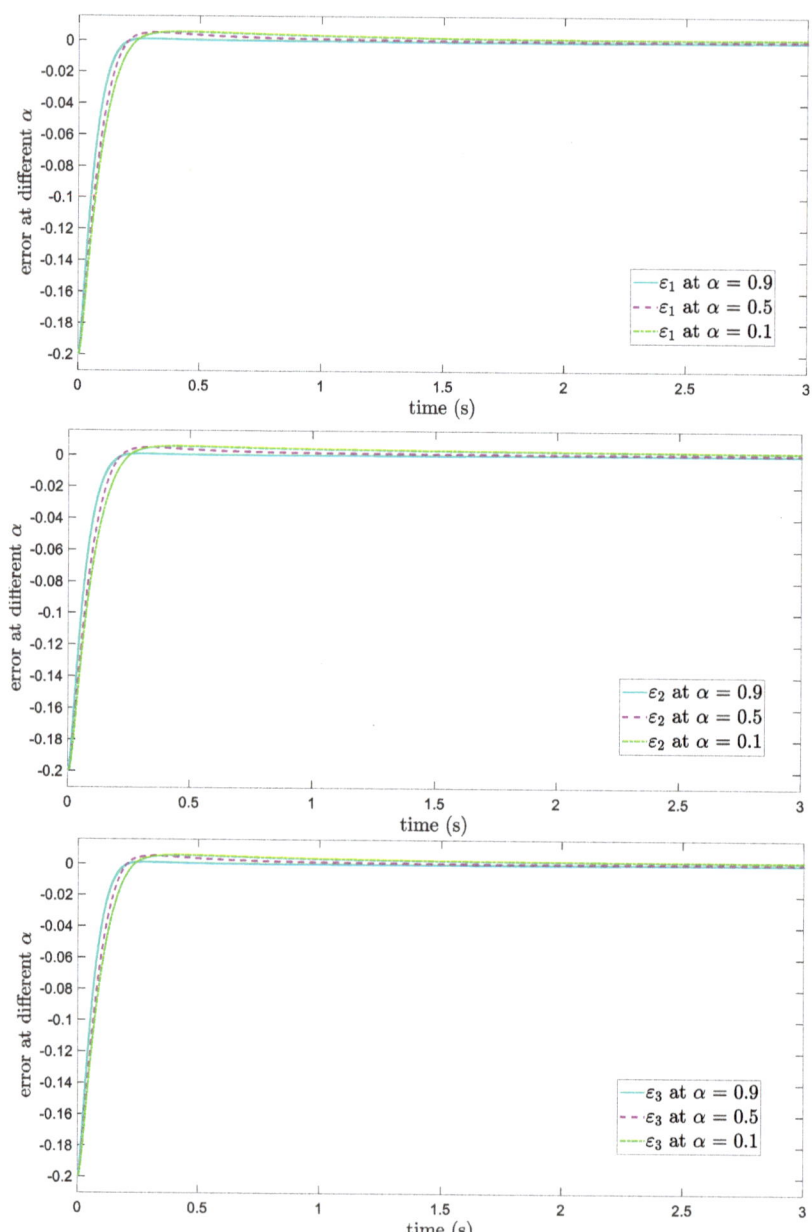

Figure 2. Tracking errors at different α values.

5.1. Case 1: Comparison for Nominal Plant

In this subsection, the proposed FoFxNTSM approach is applied to the $3-DOF$ PUMA 560 robotic manipulator with known dynamics; however, external disturbances are not taken into consideration. For (9), the FoFxNTSM parameters are set to $\delta_1 = 6$, $\delta_2 = 6$, $\delta_3 = 6$, $\beta_1 = 0.8$, $\beta_2 = 1.9$, and $\alpha = 0.9$. The suitable parameters of (19) are set as $\delta_4 = 50$, $\delta_5 = 50$, $\delta_6 = 0.01$, $\alpha_1 = 0.1$, $\varsigma_1 = 0.7$, $\varsigma_2 = 1.5$, and $\varpi = 0.1$. The initial conditions of the joint positions are chosen as $q_1(0) = -0.2$, $q_2(0) = -0.2$, and $q_3(0) = -0.2$.

The comparative results of the proposed FoFxNTSM approach and ATDENTSM on $3-DOF$ robotic manipulators are depicted in Figures 3–6, which show the joint's position performance, its tracking errors, smooth control inputs, and sliding mode surfaces, respectively.

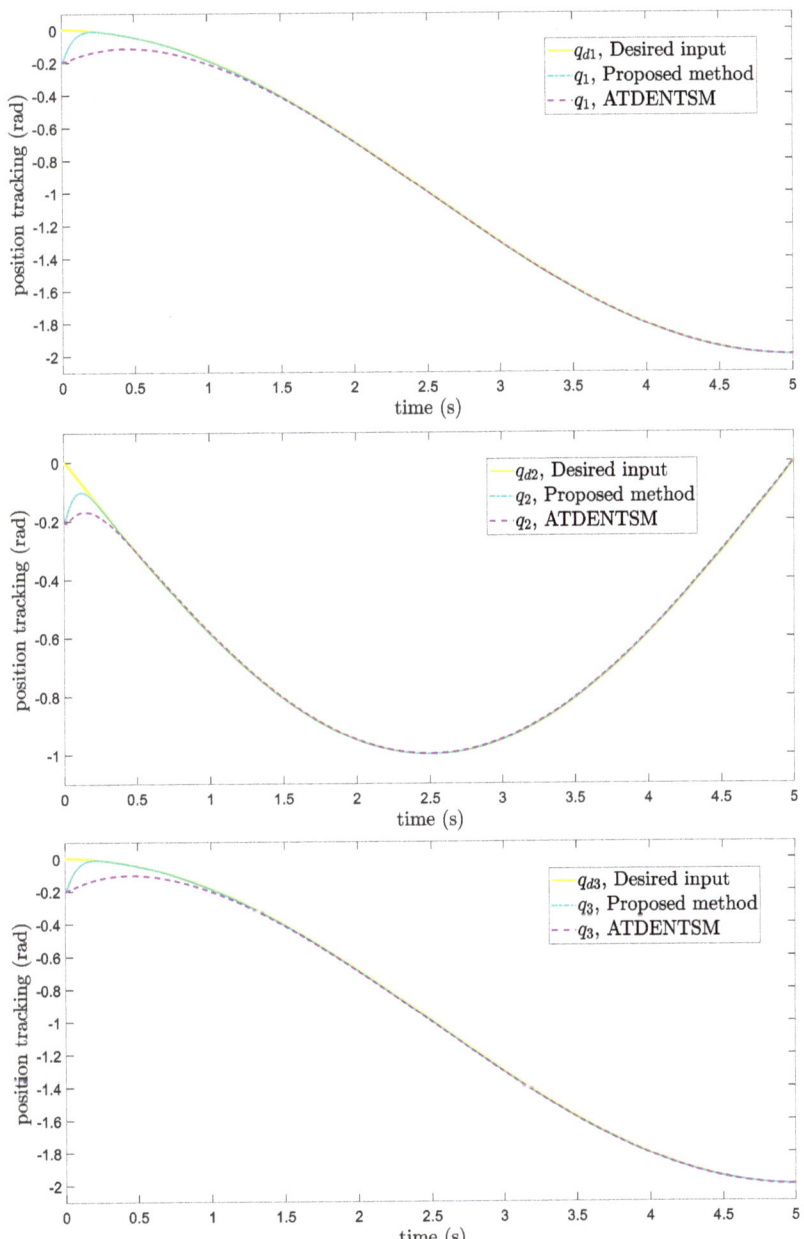

Figure 3. Position tracking.

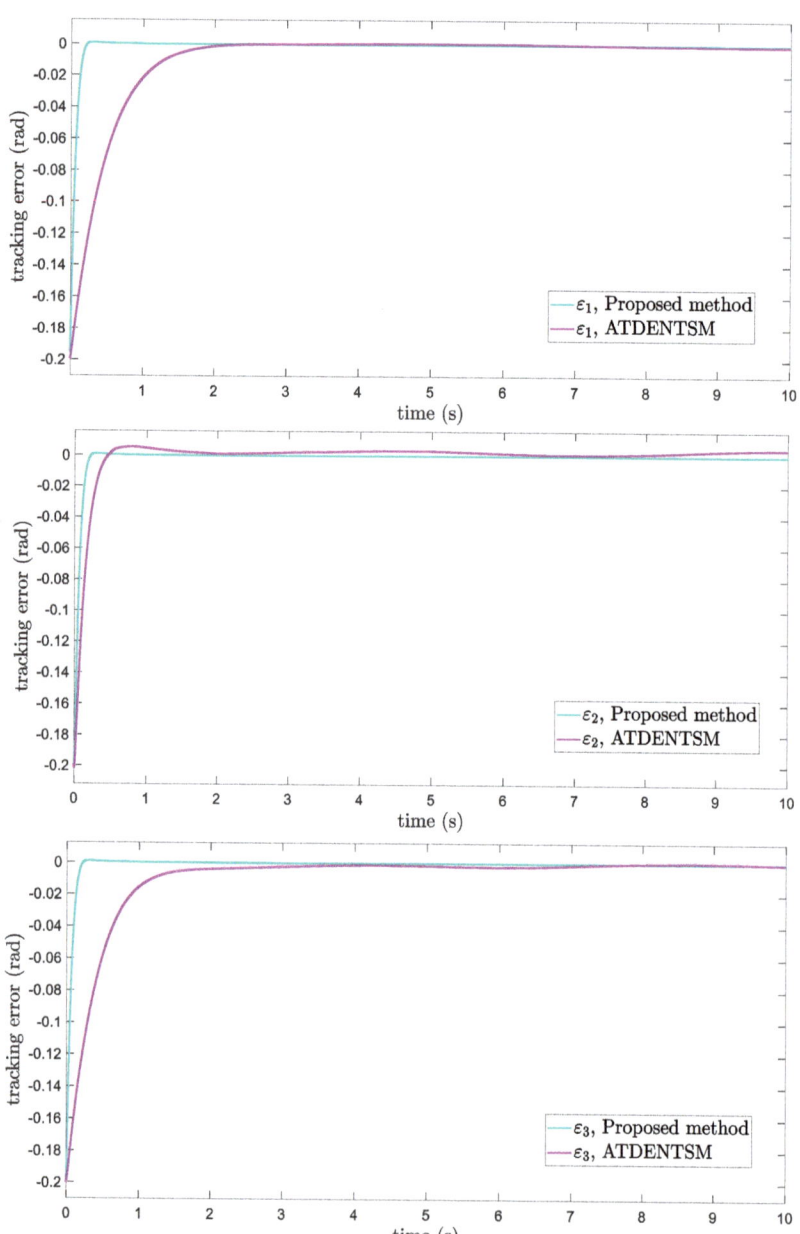

Figure 4. Tracking errors.

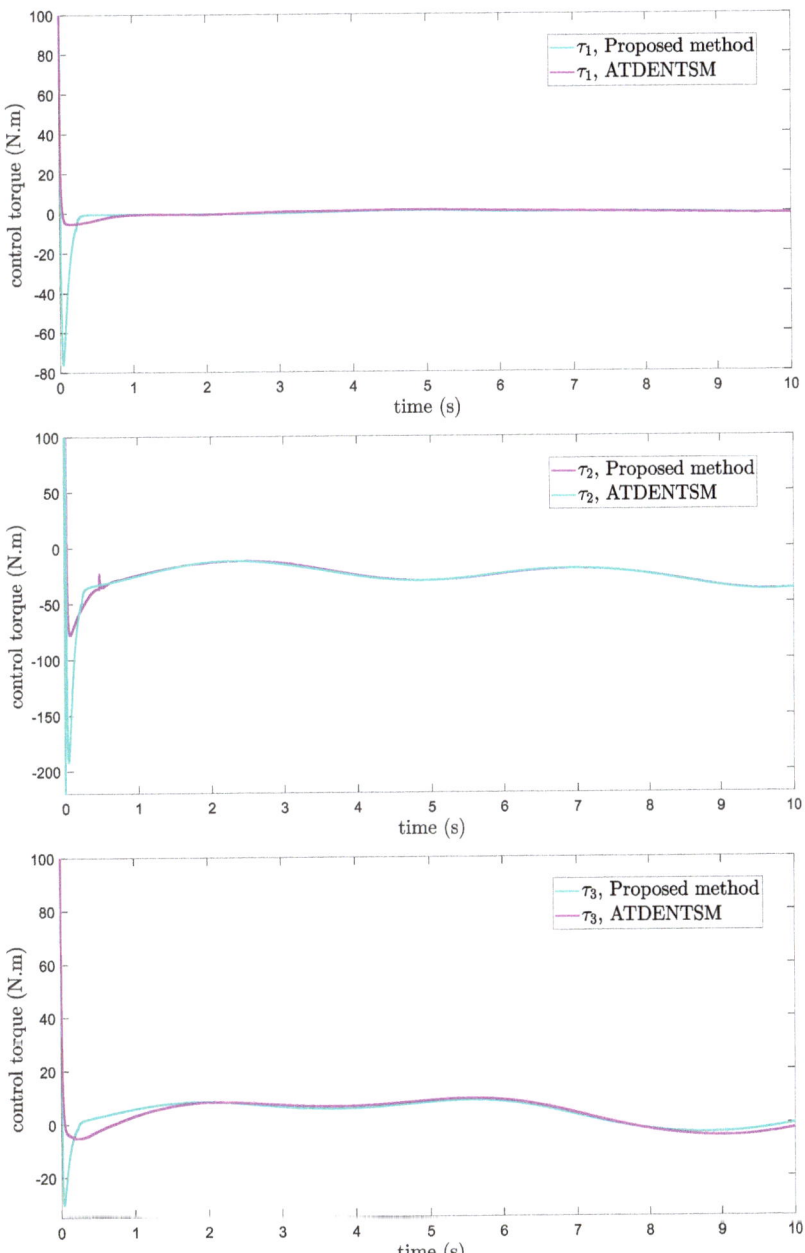

Figure 5. Control inputs.

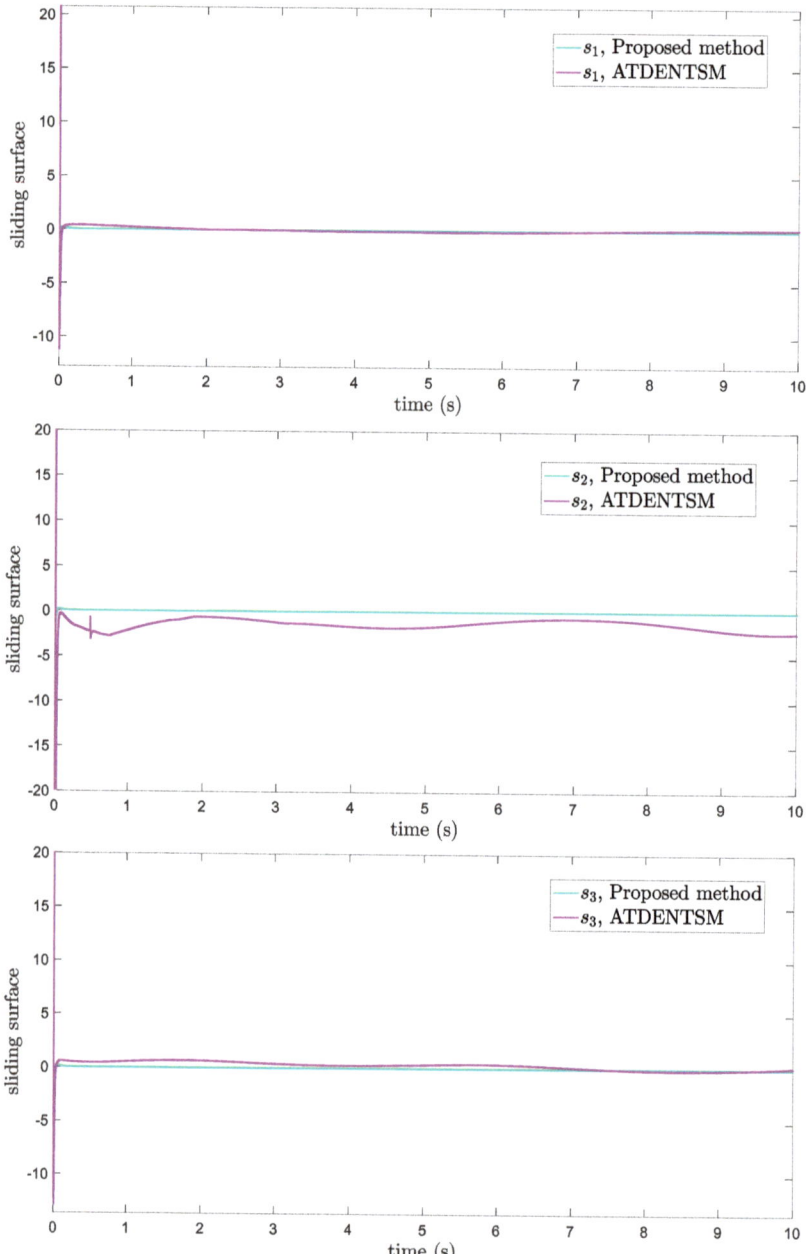

Figure 6. Sliding surfaces.

The suggested FoFxNTSM scheme has improved performance and obtains small tracking errors, rapid convergence, and chatter-free control inputs. These advantages are achieved by taking into account the high tracking performance and robustness against the system's known uncertainties.

5.2. Case 2: Comparison Under Unknown Dynamics

In this subsection, the proposed adaptive technique with the FoFxNTSM method is used to control the dynamics of the $3-DOF$ robotic manipulator in the presence of unknown uncertainties, as well as external disturbances. The parameters of (32) are set such that they are identical to those of (19), and the parameters of (34) are set such that $\gamma_1 = 0.01$, $\gamma_2 = 0.01$, and $\gamma_3 = 0.01$. Figures 7–10 present the results of comparing the proposed AFoFxNTSM scheme with ATDENTSM in terms of its performance in the face of unknown dynamics, as well as benchmark simulations of trajectories, control inputs, and sliding surfaces. Moreover, the adaptive parameter estimations of the unknown dynamics of AFoFxNTSM and ATDENTSM are given in Figures 11 and 12, respectively.

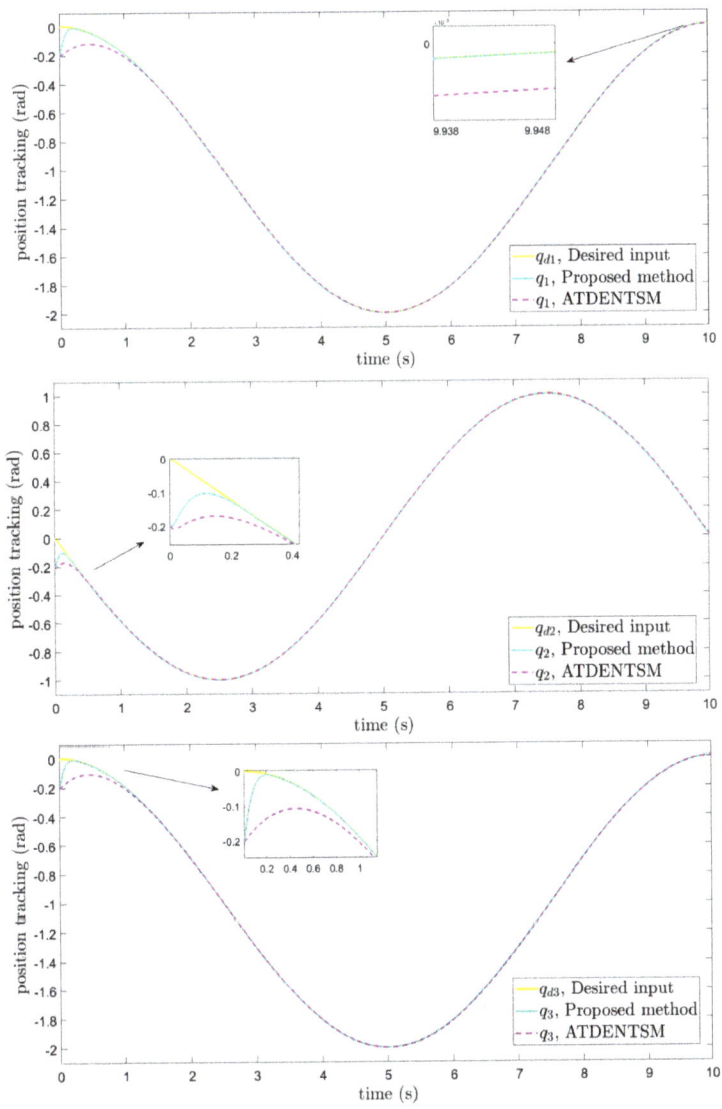

Figure 7. Position tracking method under uncertainties and disturbances.

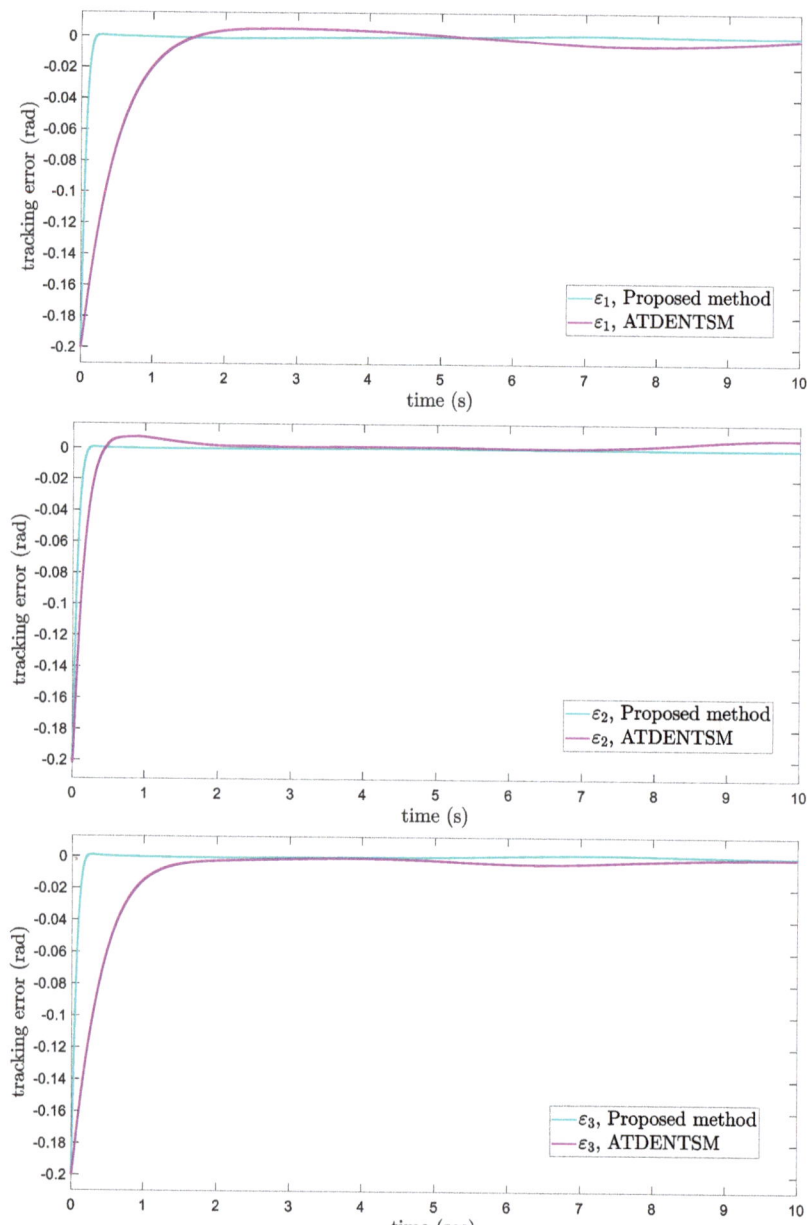

Figure 8. Tracking errors under uncertainties and disturbances.

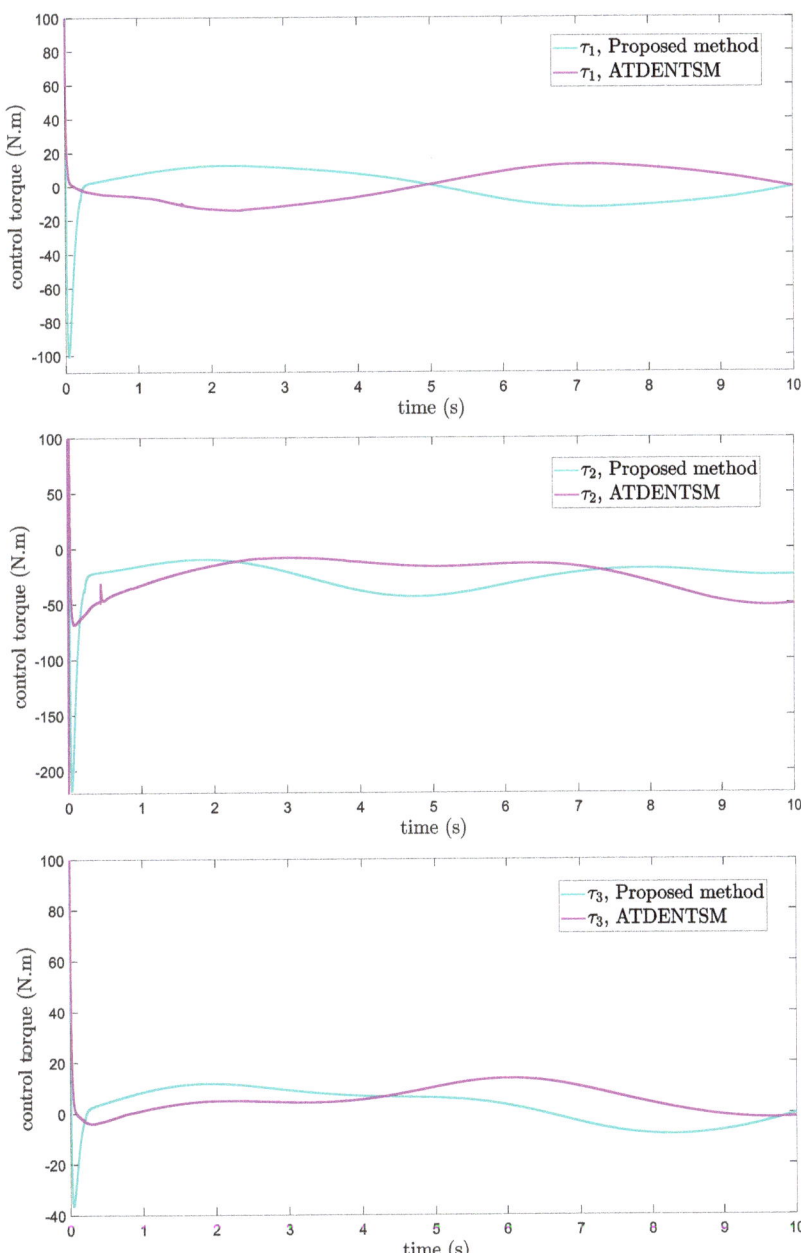

Figure 9. Control inputs under uncertainties and disturbances.

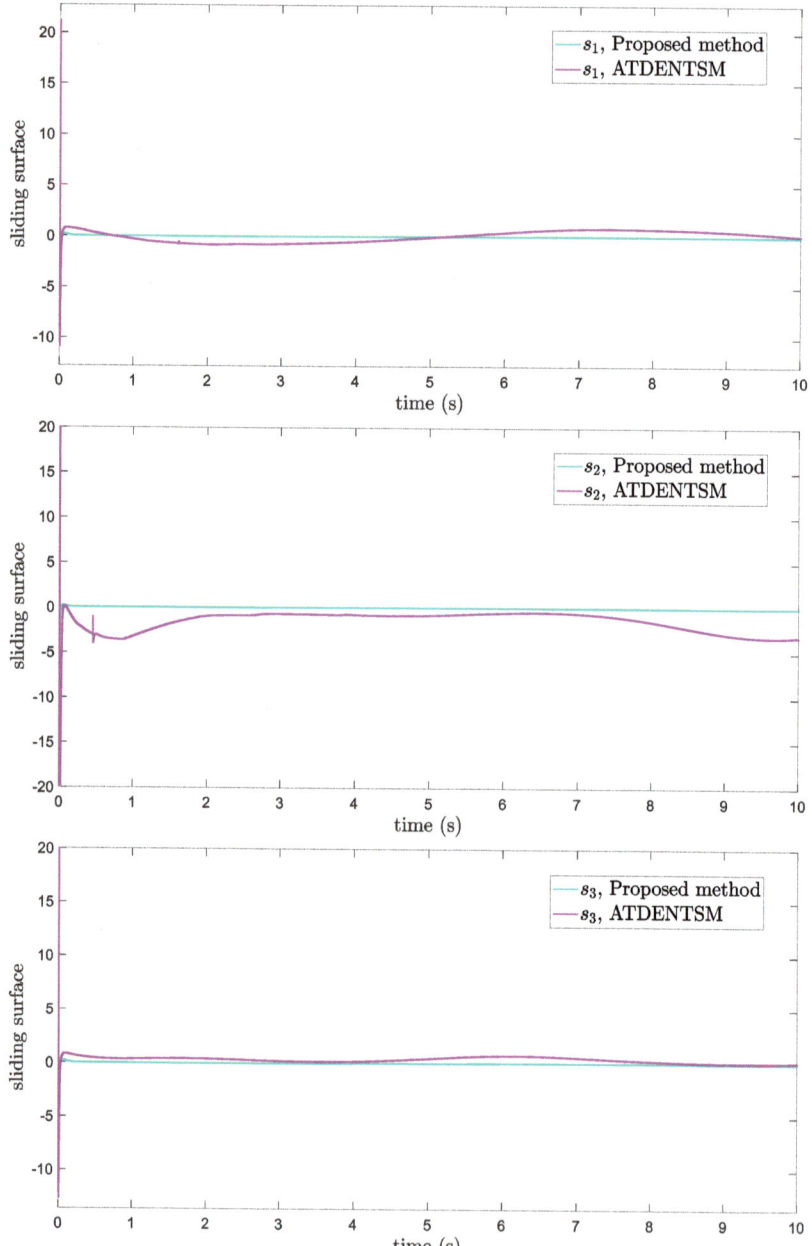

Figure 10. Sliding surfaces under uncertainties and disturbances.

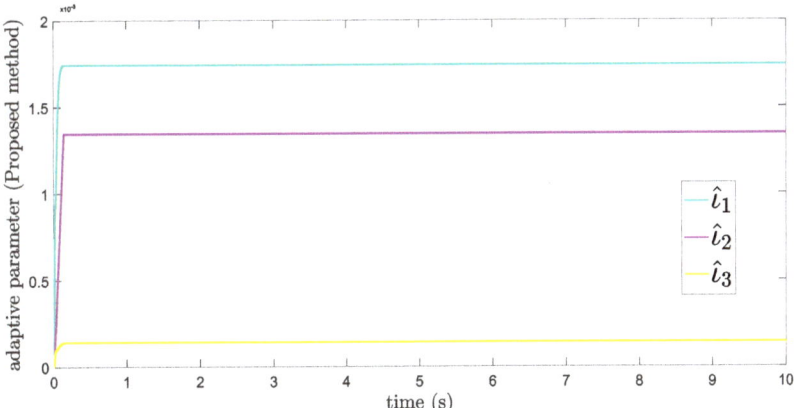

Figure 11. Adaptive parameters under uncertainties and disturbances—Proposed method.

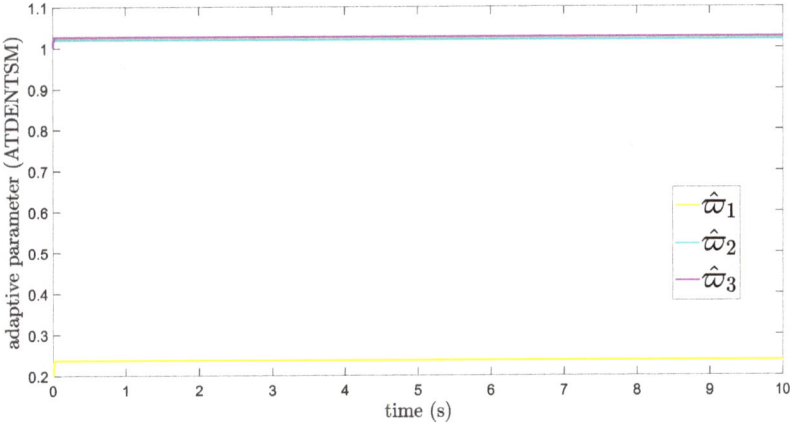

Figure 12. Adaptive parameters under uncertainties and disturbances—ATDENTSM.

The findings that are compared and obtained reveal that the AFoFxNTSM has an improved tracking performance, chatter-free control inputs, and adaptive estimation in the presence of unknown uncertainties and external disturbances. Figures 7–12 make it abundantly clear that the proposed method, when subjected to uncertainties and external disturbances, yields a superior convergence and trajectory tracking performance, whereas the ATDENTSM method demonstrates significant angular position errors and is less robust when exposed to unknown dynamics.

6. Discussion

The discussion of the simulated results of the proposed AFoFxNTSM is presented in this section. In particular, a concise discussion of the shortcomings of the suggested controller in terms of its parameters and stability analyses is included. In addition to this, potential applications of the proposed method to non-linear systems are also covered.

A comparison is made between the control strategy that has been suggested (AFoFxNTSM) and ATDENTSM, and the parameters of both systems are set in an appropriate way. Therefore, it is clear from looking at Figures 7 and 8 that the suggested controller has the least amount of tracking errors and, accordingly, the least amount of time needed to converge. In addition, the control inputs of the joints can be noticed in Figure 9, and one can see the suggested method that was provided offers the control input that is the most smooth and efficient. Figures 11 and 12 present the adaptive estimation, which demonstrates that there is no drifting problem with the adaptive rules. In addition, the root-mean-square (RMS) errors of the proposed AFoFxNTSM scheme are calculated as $\varepsilon_{1RMS} = 0.0124$, $\varepsilon_{2RMS} = 0.0125$, and $\varepsilon_{3RMS} = 0.0123$, and the RMS errors of the AT-DENTSM method are obtained as $e_{1RMS} = 0.0317$, $e_{2RMS} = 0.0189$, and $e_{3RMS} = 0.0294$. Hence, both the simulation and the quantitative analyses demonstrate that the proposed method has a superior performance.

The parameters of the suggested control technique are chosen in accordance with the range that was provided, such as $\delta_1 > 0$, $\delta_2 > 0$, $\delta_3 > 0$, $0 < \beta_1 < 1$, $\beta_2 > 1$, $0 < \alpha < 1$, $\delta_4 > 0$, $\delta_5 > 0$, $\delta_6 > 0$, $0 < \varsigma_1 < 1$, $\varsigma_2 > 1$, and $0 \leq \alpha_1 < 1$. In the event that these concerns are not considered, the closed-loop system does not continue to exhibit fixed-time stability. It is clear, based on the results of (31) and (46), that T_{s1} and T_{s2} are inversely proportional to δ_i, whereas δ_i is proportional to $\tau(t)$ in (19) and (32). Therefore, in order to attain fixed-time convergence and closed-loop system stability at the same time, the suitable values of δ_i need to be set. These values determine the stability of the system. In addition, the ranges of the other parameters are known, which makes it possible to select the suitable value in a manner that is adequate. In fact, the scope of this work could be broadened to include the consideration of non-linearities that are not smooth for the non-linear systems.

7. Conclusions

An AFoFxNTSM was proposed in order to facilitate robotic manipulator trajectory tracking in the presence of uncertainties and external disturbances. An adaptive method was used in the construction of the proposed scheme so that it could estimate the unknown bounds of uncertainties and disturbances. This method also made it possible for the FoFxNTSM to achieve fixed-time convergence and tracking performance. On the $3-DOF$ PUMA 560 robotic manipulator, the AFoFxNTSM is implemented with known and unknown dynamics to demonstrate and explain the usefulness of the suggested technique. The findings of the simulation reveal that the suggested AFoFxNTSM method, compared with ATDENTSM, is superior in terms of response time and trajectory tracking errors, and has a higher capability to reject uncertainties and disturbances.

Author Contributions: Conceptualization, S.A. and A.T.A.; Formal analysis, S.A., A.T.A. and M.T.; Funding acquisition, M.T.; Investigation, A.T.A. and M.T.; Methodology, S.A., A.T.A. and M.T.; Project administration, M.T.; Resources, S.A. and M.T.; Software, S.A.; Supervision, A.T.A.; Validation, A.T.A. and M.T.; Visualization, S.A. and A.T.A.; Writing – original draft, S.A.; Writing – review & editing, S.A., A.T.A. and M.T. All authors have read and agreed to the published version of the manuscript

Funding: This research was funded by Prince Sultan University, Riyadh, Saudi Arabia grant number: Seed-CCIS-2022-117.

Institutional Review Board Statement: Not applicable.

Informed Consent Statement: Not applicable.

Data Availability Statement: Not applicable.

Acknowledgments: The authors would like to acknowledge the support of Prince Sultan University, for paying the Article Processing Charges (APC) of this publication. Special acknowledgement to Automated Systems & Soft Computing Lab (ASSCL), Prince Sultan University, Riyadh, Saudi Arabia. In addition, the authors wish to acknowledge the editorial office and anonymous reviewers for their insightful comments, which have improved the quality of this publication.

Conflicts of Interest: The authors declare no conflict of interest.

References

1. Ahmed, S.; Wang, H.; Tian, Y. Modification to model reference adaptive control of 5-link exoskeleton with gravity compensation. In Proceedings of the Chinese Control Conference, Chengdu, China, 27 July 2016; pp. 6115–6120.
2. Hagh, Y. S.; Asl, R.M.; Cocquempot, V. A hybrid robust fault tolerant control based on adaptive joint unscented Kalman filter. *ISA Trans.* **2017**, *66*, 262–274. [CrossRef]

1. Wang, N.; Ahn, C.K. Coordinated trajectory-tracking control of a marine aerial-surface heterogeneous system. *IEEE/ASME Trans. Mechatronics* **2021**, *26*, 3198–3210. [CrossRef]
2. Zhao, D.; Li, S.; Gao, F. A new terminal sliding mode control for robotic manipulators. *Int. J. Control* **2009**, *82*, 1804–1813. [CrossRef]
3. Feng, Y.; Yu, X.; Man, Z. Non-singular terminal sliding mode control of rigid manipulators. *Automatica* **2002**, *38*, 2159–2167. [CrossRef]
4. Yang, L.; Yang, J. Nonsingular fast terminal sliding-mode control for nonlinear dynamical systems. *Int. J. Robust Nonlinear Control.* **2011**, *21*, 1865–1879. [CrossRef]
5. Moulay, E.; Lechappe, V.; Bernuau, E.; Defoort, M.; Plestan, F. Fixed-time sliding mode control with mismatched disturbances. *Automatica* **2022**, *136*, 110009 [CrossRef]
6. Ton, C.; Petersen, C. Continuous fixed-time sliding mode control for spacecraft with flexible appendages. *IFAC-PapersOnLine* **2018**, *51*, 1–5. [CrossRef]
7. Chavez-Vazquez, S.; Gomez-Aguilar, J.F.; Lavin-Delgado, J.E.; Escobar-Jimenez, R.F.; Olivares-Peregrino, V.H. Applications of fractional operators in robotics: A review. *J. Intell. Robot. Syst.* **2022**, *104*, 1–40. [CrossRef]
8. Ouannas, A.; Azar, A.T.; Ziar, T. On Inverse Full State Hybrid Function Projective Synchronization for Continuous-time Chaotic Dynamical Systems with Arbitrary Dimensions. *Differ. Equ. Dyn. Syst.* **2020**, *28*, 1045–1058. [CrossRef]
9. Ouannas, A.; Azar, A.T.; Ziar, T.; Radwan, A.G. *Generalized Synchronization of Different Dimensional Integer-order and Fractional Order Chaotic Systems. Studies in Computational Intelligence*; Springer: Berlin/Heidelberg, Germany, 2017; Volume 688, pp. 671–697.
10. Azar, A.T.; Radwan, A.G.; Vaidyanathan, S. *Mathematical Techniques of Fractional Order Systems*; Elsevier: Amsterdam, The Netherlands, 2018; ISBN 9780128135921.
11. Azar, A.T.; Radwan, A.G.; Vaidyanathan, S. *Fractional Order Systems: Optimization, Control, Circuit Realizations and Applications*; Elsevier: Amsterdam, The Netherlands, 2018; ISBN: 9780128161524.
12. Zhang, Q.; Li, Y.; Shang, Y.; Duan, B.; Cui, N.; Zhang, C. A fractional-Order kinetic battery model of lithium-Ion batteries considering a nonlinear capacity. *Electronics* **2019**, *8*, 394. [CrossRef]
13. Magin, R. Fractional calculus in bioengineering, part 1. *Crit. Rev. Biomed. Eng.* **2004**, *32*.
14. Tarasov, V.E. Mathematical economics: Application of fractional calculus. *Mathematics* **2020**, *8*, 660. [CrossRef]
15. Tapadar, A.; Khanday, F.A.; Sen, S.; Adhikary, A. Fractional calculus in electronic circuits: A review. *Fract. Order Syst.* **2022**, *1*, 441–482.
16. Radwan, A.G.; Emira, A.A.; Abdelaty, A.; Azar, A.T. Modeling and Analysis of Fractional Order DC-DC Converter. *ISA Trans.* **2018**, *82*, 184–199. [CrossRef] [PubMed]
17. Meghni, B.; Dib, D.; Azar, A.T.; Ghoudelbourk, S.; Saadoun, A. *Robust Adaptive Supervisory Fractional order Controller For optimal Energy Management in Wind Turbine with Battery Storage. Studies in Computational Intelligence*; Springer: Berlin/Heidelberg, Germany, 2017; Volume 688, pp. 165–202.
18. Ouannas, A.; Azar, A.T.; Ziar, T.; Vaidyanathan, S. *Fractional Inverse Generalized Chaos Synchronization Between Different Dimensional Systems. Studies in Computational Intelligence*; Springer: Berlin/Heidelberg, Germany, 2017; Volume 688, pp. 525–551.
19. Ouannas, A., Azar, A.T.; Ziar, T.; Vaidyanathan, S. *A New Method To Synchronize Fractional Chaotic Systems With Different Dimensions. Studies in Computational Intelligence*; Springer: Berlin/Heidelberg, Germany, 2017; Volume 688, pp. 581–611.
20. Ibraheem, G.A.R.; Azar, A.T.; Ibraheem, I.K.; Humaidi, A.J. A Novel Design of a Neural Network based Fractional PID Controller for Mobile Robots Using Hybridized Fruit Fly and Particle Swarm Optimization. *Complexity* **2020**, *2020*, 1–18. [CrossRef]
21. Gorripotu, T.S.; Samalla, H.; Jagan Mohana Rao, C.; Azar, A.T.; Pelusi, D. TLBO Algorithm optimized fractional-order PID controller for AGC of interconnected power system. In *Soft Computing in Data Analytics. Advances in Intelligent Systems and Computing*; Nayak, J., Abraham, A., Krishna, B., Chandra Sekhar, G., Das, A., Eds.; Springer: Singapore, 2019; Volume 758, pp. 847–855.
22. Daraz, A.; Malik, S.A.; Azar, A.T.; Aslam, S.; Alkhalifah, T.; Alturise, F. Optimized Fractional Order Integral-Tilt Derivative Controller for Frequency Regulation of Interconnected Diverse Renewable Energy Resources. *IEEE Access* **2022**, *10*, 43514–43527. [CrossRef]
23. Ahmed, S.; Wang, H.; Tian, Y. Fault tolerant control using fractional-order terminal sliding mode control for robotic manipulators. *Stud. Inform. Control.* **2018**, *27*, 55–64. [CrossRef]
24. Abro, G.E.M.; Zulkifli, S.A.B.; Asirvadam, V.S.; Ali, Z.A. Model-free-based single-dimension fuzzy SMC design for underactuated quadrotor UAV. *Actuators* **2021**, *10*, 191. [CrossRef]
25. Tepljakov, A.; Alagoz, B.B.; Yeroglu, C.; Gonzalez, E.; HosseinNia, S.H.; Petlenkov, E. FOPID controllers and their industrial applications: a survey of recent results. *IFAC-PapersOnLine* **2018**, *51*, 25–30. [CrossRef]
26. Fei, J.; Wang, H.; Fang, Y. Novel neural network fractional-order sliding-mode control with application to active power filter. *IEEE Trans. Syst. Man Cybern. Syst.* **2021**, *52*, 3508–3518. [CrossRef]
27. Zheng, W.; Luo, Y.; Chen, Y.; Wang, X. A simplified fractional order PID controller's optimal tuning: A case study on a PMSM speed servo. *Entropy* **2021**, *23*, 130. [CrossRef]
28. Dadras, S.; Momeni, H.R. Fractional terminal sliding mode control design for a class of dynamical systems with uncertainty. *Commun. Nonlinear Sci. Numer. Simul.* **2012**, *17*, 367–377. [CrossRef]

31. Ahmed, S.; Ahmed, A.; Mansoor, I.; Junejo, F.; Saeed, A. Output feedback adaptive fractional-order super-twisting sliding mode control of robotic manipulator. *Iran. J. Sci. Technol. Trans. Electr. Eng.* **2021**, *45*, 335–347. [CrossRef]
32. Fei, J.; Wang, Z.; Liang, X. Robust adaptive fractional fast terminal sliding mode controller for microgyroscope. *Complexity* **2020**, *2020*. [CrossRef]
33. Ni, J.; Liu, L.; Liu, C.; Hu, X. Fractional order fixed-time nonsingular terminal sliding mode synchronization and control of fractional order chaotic systems. *Nonlinear Dyn.* **2017**, *89*, 2065–2083. [CrossRef]
34. Chen, D.; Zhang, J.; Li, Z. A novel fixed-time trajectory tracking strategy of unmanned surface vessel based on the fractional sliding mode control method. *Electronics* **2022**, *11*, 726. [CrossRef]
35. Labbadi, M.; Boubaker, S.; Djemai, M.; Mekni, S.K.; Bekrar, A. Fixed-Time Fractional-Order Global Sliding Mode Control for Nonholonomic Mobile Robot Systems under External Disturbances. *Fractal Fract.* **2022**, *6*, 177. [CrossRef]
36. Huang, S.; Xiong, L.; Wang, J.; Li, P.; Wang, Z.; Ma, M. Fixed-time fractional-order sliding mode controller for multimachine power systems. *IEEE Trans. Power Syst.* **2020**, *36*, 2866–2876. [CrossRef]
37. Tao, G. Multivariable adaptive control: A survey. *Automatica* **2014**, *50*, 2737–2764. [CrossRef]
38. Wang, N.; Qian, C.; Sun, J.C.; Liu, Y.C. Adaptive robust finite-time trajectory tracking control of fully actuated marine surface vehicles. *IEEE Trans. Control. Syst. Technol.* **2015**, *24*, 1454–1462. [CrossRef]
39. Wang, N.; Su, S.F.; Han, M.; Chen, W.H. Backpropagating constraints-based trajectory tracking control of a quadrotor with constrained actuator dynamics and complex unknowns. *IEEE Trans. Syst. Man Cybern. Syst.* **2018**, *49*, 1322–1337. [CrossRef]
40. Lavretsky, E.; Wise, K.A. *Robust Adaptive Control*; Springer: London, UK, 2013; pp. 317–353.
41. Tian, X.; Fei, S. Robust control of a class of uncertain fractional-order chaotic systems with input nonlinearity via an adaptive sliding mode technique. *Entropy* **2014**, *16*, 729–746. [CrossRef]
42. Zhang, X.; Quan, Y. Adaptive fractional-order non-singular fast terminal sliding mode control based on fixed time disturbance observer for manipulators. *IEEE Access* **2022**, *10*, 76504–76511. [CrossRef]
43. Lopes, A.M.; Machado, J.A.T. A review of fractional order entropies. *Entropy* **2020**, *22*, 1374. [CrossRef]
44. Zhai, J.; Li, Z. Fast-exponential sliding mode control of robotic manipulator with super-twisting method. *IEEE Trans. Circuits Syst. II Express Briefs* **2021**, *69*, 489–493. [CrossRef]
45. Yin, C.; Huang, X.; Chen, Y.; Dadras, S.; Zhong, S.M.; Cheng, Y. Fractional-order exponential switching technique to enhance sliding mode control. *Appl. Math. Model.* **2017**, *44*, 705–726. [CrossRef]
46. Ahmed, S. Robust model reference adaptive control for five-link robotic exoskeleton. *Int. J. Model. Identif. Control.* **2021**, *39*, 324–331. [CrossRef]
47. Han, Z.; Zhang, K.; Yang, T.; Zhang, M. Spacecraft fault-tolerant control using adaptive non-singular fast terminal sliding mode. *IET Control Theory Appl.* **2010**, *10*, 1991–1999. [CrossRef]
48. Armstrong, B.; Khatib, O.; Burdick, J. April. The explicit dynamic model and inertial parameters of the PUMA 560 arm. In Proceedings of the 1986 IEEE International Conference on Robotics and Automation, San Francisco, CA, USA, 7–10 April 1986; Volume 3, pp. 510–518.
49. Ahmed, S.; Wang, H.; Tian, Y. Adaptive high-order terminal sliding mode control based on time delay estimation for the robotic manipulators with backlash hysteresis. *IEEE Trans. Syst. Man Cybern. Syst.* **2019**, *51*, 1128–1137. [CrossRef]

Article

Adaptive Fixed-Time Neural Networks Control for Pure-Feedback Non-Affine Nonlinear Systems with State Constraints

Yang Li [1], Quanmin Zhu [2], Jianhua Zhang [1,*] and Zhaopeng Deng [1]

1. School of Information and Control Engineering, Qingdao University of Technology, Qingdao 266525, China; yang_li@qut.edu.cn (Y.L.); dengzhaopeng@qut.edu.cn (Z.D.)
2. Department of Engineering Design and Mathematics, University of the West of England, Coldharbour Lane, Bristol BS16 1QY, UK; quan.zhu@uwe.ac.uk
* Correspondence: jianhuazhang@qut.edu.cn

Abstract: A new fixed-time adaptive neural network control strategy is designed for pure-feedback non-affine nonlinear systems with state constraints according to the feedback signal of the error system. Based on the adaptive backstepping technology, the Lyapunov function is designed for each subsystem. The neural network is used to identify the unknown parameters of the system in a fixed-time, and the designed control strategy makes the output signal of the system track the expected signal in a fixed-time. Through the stability analysis, it is proved that the tracking error converges in a fixed-time, and the design of the upper bound of the setting time of the error system only needs to modify the parameters and adaptive law of the controlled system controller, which does not depend on the initial conditions.

Keywords: adaptive control; neural network control; nonlinear constraint systems; non-affine nonlinear systems; pure feedback

Citation: Li, Y.; Zhu, Q.; Zhang, J.; Deng, Z. Adaptive Fixed-Time Neural Networks Control for Pure-Feedback Non-Affine Nonlinear Systems with State Constraints. *Entropy* 2022, 24, 737. https://doi.org/10.3390/e24050737

Academic Editor: Katalin M. Hangos

Received: 9 April 2022
Accepted: 20 May 2022
Published: 22 May 2022

Publisher's Note: MDPI stays neutral with regard to jurisdictional claims in published maps and institutional affiliations.

Copyright: © 2022 by the authors. Licensee MDPI, Basel, Switzerland. This article is an open access article distributed under the terms and conditions of the Creative Commons Attribution (CC BY) license (https://creativecommons.org/licenses/by/4.0/).

1. Introduction

In recent years, great breakthroughs have been made in the research of adaptive trajectory tracking control for uncertain nonlinear systems [1–3]. When solving such problems, neural network technology has become the key technology [4–6]. Combining neural network technology with backstepping control and adaptive control, the results have been widely used in different types of nonlinear systems such as strict feedback and pure feedback [7–9]. With the development of increasing power integrators, great progress has been made in the research of non-affine nonlinear systems. In recent years, the problems studied include output feedback stability, state output constraints, etc. Many methods have been introduced to solve these problems, such as backstepping technology, adaptive technology, and neural network control [10–12]. For nonlinear systems with time delays, the authors of reference [13] designed the control strategy by combining adaptive neural network and backstepping technology, and then the neural network technology based on adaptive backstepping was developed and applied [14–16].

With the development of society, the accuracy requirements of industrial control systems for convergence time are increasing. For example, in antimissile control systems, aircraft attitude control systems, and robot control systems, the purpose of controller design is to realize the stability of the controlled system and maintain stability in finite time (for example, in antimissile control systems, there is no need for control after missile explosion). For nonlinear systems with uncertainties, researchers have combined fixed-time controls with adaptive neural network technology to produce many excellent control schemes [17–19].

Researchers combine neural networks with adaptive control for online identification of complex nonlinear objects. In the design of these control systems, neural networks are generally used to approximate the uncertain nonlinear terms of the system, and neural networks are effective in compact sets [20–22]. In recent years, some fixed-time control methods based on the neural networks have been developed [23–25]. The author of reference [26] studies the control method of unknown nonlinear systems based on free model control. Based on Lyapunov functional and analysis technology, combined with advanced control algorithms, sufficient conditions for the master–slave memristor systems to realize timing synchronization are established. The authors of reference [27] extended the method to time-varying delay discontinuous fuzzy inertial neural network fixed-time synchronous control.

Although the research on fixed-time adaptive neural network control has produced a series of research results, there are still many problems to be solved in the existing control strategies, such as system constraints. In reference [28–30], the control problem of constrained nonlinear systems is discussed. For the control problem of systems with state constraints, the difficulty of constraints can be solved by using the boundary Lyapunov function. However, for the control problem of constrained non-affine nonlinear systems, the control strategies in the above literature cannot be used directly, and the research results based on fixed-time control are relatively few.

In summary, when there are state constraints in non-affine nonlinear systems, how to combine the adaptive neural network control and backstepping control to design effective control strategies so that the system can achieve the expected performance in fixed time, with the setting time not depending on the initial state of the system, is a problem. To solve this problem, some control problems have not been solved, such as for pure-feedback non-affine nonlinear systems, how to combine the backstepping method with Lyapunov function theory to design a fixed-time adaptive neural network tracking control strategy, so that the system output can track the desired signal and maintain fixed-time stability, the control performance can be guaranteed without initial conditions, and all state variables are bounded to a fixed region.

This article consists of the following parts. In Section 2, a constrained nonlinear system mathematical description of the problem is presented. In Section 3, firstly, the novel fixed-time stability theorem for constrained nonlinear systems is proposed, secondly, the adaptive neural network fixed-time tracking control scheme for constrained nonlinear systems is presented. In Section 4, the performances of the tracking control scheme are illustrated by a simulation example. In Section 5, some conclusions of the article are summarized.

2. Problem Formation and Preliminaries

Based on backstepping technology, combined with an adaptive neural network and fixed-time control, the tracking control of pure-feedback non-affine nonlinear interconnected systems was studied. Consider pure-feedback nonlinear systems:

$$\begin{cases} \dot{x}_i(t) = f_i(\overline{x}_{i+1}(t)), i = 1, 2, \ldots, n-1 \\ \dot{x}_n(t) = f_n(\overline{x}_n(t), u(t)) \\ y(t) = x_1(t) \end{cases} \quad (1)$$

where $x = \begin{pmatrix} x_1 & x_2 & \cdots & x_n \end{pmatrix}^T \in \Re^n$, $u \in \Re$, $y \in \Re$, indicate the state, control and output, respectively, $f_i(\cdot), i = 1, 2, \ldots, n$ are nonlinear smooth functions, $y_d \in \Re$ is desired trajectory.

Remark 1. *Based on the existing algorithms, this article attempts to further design a novel neural network adaptive control algorithm. The control objective of the algorithm is the output of the pure-feedback non-affine nonlinear system that can track the desired signal and maintain fixed-time stability. The designed upper bound of the setting time does not rely on the initial parameters, only by adjusting the parameters of the controller.*

Lemma 1 [6]. *For $x_i \in R$ and $x_i \geq 0$, $i = 1, 2, \cdots, n$, $0 < p < 1$, $q > 1$, then*

$$\left(\sum_{i=1}^{n} x_i\right)^p \leq \sum_{i=1}^{n} x_i^p \leq n^{1-p}\left(\sum_{i=1}^{n} x_i\right)^p \tag{2}$$

$$n^{1-q}\left(\sum_{i=1}^{n} x_i\right)^q \leq \sum_{i=1}^{n} x_i^q \leq \left(\sum_{i=1}^{n} x_i\right)^q \tag{3}$$

3. Main Results

The control algorithm was designed for the system (1). The objective of the control was to propose a new adaptive fixed-time neural network tracking control algorithm for the pure-feedback nonlinear system. Adaptive neural network technology is used to solve the uncertainty of the unknown system. Under the proposed control scheme, through the Lyapunov stability analysis, the closed system is fixed-time stability.

For a nonlinear system (1), combine homeomorphism mapping and backstepping control to design constraint control, in the first step, consider system state

$$z_1 = x_1 - y_d \tag{4}$$

Design homeomorphism mapping

$$\zeta_1 = \text{arctanh}\left(\frac{z_1}{k_{b1}}\right) \tag{5}$$

where $k_{b1} > 0$ is the bound of z_1 and satisfy the $|z_1| < k_{b1}$, then the system can obtain

$$\begin{aligned}\dot{\zeta}_1 &= \frac{k_{b1}}{k_{b1}^2 - z_1^2}\dot{z}_1 \\ &= \frac{k_{b1}}{k_{b1}^2 - z_1^2}(f_1(x_1, x_2) - \dot{y}_d)\end{aligned} \tag{6}$$

Choose the NN to approximate the nonlinear system $f_1(\bar{x}_2), \bar{x}_2 \in \Omega_1 \subset \Re^2$ and Ω_1 is compact set

$$\frac{k_{b1}}{k_{b1}^2 - z_1^2}(f_1(x_1, x_2) - \dot{y}_d) = W_1^{T*}\Psi_1(\bar{x}_2) + \varepsilon_1 \tag{7}$$

where $\|W_1^*\| = \theta_1$, $\hat{\theta}_1$ is estimation of θ_1 and $\tilde{\theta}_1 = \hat{\theta}_1 - \theta_1$, then we have

$$W_1^{*T}\Psi_1(\bar{x}_2) \leq \theta_1 \|\Psi_1(\bar{x}_2)\| \tag{8}$$

Define a Lyapunov functional candidate as

$$V_1 = \frac{1}{2}\zeta_1^2 + \frac{1}{2\mu_1}\tilde{\theta}_1^2 \tag{9}$$

take the time derivative (9) along the trajectory of (6) as

$$\begin{aligned}\dot{V}_1 &= \zeta_1 \frac{k_{b1}}{k_{b1}^2 - z_1^2}(f_1(x_1, x_2) - \dot{y}_d) + \frac{1}{\mu_1}\tilde{\theta}_1\dot{\hat{\theta}}_1 \\ &= \zeta_1(W_1^{*T}\Psi_1(\bar{x}_2) + \varepsilon_1) + \frac{1}{\mu_1}\tilde{\theta}_1\dot{\hat{\theta}}_1\end{aligned} \tag{10}$$

Choose the virtual control law

$$\zeta_2 = k_1\zeta_1 + k_{p1}\zeta_1^p + k_{q1}\zeta_1^q + \text{sign}(\zeta_1)\hat{\theta}_1\|\Psi_1(\bar{x}_2)\| \tag{11}$$

where $k_1 > \frac{1}{2}, k_{p1} > 0, k_{q1} > 0, 0 < p < 1, q > 1$, based on homeomorphism mapping

$$z_2 = k_{b2}\tanh(\xi_2) \tag{12}$$

where $k_{b2} > 0$ is the bound of z_2 and satisfies the $|z_2| < k_{b2}$, and

$$\alpha_1 = x_2 - z_2 \tag{13}$$

then we have

$$\begin{aligned}\dot{V}_1 &= \xi_1 W_1^{*T}\Psi_1(\bar{x}_2) + \xi_1\varepsilon_1 - k_1\xi_1^2 - k_{p1}\xi_1^{p+1} - k_{q1}\xi_1^{q+1} \\ &\quad - \hat{\theta}_1|\xi_1|\|\Psi_1(\bar{x}_2)\| + \xi_1\xi_2 + \frac{1}{\mu_1}\tilde{\theta}_1\dot{\hat{\theta}}_1\end{aligned} \tag{14}$$

when $\|W_1^*\| = \theta_1$, we have

$$\xi_1 W_1^{*T}\Psi_1(\bar{x}_2) \le \theta_1|\xi_1|\|\Psi_1(\bar{x}_2)\| \tag{15}$$

and

$$\xi_1\varepsilon_1 \le \frac{1}{2}\xi_1^2 + \frac{1}{2}\varepsilon_1^2 \tag{16}$$

then we have

$$\begin{aligned}\dot{V}_1 &= -\tilde{\theta}_1|\xi_1|\|\Psi_1(\bar{x}_2)\| + \frac{1}{2}\xi_1^2 + \frac{1}{2}\varepsilon_1^2 - k_1\xi_1^2 - k_{p1}\xi_1^{p+1} - k_{q1}\xi_1^{q+1} \\ &\quad + \xi_1\xi_2 + \frac{1}{\mu_1}\tilde{\theta}_1\dot{\hat{\theta}}_1\end{aligned} \tag{17}$$

where

$$\tilde{\theta}_1 = \hat{\theta}_1 - \theta_1 \tag{18}$$

Choose the NN adaptive law as

$$\dot{\hat{\theta}}_1 = \mu_1\left(|\xi_1|\|\Psi_1(\bar{x}_2)\| - \rho_{p1}\hat{\theta}_1^p - \rho_{q1}\hat{\theta}_1^q\right), \hat{\theta}_1(0) = 0 \tag{19}$$

where $\mu_1 > 0, \rho_{p1} > 0, \rho_{q1} > 0$, then we have

$$\begin{aligned}\dot{V}_1 &= -\tilde{\theta}_1|\xi_1|\|\Psi_1(\bar{x}_2)\| + \frac{1}{2}\xi_1^2 + \frac{1}{2}\varepsilon_1^2 - k_1\xi_1^2 - k_{p1}\xi_1^{p+1} - k_{q1}\xi_1^{q+1} \\ &\quad + \xi_1\xi_2 + \tilde{\theta}_1\left(|\xi_1|\|\Psi_1(\bar{x}_2)\| - \rho_{p1}\hat{\theta}_1^p - \rho_{q1}\hat{\theta}_1^q\right) \\ &= -\left(k_1 - \frac{1}{2}\right)\xi_1^2 - k_{p1}\xi_1^{p+1} - k_{q1}\xi_1^{q+1} + \xi_1\xi_2 \\ &\quad - \rho_{p1}\tilde{\theta}_1\hat{\theta}_1^p - \rho_{q1}\tilde{\theta}_1\hat{\theta}_1^q + \frac{1}{2}\varepsilon_1^2\end{aligned} \tag{20}$$

based on inequalities from [7], the following hold:

$$\begin{aligned}-\rho_{p1}\tilde{\theta}_1\hat{\theta}_1^p &\le -\varsigma_{p1}\tilde{\theta}_1^{p+1} + v_{p1}\theta_1^{p+1} \\ -\rho_{q1}\tilde{\theta}_1\hat{\theta}_1^q &\le -\varsigma_{q1}\tilde{\theta}_1^{q+1} + v_{q1}\theta_1^{q+1}\end{aligned} \tag{21}$$

where $\rho_{p1}, \varsigma_{p1}, v_{p1}, \rho_{q1}, \varsigma_{q1}, v_{q1} > 0$, therefore, we have

$$\begin{aligned}\dot{V}_1 &\le -\left(k_1 - \frac{1}{2}\right)\xi_1^2 - k_{p1}\xi_1^{p+1} - k_{q1}\xi_1^{q+1} + \xi_1\xi_2 - \varsigma_{p1}\tilde{\theta}_1^{p+1} + v_{p1}\theta_1^{p+1} \\ &\quad - \varsigma_{q1}\tilde{\theta}_1^{q+1} + v_{q1}\theta_1^{q+1} + \frac{1}{2}\varepsilon_1^2 \\ &\le -k_{p1}\xi_1^{p+1} - k_{q1}\xi_1^{q+1} - \varsigma_{p1}\tilde{\theta}_1^{p+1} - \varsigma_{q1}\tilde{\theta}_1^{q+1} + \xi_1\xi_2 + \delta_1\end{aligned} \tag{22}$$

where

$$\delta_1 = \frac{1}{2}\varepsilon_1^2 + v_{p1}\theta_1^{p+1} + v_{q1}\theta_1^{q+1} \tag{23}$$

The ith step $2 \leq i \leq n$, consider system state

$$z_i = x_i - \alpha_{i-1} \tag{24}$$

Design homeomorphism mapping

$$\varsigma_i = \text{arctanh}\left(\frac{z_i}{k_{bi}}\right) \tag{25}$$

where $k_{bi} > 0$ is the bound of z_i and satisfies the $|z_i| < k_{bi}$, then the system can obtain

$$\begin{aligned}\dot{\varsigma}_i &= \frac{k_{bi}}{k_{bi}^2 - z_i^2}\dot{z}_i \\ &= \frac{k_{bi}}{k_{bi}^2 - z_i^2}(f_i - \dot{\alpha}_{i-1})\end{aligned} \tag{26}$$

The neural network is constructed as $f_i, \bar{x}_{i+1} \in \Omega_i \subset \Re^{i+1}$ and Ω_i is compact set

$$\frac{k_{bi}}{k_{bi}^2 - z_i^2}(f_i - \dot{\alpha}_{i-1}) = W_i^{T*}\Psi_i(\bar{x}_{i+1}) + \varepsilon_i \tag{27}$$

where $\|W_i^*\| = \theta_i$, $\hat{\theta}_i$ is estimation of θ_i and $\tilde{\theta}_i = \hat{\theta}_i - \theta_i$, then we have

$$W_i^{*T}\Psi_1(\bar{x}_{i+1}) \leq \theta_i \|\Psi_i(\bar{x}_{i+1})\| \tag{28}$$

Define a Lyapunov functional candidate as

$$V_i = \frac{1}{2}\varsigma_i^2 + \frac{1}{2\mu_i}\tilde{\theta}_i^2 \tag{29}$$

Take the time derivative (29) along the trajectory of (26) as

$$\begin{aligned}\dot{V}_i &= \varsigma_i \frac{k_{bi}}{1-z_i^2}(f_i - \dot{\alpha}_{i-1}) + \frac{1}{\mu_i}\tilde{\theta}_i\dot{\hat{\theta}}_i \\ &= \varsigma_i(W_i^{*T}\Psi_i(\bar{x}_{i+1}) + \varepsilon_i) + \frac{1}{\mu_i}\tilde{\theta}_i\dot{\hat{\theta}}_i\end{aligned} \tag{30}$$

The virtual control signal is constructed as

$$\varsigma_{i+1} = \varsigma_{i-1} + k_i\varsigma_i + k_{pi}\varsigma_i^p + k_{qi}\varsigma_i^q + \text{sign}(\varsigma_i)\hat{\theta}_i\|\Psi_i(\bar{x}_{i+1})\| \tag{31}$$

where $k_i > \frac{1}{2}, k_{pi} > 0, k_{qi} > 0, 0 < p < 1, q > 1$, based on homeomorphism mapping

$$z_{i+1} = k_{bi+1}\tanh(\varsigma_{i+1}) \tag{32}$$

where $k_{bi+1} > 0$ is the bound of z_{i+1} and satisfies the $|z_{i+1}| < k_{bi+1}$ where

$$z_{i+1} = x_{i+1} - \alpha_i \tag{33}$$

and assume $x_{n+1} = u$, then we have

$$\begin{aligned}\dot{V}_i &= \varsigma_i W_i^{*T}\Psi_i(\bar{x}_{i+1}) + \varsigma_i\varepsilon_i - \varsigma_{i-1}\varsigma_i - k_i\varsigma_i^2 - k_{pi}\varsigma_i^{p+1} - k_{qi}\varsigma_i^{q+1} \\ &\quad - \hat{\theta}_i|\varsigma_i|\|\Psi_i(\bar{x}_{i+1})\| + \varsigma_i\varsigma_{i+1} + \frac{1}{\mu_i}\tilde{\theta}_i\dot{\hat{\theta}}_i\end{aligned} \tag{34}$$

when $\|W_i^*\| = \theta_i$, we have

$$\varsigma_i W_i^{*T}\Psi_i(\bar{x}_{i+1}) \leq \theta_i|\varsigma_i|\|\Psi_i(\bar{x}_{i+1})\| \tag{35}$$

and
$$\zeta_i \varepsilon_i \leq \frac{1}{2}\zeta_i^2 + \frac{1}{2}\varepsilon_i^2 \tag{36}$$

then we have
$$\begin{aligned}\dot{V}_i &= -\tilde{\theta}_i|\zeta_i|\|\Psi_i(\overline{x}_{i+1})\| + \frac{1}{2}\zeta_i^2 + \frac{1}{2}\varepsilon_i^2 - k_1\zeta_i^2 - k_{pi}\zeta_i^{p+1} - k_{qi}\zeta_i^{q+1} \\ &\quad - \zeta_{i-1}\zeta_i + \zeta_i\zeta_{i+1} + \frac{1}{\mu_i}\tilde{\theta}_i\dot{\hat{\theta}}_i\end{aligned} \tag{37}$$

where
$$\tilde{\theta}_i = \hat{\theta}_i - \theta_i \tag{38}$$

The NN adaptive signal is constructed as
$$\dot{\hat{\theta}}_i = \mu_i\left(|\zeta_i|\|\Psi_i(\overline{x}_{i+1})\| - \rho_{pi}\hat{\theta}_i^p - \rho_{qi}\hat{\theta}_i^q\right), \hat{\theta}_i(0) = 0 \tag{39}$$

where $\mu_i > 0, \rho_{pi} > 0, \rho_{qi} > 0$, then we have
$$\begin{aligned}\dot{V}_i &= -\tilde{\theta}_i|\zeta_i|\|\Psi_i(\overline{x}_{i+1})\| + \frac{1}{2}\zeta_i^2 + \frac{1}{2}\varepsilon_i^2 - k_i\zeta_i^2 - k_{pi}\zeta_i^{p+1} - k_{qi}\zeta_i^{q+1} \\ &\quad - \zeta_{i-1}\zeta_i + \zeta_i\zeta_{i+1} + \tilde{\theta}_i\left(|\zeta_i|\|\Psi_i(\overline{x}_{i+1})\| - \rho_{pi}\hat{\theta}_i^p - \rho_{qi}\hat{\theta}_i^q\right) \\ &= -\left(k_i - \frac{1}{2}\right)\zeta_i^2 - k_{pi}\zeta_i^{p+1} - k_{qi}\zeta_i^{q+1} - \zeta_{i-1}\zeta_i + \zeta_i\zeta_{i+1} \\ &\quad - \rho_{pi}\tilde{\theta}_i\hat{\theta}_i^p - \rho_{qi}\tilde{\theta}_i\hat{\theta}_i^q + \frac{1}{2}\varepsilon_i^2\end{aligned} \tag{40}$$

Based on inequalities from [7], the following hold:
$$\begin{aligned}-\rho_{pi}\tilde{\theta}_i\hat{\theta}_i^p &\leq -\varsigma_{pi}\tilde{\theta}_i^{p+1} + v_{pi}\theta_i^{p+1} \\ -\rho_{qi}\tilde{\theta}_i\hat{\theta}_i^q &\leq -\varsigma_{qi}\tilde{\theta}_i^{q+1} + v_{qi}\theta_i^{q+1}\end{aligned} \tag{41}$$

where $\rho_{pi}, \varsigma_{pi}, v_{pi}, \rho_{qi}, \varsigma_{qi}, v_{qi} > 0$, therefore we have
$$\begin{aligned}\dot{V}_i &\leq -\left(k_i - \frac{1}{2}\right)\zeta_i^2 - k_{pi}\zeta_i^{p+1} - k_{qi}\zeta_i^{q+1} - \zeta_{i-1}\zeta_i + \zeta_i\zeta_{i+1} - \varsigma_{pi}\tilde{\theta}_i^{p+1} \\ &\quad + v_{pi}\theta_i^{p+1} - \varsigma_{qi}\tilde{\theta}_i^{q+1} + v_{qi}\theta_i^{q+1} + \frac{1}{2}\varepsilon_i^2 \\ &\leq -\zeta_{i-1}\zeta_i - k_{pi}\zeta_i^{p+1} - k_{qi}\zeta_i^{q+1} - \varsigma_{pi}\tilde{\theta}_i^{p+1} - \varsigma_{qi}\tilde{\theta}_i^{q+1} + \zeta_i\zeta_{i+1} + \delta_i\end{aligned} \tag{42}$$

where
$$\delta_i = \frac{1}{2}\varepsilon_i^2 + v_{pi}\theta_i^{p+1} + v_{qi}\theta_i^{q+1} \tag{43}$$

The $n + 1$th step, this is the most important step.
$$z_{n+1} = u - \alpha_n \tag{44}$$

Based on system
$$\dot{z}_{n+1} = v - \dot{\alpha}_n \tag{45}$$

Design homeomorphism mapping
$$\zeta_{n+1} = \operatorname{arctanh}\left(\frac{z_{n+1}}{k_{bn+1}}\right) \tag{46}$$

where $k_{bn+1} > 0$ is the bound of z_{n+1} and satisfies the $|z_{n+1}| < k_{bn+1}$, then the system can obtain
$$\begin{aligned}\dot{\zeta}_{n+1} &= \frac{k_{bn+1}}{k_{bn+1}^2 - z_{n+1}^2}\dot{z}_{n+1} \\ &= \frac{k_{bn+1}}{k_{bn+1}^2 - z_{n+1}^2}(v - \dot{\alpha}_n)\end{aligned} \tag{47}$$

The neural network is constructed as $\dot{\alpha}_n, \bar{x}_{n+1} = (x, u) \in \Omega_{n+1} \subset \Re^{n+1}$ and Ω_{n+1} is compact set

$$\frac{k_{bn+1}}{k_{bn+1}^2 - z_{n+1}^2} \dot{\alpha}_n = W_{n+1}^{T*} \Psi_{n+1}(\bar{x}_{n+1}) + \varepsilon_{n+1} \quad (48)$$

where $\|W_{n+1}^*\| = \theta_{n+1}$, $\hat{\theta}_{n+1}$ is an estimation of θ_{n+1} and $\tilde{\theta}_{n+1} = \hat{\theta}_{n+1} - \theta_{n+1}$, then we have

$$W_{n+1}^{*T} \Psi_{n+1}(\bar{x}_{n+1}) \leq \theta_{n+1} \|\Psi_{n+1}(\bar{x}_{n+1})\| \quad (49)$$

Define a Lyapunov functional candidate as

$$V_{n+1} = \frac{1}{2}\zeta_{n+1}^2 + \frac{1}{2\mu_{n+1}}\tilde{\theta}_{n+1}^2 \quad (50)$$

Take the time derivative (9) along the trajectory of (6) as

$$\begin{aligned}\dot{V}_{n+1} &= \zeta_{n+1}\frac{k_{bn+1}}{k_{bn+1}^2-z_{n+1}^2}(v-\dot{\alpha}_i)+\frac{1}{\mu_{n+1}}\tilde{\theta}_{n+1}\dot{\hat{\theta}}_{n+1}\\ &= \zeta_{n+1}\frac{k_{bn+1}}{k_{bn+1}^2-z_{n+1}^2}v-\zeta_{n+1}\left(W_{n+1}^{*T}\Psi_{n+1}(\bar{x}_{n+1})+\varepsilon_{n+1}\right)\\ &\quad +\frac{1}{\mu_{n+1}}\tilde{\theta}_{n+1}\dot{\hat{\theta}}_{n+1}\end{aligned} \quad (51)$$

Choose the control

$$v = \frac{-\zeta_n - k_{n+1}\zeta_{n+1} - k_{pn+1}\zeta_{n+1}^p - k_{qn+1}\zeta_{n+1}^q - \operatorname{sign}(\zeta_{n+1})\hat{\theta}_{n+1}\|\Psi_{n+1}(\bar{x}_{n+1})\|}{\frac{k_{bn+1}}{k_{bn+1}^2-z_{n+1}^2}} \quad (52)$$

where $k_{n+1} > \frac{1}{2}, k_{pn+1} > 0, k_{qn+1} > 0, 0 < p < 1, q > 1$, then

$$\begin{aligned}\dot{V}_{n+1} &= -\zeta_{n+1}\left(W_{n+1}^{*T}\Psi_{n+1}(\bar{x}_{n+1})+\varepsilon_{n+1}\right)+\frac{1}{\mu_{n+1}}\tilde{\theta}_{n+1}\dot{\hat{\theta}}_{n+1}-\zeta_n\zeta_{n+1}-k_{n+1}\zeta_{n+1}^2\\ &\quad -k_{pn+1}\zeta_{n+1}^{p+1}-k_{qn+1}\zeta_{n+1}^{q+1}-\hat{\theta}_{n+1}|\zeta_{n+1}|\|\Psi_{n+1}(\bar{x}_{n+1})\|\end{aligned} \quad (53)$$

when $\|W_{n+1}^*\| = \theta_{n+1}$, we have

$$\zeta_{n+1}W_{n+1}^{*T}\Psi_{n+1}(\bar{x}_{n+1}) \leq \theta_{n+1}|\zeta_{n+1}|\|\Psi_{n+1}(\bar{x}_{n+1})\| \quad (54)$$

and

$$\zeta_{n+1}\varepsilon_{n+1} \leq \frac{1}{2}\zeta_{n+1}^2 + \frac{1}{2}\varepsilon_{n+1}^2 \quad (55)$$

then we have

$$\begin{aligned}\dot{V}_{n+1} &= -\tilde{\theta}_{n+1}|\zeta_{n+1}|\|\Psi_{n+1}(\bar{x}_{n+1})\|+\frac{1}{2}\zeta_{n+1}^2+\frac{1}{2}\varepsilon_{n+1}^2+\frac{1}{\mu_{n+1}}\tilde{\theta}_{n+1}\dot{\hat{\theta}}_{n+1}\\ &\quad -\zeta_n\zeta_{n+1}-k_{n+1}\zeta_{n+1}^2-k_{pn+1}\zeta_{n+1}^{p+1}-k_{qn+1}\zeta_{n+1}^{q+1}\end{aligned} \quad (56)$$

where

$$\tilde{\theta}_{n+1} = \hat{\theta}_{n+1} - \theta_{n+1} \quad (57)$$

choose the NN adaptive law as

$$\dot{\hat{\theta}}_{n+1} = \mu_{n+1}\left(|\zeta_{n+1}|\|\Psi_{n+1}(\bar{x}_{n+1})\| - \rho_{pn+1}\hat{\theta}_{n+1}^p - \rho_{qn+1}\hat{\theta}_{n+1}^q\right), \hat{\theta}_{n+1}(0) = 0 \quad (58)$$

where $\mu_{n+1} > 0, \rho_{pn+1} > 0, \rho_{qn+1} > 0$, then we have

$$\begin{aligned}\dot{V}_{n+1} &= -\tilde{\theta}_{n+1}|\xi_{n+1}|\|\Psi_{n+1}(\bar{x}_{n+1})\| + \tfrac{1}{2}\xi_{n+1}^2 + \tfrac{1}{2}\varepsilon_{n+1}^2 \\ &+ \tilde{\theta}_{n+1}\left(|\xi_{n+1}|\|\Psi_{n+1}(\bar{x}_{n+1})\| - \rho_{pn+1}\hat{\theta}_{n+1}^p - \rho_{qn+1}\hat{\theta}_{n+1}^q\right) \\ &- \zeta_n\xi_{n+1} - k_{n+1}\xi_{n+1}^2 - k_{pn+1}\xi_{n+1}^{p+1} - k_{qn+1}\xi_{n+1}^{q+1}\end{aligned} \quad (59)$$

based on inequalities from [7], the following hold:

$$\begin{aligned}-\rho_{pn+1}\tilde{\theta}_{n+1}\hat{\theta}_{n+1}^p &\leq -\varsigma_{pn+1}\tilde{\theta}_{n+1}^{p+1} + v_{pn+1}\theta_{n+1}^{p+1} \\ -\rho_{qn+1}\tilde{\theta}_{n+1}\hat{\theta}_{n+1}^q &\leq -\varsigma_{qn+1}\tilde{\theta}_{n+1}^{q+1} + v_{qn+1}\theta_{n+1}^{q+1}\end{aligned} \quad (60)$$

where $\rho_{pn+1}, \varsigma_{pn+1}, v_{pn+1}, \rho_{qn+1}, \varsigma_{qn+1}, v_{qn+1} > 0$, therefore, we have

$$\begin{aligned}\dot{V}_{n+1} &\leq -\left(k_{n+1} - \tfrac{1}{2}\right)\xi_{n+1}^2 + \tfrac{1}{2}\varepsilon_{n+1}^2 - \varsigma_{pn+1}\tilde{\theta}_{n+1}^{p+1} + v_{pn+1}\theta_{n+1}^{p+1} \\ &- \varsigma_{qn+1}\tilde{\theta}_{n+1}^{q+1} + v_{qn+1}\theta_{n+1}^{q+1} - \zeta_n\xi_{n+1} - k_{pn+1}\xi_{n+1}^{p+1} - k_{qn+1}\xi_{n+1}^{q+1} \\ &\leq -\zeta_n\xi_{n+1} - k_{pn+1}\xi_{n+1}^{p+1} - k_{qn+1}\xi_{n+1}^{q+1} - \varsigma_{pn+1}\tilde{\theta}_{n+1}^{p+1} - \varsigma_{qn+1}\tilde{\theta}_{n+1}^{q+1} + \delta_{n+1}\end{aligned} \quad (61)$$

where

$$\delta_{n+1} = \frac{1}{2}\varepsilon_{n+1}^2 + v_{pn+1}\theta_{n+1}^{p+1} + v_{qn+1}\theta_{n+1}^{q+1} \quad (62)$$

Theorem 1. *Consider the non-affine pure-feedback nonlinear system (1), based on the homeomorphism mapping and adaptive fixed-time neural network control scheme, choose the virtual control law as (8), (27), the adaptive fixed-time law (16) as (35), and the actual controller as (47). The tracking error system is practical fixed-time stability, and the upper bound of the settling time T is independent of the initial parameters. The settling time T satisfies*

$$T \leq T_{\max} = \frac{2^{\frac{3-p}{2}}}{k_p(1-p)} + \frac{2}{k_q(q-1)} \quad (63)$$

Proof. Select the following Lyapunov function

$$V = \sum_{i=1}^{n+1} V_i \quad (64)$$

then it has

$$\dot{V} \leq -\sum_{i=1}^{n+1}\left(k_{pi}\xi_i^{p+1} + \varsigma_{pi}\tilde{\theta}_i^{p+1}\right) - \sum_{i=1}^{n+1}\left(k_{qi}\xi_i^{q+1} + \varsigma_{qi}\tilde{\theta}_i^{q+1}\right) + \sum_{i=1}^{n+1}\delta_i \quad (65)$$

Based on Lemma 1

$$\begin{aligned}\sum_{i=1}^{n+1}\left(k_{pi}\xi_i^{p+1} + \varsigma_{pi}\tilde{\theta}_i^{p+1}\right) &\geq k_p\left(\sum_{i=1}^{n+1}\left(\tfrac{\xi_i^2}{2} + \tfrac{1}{2\mu_i}\tilde{\theta}_i^2\right)\right)^{\frac{p+1}{2}} \\ \sum_{i=1}^{n+1}\left(k_{qi}\xi_i^{q+1} + \varsigma_{qi}\tilde{\theta}_i^{q+1}\right) &\geq k_q\left(\sum_{i=1}^{n+1}\left(\tfrac{\xi_i^2}{2} + \tfrac{1}{2\mu_i}\tilde{\theta}_i^2\right)\right)^{\frac{p+1}{2}}\end{aligned} \quad (66)$$

where

$$\begin{aligned}k_p &= \min\left(2^{\frac{p+1}{2}}k_{pi}, 2^{\frac{p+1}{2}}\mu_i^{\frac{p+1}{2}}\varsigma_{pi}\right), i = 1, 2, 3 \cdots n+1 \\ k_q &= \min\left(2(n+1)^{\frac{1-q}{2}}k_{qi}, 2(n+1)^{\frac{1-q}{2}}\mu_i^{\frac{q+1}{2}}\varsigma_{qi}\right), i = 1, 2, 3 \cdots n+1 \\ \delta &= \sum_{i=1}^{n+1}\delta_i\end{aligned} \quad (67)$$

$$\dot{V} \leq -k_p V^{\frac{p+1}{2}} - k_q V^{\frac{q+1}{2}} + \delta \qquad (68)$$

based on Lemma in [6], the system is practically fixed-time stability. □

Remark 2. A new adaptive neural network control strategy is designed. The control objective is to drive the output signal of the error system to track the expected signal in a fixed-time. The neural network is used to approximate the unknown function of the system and design a fixed-time adaptive law to update the weight of the neural network. Without considering the initial conditions, the setting time can be designed by selecting the controller parameters. Based on the fixed-time stability theory, it is proved that the controller can realize the fixed-time stability of the closed-loop system.

Remark 3. The control deviation is obtained from the given value and the actual output value of the system, the fixed-time adaptive laws are designed by the homeomorphic mapping of the deviation, and the neural network weights are trained through the adaptive rate to form the control signal, to change the regulation quality of the system. This forms a fixed-time adaptive neural network control system, and its control structure is shown in Figure 1.

Remark 4. Programming according to the control algorithm described in equation to Equations (4), (23), (43) and the program block diagram is shown in Figure 2

Step 1: Calculate the control deviation z_i by value and output value.
Step 2: Calculate ζ_i according to the principle of homeomorphic mapping.
Step 3: Design the fixed-time adaptive laws to train the weights of the neural network.
Step 4: Design the neural network to estimate the nonlinear system.
Step 5: Repeat Step 1 to Step 4 when $i \leq n+1$.
Step 6: The control variables are determined based on backstepping control.

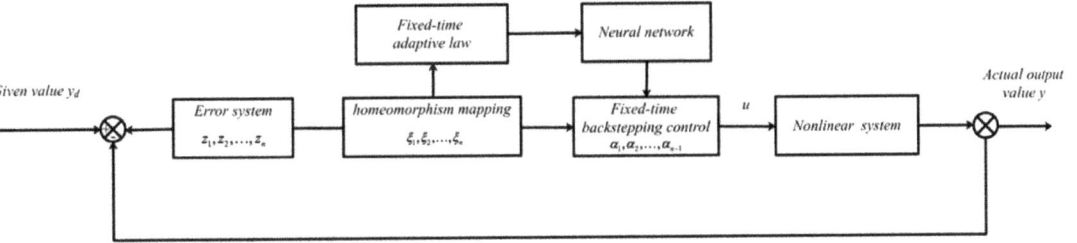

Figure 1. Fixed-time adaptive neural network control system.

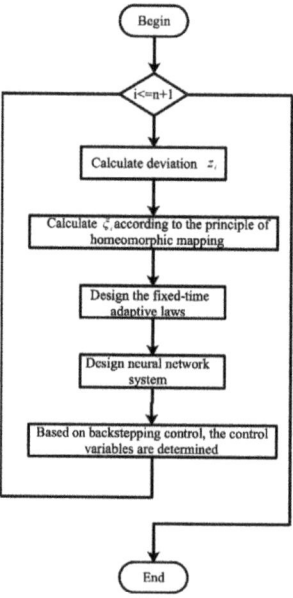

Figure 2. Fixed-time adaptive neural network control algorithm.

4. Numerical Examples

This section gives two examples to show the effectiveness of the proposed control scheme.

A. Mathematical example

The nonlinear dynamics is

$$\begin{aligned}\dot{x}_1 &= x_2 \sin(x_1) + (3 + x_1^2)(x_2 + x_2^3) \\ \dot{x}_2 &= 2x_1 x_2 (\sin(x_1) + x_2^2) + x_3 + \frac{x_3^3}{7} \\ \dot{x}_3 &= x_1 + x_1 x_2 + x_3 + u \\ y &= x_1 \end{aligned} \qquad (69)$$

Consider the system state

$$z_1 = y - y_d \qquad (70)$$

Choose the homeomorphism mapping

$$\zeta_1 = \operatorname{arctanh}(z_1) \qquad (71)$$

and adaptive functions have the following form:

$$\zeta_2 = \zeta_1 + \zeta_1^{\frac{3}{5}} + \zeta_1^{\frac{5}{3}} + sign(\zeta_1)\hat{\theta}_1 \|\Psi_1(\bar{x}_2)\| \qquad (72)$$

and controller has the following form:

$$u = \left(1 - z_3^2\right)\left(-\zeta_2 - \zeta_3 - \zeta_3^{\frac{3}{5}} - \zeta_3^{\frac{5}{3}} - sign(\zeta_3)\hat{\theta}_3 \|\Psi_3(\bar{x}_3)\|\right) \qquad (73)$$

where $y_d = \sin(t)$ being the desired signal. Select the initial parameters as $x = (1, 0, 0)^T$, and the neural network parameters chosen zeros.

The simulation results are shown in Figures 3–6. Figure 3 depicts the tracking curve of the given value and output value. It can be seen from the figure that the tracking error can be sufficiently small in fixed-time and the system output is bounded. Figure 4 shows that the system state is bounded and can converge to zero in fixed time. Figure 5 depicts the tracking errors' tracking curve, which shows that the tracking errors are bounded. Because $\tanh(\xi_i) = z_i, i = 1,2,3$, therefore, the system states $z_i, i = 1,2,3$ are bounded with $|z_i| < 1$. Figure 6 shows the time response of the output, the output is bounded, and its value is constant after a fixed time.

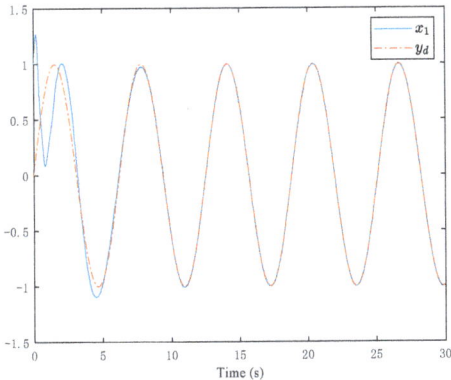

Figure 3. Trajectories of the output and the desired signal.

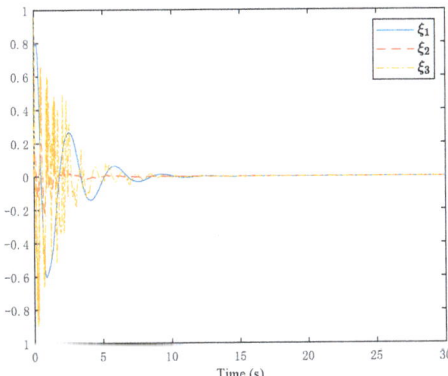

Figure 4. Trajectories of the homeomorphism mapping states.

B. Robot model

Consider a robot model [31] is

$$M_r \ddot{q}_r + \frac{1}{2} m_r g l_r \sin(q_r) = \tau_r \tag{74}$$

where q_r is angle displacement, g and M_r are the gravitational acceleration and moment of inertia, respectively, and m_r is the mass of link and l_r represents its length, τ_r is the considered input torque. If $x_1 = q_r, x_2 = \dot{q}_r$, and $u = \tau_r$, the dynamic system can be transformed as follows:

$$\begin{cases} \dot{x}_1 = x_2 \\ \dot{x}_2 = -\frac{mgl_r}{2M_r} \sin(x_1) + \frac{1}{M_r} u \end{cases} \tag{75}$$

For simulation process, the neural networks adaptive fixed-time control, the $y_d = 0.1\sin(t)$ being the desired signal.

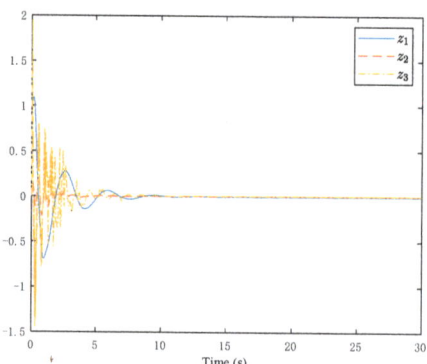

Figure 5. Trajectories of the system states.

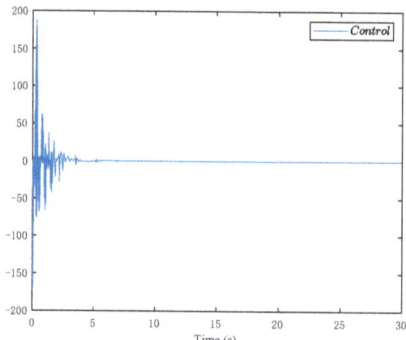

Figure 6. Trajectories of the controller.

The simulation results are shown in Figures 7 and 8. Figure 7 depicts the tracking curve of the given value and output value. It can be seen from the figure that the tracking error can be sufficiently small in fixed-time and the system output is bounded. Figure 8 shows the time response of the control input.

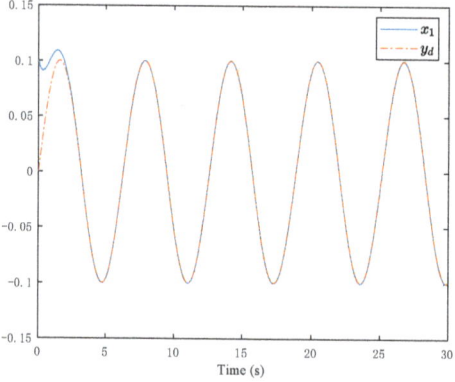

Figure 7. Trajectories of the output and the desired signal.

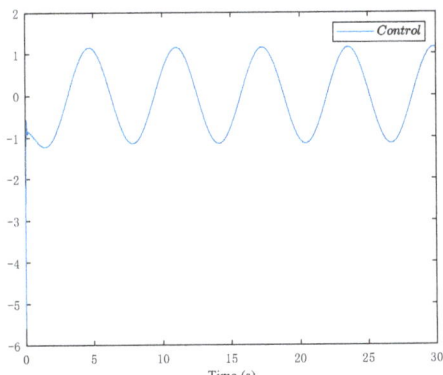

Figure 8. Trajectories of the controller.

5. Conclusions

So far, great breakthroughs have been made in the research of adaptive neural network tracking controls for nonlinear systems, but there are still some control problems to be solved. In this paper, a new fixed-time adaptive neural network tracking control strategy is designed for pure-feedback non-affine nonlinear constrained systems. Based on the backstepping control technology, the fixed-time adaptive neural network function of the error system is designed. The setting time by the control parameters and adaptive law gain parameters, that is, the control performance can be guaranteed without initial conditions, which is more practical than the control algorithm based on Lyapunov stability theory.

Author Contributions: Conceptualization, Y.L. and J.Z.; methodology, Y.L.; software, Z.D.; validation, Y.L., J.Z. and Z.D.; formal analysis, J.Z.; investigation, Y.L.; resources, Q.Z.; data curation, Q.Z.; writing—original draft preparation, Y.L.; writing—review and editing, Y.L.; visualization, Y.L.; supervision, Y.L.; project administration, Y.L.; funding acquisition, Y.L. All authors have read and agreed to the published version of the manuscript.

Funding: This work was supported in part by the National Natural Science Foundation of China under Grant 62001263.

Informed Consent Statement: Not applicable.

Conflicts of Interest: The authors declare no conflict of interest.

References

1. Fang, J.S.; Tsai, J.S.; Yan, J.J.; Chiang, L.H.; Guo, S.M. H-infinity Synchronization for Fuzzy Markov Jump Chaotic Systems with Piecewise-Constant Transition Probabilities Subject to PDT Switching Rule. *IEEE Trans. Fuzzy Syst.* **2021**, *29*, 3082–3092.
2. Zhou, Q.; Du, P.; Li, H.; Lu, R.; Yang, J. Adaptive Fixed-Time Control of Error-Constrained Pure-Feedback Interconnected Nonlinear Systems. *IEEE Trans. Syst. Man Cybern. Syst.* **2021**, *51*, 6369–6380. [CrossRef]
3. Li, R.; Zhu, Q.; Narayan, P.; Yue, A.; Yao, Y.; Deng, M. U-Model-Based Two-Degree-of-Freedom Internal Model Control of Nonlinear Dynamic Systems. *Entropy* **2021**, *23*, 169. [CrossRef] [PubMed]
4. Luo, X.; Ge, S.; Wang, J.; Guan, X. Time Delay Estimation-based Adaptive Sliding-Mode Control for Nonholonomic Mobile Robots. *Int. J. Appl. Math. Control Eng.* **2018**, *1*, 1–8.
5. Du, P.; Liang, H.; Zhao, S.; Ahn, C.K. Neural-Based Decentralized Adaptive Finite-Time Control for Nonlinear Large-Scale Systems with Time-Varying Output Constraints. *IEEE Trans. Syst. Man Cybern. Syst.* **2021**, *51*, 3136–3147. [CrossRef]
6. Zhang, J.; Li, Y.; Fei, W. Neural Network-Based Nonlinear Fixed-Time Adaptive Practical Tracking Control for Quadrotor Unmanned Aerial Vehicles. *Complexity* **2020**, *2020*, 13. [CrossRef]
7. Li, Y.; Zhang, J.; Ye, X.; Chin, C. Adaptive Fixed-Time Control of Strict-Feedback High-Order Nonlinear Systems. *Entropy* **2021**, *23*, 963. [CrossRef]
8. Li, Y.; Zhang, J.; Xu, X.; Chin, C.S. Adaptive Fixed-Time Neural Network Tracking Control of Nonlinear Interconnected Systems. *Entropy* **2021**, *23*, 1152. [CrossRef]
9. Ni, J.; Liu, L.; Liu, C.; Hu, X.; Li, S. Fast Fixed-Time Nonsingular Terminal Sliding Mode Control and Its Application to Chaos Suppression in Power System. *IEEE Trans. Circuits Syst. II Express Briefs* **2017**, *64*, 151–155. [CrossRef]

10. Zuo, Z.Y. Nonsingular fixed-time consensus tracking for second-order multi-agent networks. *Automatica* **2015**, *54*, 305–309. [CrossRef]
11. Ngo, K.; Mahony, R.; Jiang, Z.-P. Integrator Backstepping using Barrier Functions for Systems with Multiple State Constraints. In Proceedings of the 44th IEEE Conference on Decision and Control, Seville, Spain, 12–15 December 2005.
12. Tee, K.P.; Ge, S.S.; Tay, E.H. Barrier Lyapunov Functions for the control of output-constrained nonlinear systems. *Automatica* **2009**, *45*, 918–927. [CrossRef]
13. Tee, K.P.; Ren, B.; Ge, S.S. Control of nonlinear systems with time-varying output constraints. *Automatica* **2011**, *47*, 2511–2516. [CrossRef]
14. Zhang, J.; Zhu, Q.; Li, Y. Convergence Time Calculation for Supertwisting Algorithm and Application for Nonaffine Nonlinear Systems. *Complexity* **2019**, *2019*, 6235190. [CrossRef]
15. Zhang, J.; Zhu, Q.; Li, Y.; Wu, X. Homeomorphism Mapping Based Neural Networks for Finite Time Constraint Control of a Class of Nonaffine Pure-Feedback Nonlinear Systems. *Complexity* **2019**, *2019*, 9053858. [CrossRef]
16. Li, Y.; Zhang, J.; Wu, Q. *Acknowledgments, Adaptive Sliding Mode Neural Network Control for Nonlinear Systems*; Elsevier: Amsterdam, The Netherlands; Academic Press: Cambridge, MA, USA, 2019.
17. Jung, S.; Kim, S.S. Control experiment of a wheel-driven mobile inverted pendulum using neural network. *IEEE Trans. Control Syst. Technol.* **2008**, *16*, 297–303. [CrossRef]
18. Li, Z.J.; Yang, C.G. Neural-Adaptive Output Feedback Control of a Class of Transportation Vehicles Based on Wheeled Inverted Pendulum Models. *IEEE Trans. Control Syst. Technol.* **2012**, *20*, 1583–1591. [CrossRef]
19. Ye, W.; Li, Z.; Yang, C.; Sun, J.; Su, C.-Y.; Lu, R. Vision-Based Human Tracking Control of a Wheeled Inverted Pendulum Robot. *IEEE Trans. Cybern.* **2016**, *46*, 2423–2434. [CrossRef]
20. Noh, J.S.; Lee, G.H.; Jung, S. Position Control of a Mobile Inverted Pendulum System Using Radial Basis Function Network. *Int. J. Control Syst. Technol.* **2010**, *8*, 157–162. [CrossRef]
21. Ping, Z.W. Tracking problems of a spherical inverted pendulum via neural network enhanced design. *Neurocomputing* **2013**, *106*, 137–147. [CrossRef]
22. Ping, Z.; Hu, H.; Huang, Y.; Ge, S.; Lu, J.-G. Discrete-Time Neural Network Approach for Tracking Control of Spherical Inverted Pendulum. *IEEE Trans. Syst. Man Cybern. Syst.* **2020**, *50*, 2989–2995. [CrossRef]
23. Liu, C.; Ping, Z.; Huang, Y.; Lu, J.-G.; Wang, H. Position control of spherical inverted pendulum via improved discrete-time neural network approach. *Nonlinear Dyn.* **2020**, *99*, 2867–2875. [CrossRef]
24. Consolini, L.; Tosques, M. On the exact tracking of the spherical inverted pendulum via an homotopy method. *Syst. Control Lett.* **2009**, *58*, 1–6. [CrossRef]
25. Ping, Z.W.; Huang, J. Approximate output regulation of spherical inverted pendulum by neural network control. *Neurocomputing* **2012**, *85*, 38–44. [CrossRef]
26. Zhu, Q. Complete model-free sliding mode control (CMFSMC). *Sci. Rep.* **2021**, *11*, 22565. [CrossRef]
27. Dong, S.; Zhu, H.; Zhong, S.; Shi, K.; Liu, Y. New study on fixed-time synchronization control of delayed inertial memristive neural networks. *Appl. Math. Comput.* **2021**, *399*, 126035. [CrossRef]
28. Zhang, J.; Zhu, Q.; Wu, X.; Li, Y. A generalized indirect adaptive neural networks backstepping control procedure for a class of non-affine nonlinear systems with pure-feedback prototype. *Neurocomputing* **2013**, *121*, 131–139. [CrossRef]
29. Zhang, J.; Li, Y.; Fei, W.; Wu, X. U-Model Based Adaptive Neural Networks Fixed-Time Backstepping Control for Uncertain Nonlinear System. *Math. Probl. Eng.* **2020**, *2020*, 7. [CrossRef]
30. Wang, Z.; Wang, J. A Practical Distributed Finite-Time Control Scheme for Power System Transient Stability. *IEEE Trans. Power Syst.* **2020**, *35*, 3320–3331. [CrossRef]
31. Zhou, S.; Song, Y.; Wen, C. Event-Triggered Practical Prescribed Time Output Feedback Neuroadaptive Tracking Control under Saturated Actuation. *IEEE Trans. Neural Netw. Learn. Syst.* **2021**, *5*, 1–11. [CrossRef]

H_∞ Observer Based on Descriptor Systems Applied to Estimate the State of Charge

Shengya Meng [1], Shihong Li [2,*], Heng Chi [1], Fanwei Meng [1] and Aiping Pang [3]

1. School of Control Engineering, Northeastern University at Qinhuangdao, Qinhuangdao 066004, China; 2001944@stu.neu.edu.cn (S.M.); 2071918@stu.neu.edu.cn (H.C.); mengfanwei@neuq.edu.cn (F.M.)
2. School of Automation, Aviation University Air Force, Changchun 130000, China
3. College of Electrical Engineering, Guizhou University, Guiyang 550025, China; appang@gzu.edu.cn
* Correspondence: b20210201@st.nuc.edu.cn

Abstract: This paper proposes an H_∞ observer based on descriptor systems to estimate the state of charge (SOC). The battery's open-current voltage is chosen as a generalized state variable, thereby avoiding the artificial derivative calculation of the algebraic equation for the SOC. Furthermore, the observer's dynamic performance is saved. To decrease the impacts of the uncertain noise and parameter perturbations, nonlinear H_∞ theory is implemented to design the observer. The sufficient conditions for the H_∞ observer to guarantee the disturbance suppression performance index are given and proved by the Lyapunov stability theory. This paper systematically gives the design steps of battery SOC H_∞ observers. The simulation results highlight the accuracy, transient performance, and robustness of the presented method.

Keywords: descriptor systems; SOC estimation; H_∞ observer; disturbance suppression performance

1. Introduction

Over the past few years, renewable energy vehicles (REVs) have become a mainstream consumer option, so related research about REV batteries has been of great interest [1]. The state of charge (SOC) is a percentage of the remaining capacity to the actual capacity of the battery, which is a vital indicator to evaluate battery performance [2,3]. Accurately tracking the SOC can dramatically avoid battery overcharge or overdischarge, thereby extending the battery life. However, due to a series of complex electrochemical reactions inside the battery, it is often impossible to obtain the SOC directly through the sensors. In other words, SOC can only be estimated by the measurable electrical signals and battery parameters. Even worse, battery parameters are affected by external factors such as temperature, battery age, and noise in electrical signals [4]. Accordingly, the SOC observer needs to provide sufficient estimation accuracy even in noise and parameter perturbations, which is a daunting task.

A variety of algorithms are proposed to estimate SOC, such as the coulomb counting method (CCM), open-circuit voltage method (OCVM), Kalaman filter (KF), sliding-mode observer (SMO), H_∞ observer, neural network algorithm, proportional-integral (PI) observer, and adaptive observer [5–7]. The CCM estimates SOC by continuously measuring and integrating the current in time. The main drawbacks of CCM are two-fold: the first is that CCM highly depends on the initial value of observers, and the second is that it is known as an open-loop method whose estimation value will drift in the long term [8]. Alternatively, because of the one-to-one correspondence (as shown in Figure 1) between the SOC and the open-circuit voltage (OCV), the OCVM estimates the SOC by measuring the OCV of the battery without load. However, this technique fails to estimate SOC online. Due to their drawbacks, CCM and OCVM are never utilized separately in practical applications [9]. The KF and SMO are widely employed in the field of SOC estimation [10,11]. Nevertheless, due to the assumption of a noise signal Gaussian, the KF falls short when the system has noise or unmodeled dynamics [12,13]. The SMO is commonly used for SOC estimation due to its robustness. In [14], an OCV–SOC formula was modeled by the Nernst equation, and a SMO was proposed to estimate SOC; simulation results validate its accuracy. However, the estimate error of SOC may fluctuate because of the discontinuous input. A new SMO, based on the two-circuit model presented in [15], exhibits good performance. However, without accurate initial states, the SMO in [15] takes longer to track the true SOC.

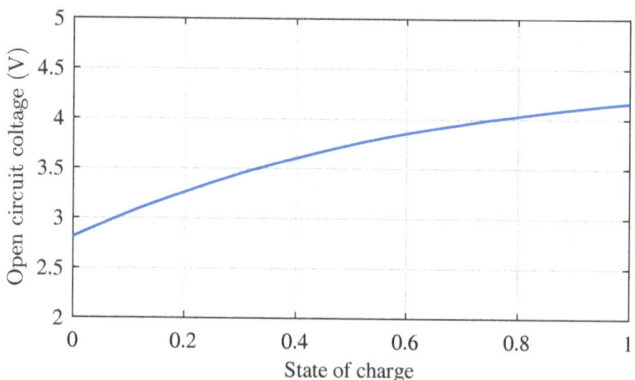

Figure 1. The relationship between SOC and OCV.

The influence of possible error sources on the SOC observation was analyzed in [16], and the results show that measurement noise and modeling errors are the main factors that limit the observation accuracy. The H_∞ observer is a promising tool to handle unknown noise and modeling errors, and its effectiveness under a variety of operating conditions has been confirmed by experiments [17–20]. Based on the OCV–SOC formula, an H_∞-switched observer was presented in [21]; the experimental results confirm that, compared with the KF, both the accuracy and robustness of SOC estimation are improved by its use.

Regardless of the above approaches, it is impossible to ignore the piecewise nonlinear function of OCV versus SOC shown in Figure 1. In the battery model, the SOC fails to be expressed explicitly in the state equation, which brings difficulties to the design observer. In [22,23], the piecewise nonlinear function was linearized and differentiated before the observer design. However, the differential operation produced two problems:

1. The derivation of the piecewise function increased the order of the observer, which did not match the original system, and the observer error was not converged potentially;
2. The derivation of the current was ignored completely, so the dynamic performance of the observer became worse.

There are both differential equations and algebraic equations in battery systems. Such systems are also called descriptor systems, singular systems, or differential-algebraic systems [24,25]. To avoid the differentiation of the OCV–SOC formula, it is feasible to design the observer after modeling the battery as a descriptor system. Various methods are developed to design observers for descriptor systems [26].

The main objective of this paper is to design a noncomplex observer to estimate SOC accurately. To balance accuracy and complexity, this paper innovatively models the battery as a descriptor system. The H_∞ theory is applied to design the observer to improve disturbance suppression performance. Compared with the traditional SOC estimation method, the method proposed in this paper can accurately estimate the SOC online, and does not require an accurate initial value. The designed observer exhibits good robustness in the presence of noise.

This paper is organized as follows. In Section 2, for the equivalent circuit model, a descriptor system with state variable OCV is established. In Section 3, the H_∞ observer is proposed. The sufficient conditions to solve the observer are given and proved. In Section 4, several simulation experiments verify the accuracy and robustness of the proposed method. Section 5 summarizes the contribution of this paper.

Notations: M^+ is the generalized inverse of matrix M, satisfying $MM^+M = M$. I denotes an identity matrix with appropriate dimensions. $\mathbf{0}$ is the zero matrix with appropriate dimensions.

2. Battery Model

A resistance–capacitance (RC) equivalent circuit model is used to build a dynamic model of the battery, as shown in Figure 2, where the variable s represents the SOC. C_N is the nominal capacity of the battery. v_{oc} represents the OCV, which is the function of SOC. v_c is the voltage across the polarized capacitor C_c. R_e and R_c represent the conduction resistance and the diffusion resistance, respectively; i_e and i_c are the currents of the two branches; R_t is the terminal resistance; v_t is the measurable terminal voltage, and i is the charge and discharge current.

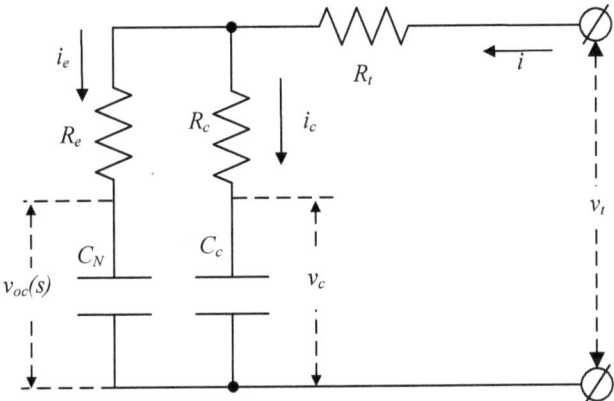

Figure 2. RC equivalent circuit model of the battery.

From the definition of SOC, the dynamic relationship of SOC is $\dot{s} = \dfrac{i_e}{C_N}$. In addition, the aforementioned nonlinear function v_{oc} can be reasonably approximated as $v_{oc} = k_1 s + k_2 + \Delta f_1$, where Δf_1 is the nonlinearity of the OCV–SOC relationship. The two constants k_1 and k_2 can be determined by fitting the curve in Figure 1. From Kirchhoff's law and Figure 2, the dynamic equations of the battery are:

$$\begin{aligned}
\dot{s} &= \frac{-v_{oc} + v_c}{C_N(R_e + R_c)} + \frac{iR_c}{C_N(R_e + R_c)} + \Delta f_2, \\
\dot{v}_c &= \frac{v_{oc} - v_c}{C_c(R_e + R_c)} + \frac{iR_e}{C_c(R_e + R_c)} + \Delta f_3, \\
v_t &= \frac{R_c v_c + R_e v_{oc}}{R_e + R_c} + \left(\frac{R_e R_c}{R_e + R_c} + R_t\right)i,
\end{aligned} \quad (1)$$

where Δf_2 and Δf_3 are the uncertainties caused by modeling accuracy.

In this model, $s \in (0,1)$ is the independent variable of the v_{oc}, which essentially introduces a piecewise algebraic constraint. To solve this piecewise algebraic system state estimation problem, Refs. [22,23] ignore the change of the current to derive the v_{oc} and v_t, respectively, and model the system as a third-order system which is primordially two-order. In the above modeling process, the derivation operation increases the order of the system, and it is doubtful whether the observer error converges. In the actual application of batteries, especially in the course of REVs, the current of the battery is constantly changing. Therefore, it is obviously unreasonable to completely ignore the derivative of the current.

Motivated by these considerations, this paper regards the OCV–SOC function as an algebraic constraint between state variables, thereby modeling the system battery as a descriptor system. $x = \begin{bmatrix} s^T & v_c^T & v_{oc}^T \end{bmatrix}^T$ is identified as a state variable, $u = \begin{bmatrix} i^T & 1 \end{bmatrix}^T$, and $\omega = \begin{bmatrix} \Delta f_1^T & \Delta f_2^T & \Delta f_3^T \end{bmatrix}^T$; then, the battery is modeled as a descriptor system (2) with $n = 3$ dimensions.

$$E\dot{x} = Ax + Bu + D_1\omega,$$
$$y = Cx + Du, \tag{2}$$

where:

$$E = \begin{bmatrix} 1 & 0 & 0 \\ 0 & 1 & 0 \\ 0 & 0 & 0 \end{bmatrix}, \quad A = \begin{bmatrix} 0 & \dfrac{1}{C_N(R_e+R_c)} & \dfrac{-1}{C_N(R_e+R_c)} \\ 0 & \dfrac{-1}{C_c(R_e+R_c)} & \dfrac{1}{C_c(R_e+R_c)} \\ k_1 & 0 & -1 \end{bmatrix},$$

$$B = \begin{bmatrix} \dfrac{R_c}{C_N(R_e+R_c)} & 0 \\ \dfrac{R_e}{C_c(R_e+R_c)} & 0 \\ 0 & k_2 \end{bmatrix}, \quad C = \begin{bmatrix} 0 & \dfrac{R_c}{R_e+R_c} & \dfrac{R_e}{R_e+R_c} \end{bmatrix},$$

$$D = \begin{bmatrix} \dfrac{R_eR_c}{R_e+R_c} + R_t & 0 \end{bmatrix}, \quad D_1 = \begin{bmatrix} 0 & 1 & 0 \\ 0 & 0 & 1 \\ 1 & 0 & 0 \end{bmatrix}.$$

Before the observer design, assume that the descriptor system (2) satisfies Assumption 1.

Assumption 1.

$$\text{rank} \begin{bmatrix} E & A \\ 0 & E \\ 0 & C \end{bmatrix} = n + \text{rank}\, E, \tag{3}$$

where n is the number of state variables.

Under Assumption 1, the descriptor system (2) is impulse observable, which guarantees there exists an observer to track the states. Actually, this assumption is not strict and easy to achieve in battery models.

3. H_∞ Observer

Design an H_∞ observer described as follows:

$$\dot{z} = Hz + J\bar{y} + Mu,$$
$$\hat{x} = Pz - Q\Phi Bu + R\bar{y}, \tag{4}$$

where $z \in R^r$ is the state variable of the observer, $\hat{x} \in R^n$ is the estimated value of the battery state, $H, J, M, P, Q,$ and R are all unknown matrices with appropriate dimensions, and $\bar{y} = y - Du$ is the virtual output. Φ satisfies $\Phi E = 0$.

The H_∞ observer design target can be expressed as designing a stable observer (4) to satisfy that:

1. With $\omega = 0$, the estimate error $e = x - \hat{x}$ is asymptotically stable;
2. With $\omega \neq 0$, for a prescribed level of noise $\gamma > 0$, $\|e\|_{L_2} < \gamma \|\omega\|_{L_2}$ will be satisfied.

Define the error $\delta = z - NEx$, where N is of appropriate dimensions. Then, one has:

$$\begin{aligned}\dot{\delta} &-\dot{z}-NE\dot{x}\\&=H\delta+(HNE+JC-NA)x+(M-NB)u-ND_1\omega,\\ e&=\hat{x}-x\\&=P\delta+(PNE+Q\Phi A+RC-I_n)x+Q\Phi D_1\omega,\end{aligned} \tag{5}$$

where $\Phi E = 0$ is applied. Under Assumption 1, to make the error system (5) be a homogeneous linear differential equation for δ, the observer (4) should satisfy:

$$\begin{bmatrix} H & \psi & J \\ P & Q & R \end{bmatrix} \begin{bmatrix} N'E \\ \Phi A \\ C \end{bmatrix} = \begin{bmatrix} N'A \\ I_n \end{bmatrix}, \tag{6}$$

$$M = NB, \tag{7}$$

where ψ is an arbitrary matrix of appropriate dimension and $N' = N + \psi\Phi$.

To facilitate the analysis, define $\varphi_1 = -ND_1$ and $\varphi_2 = Q\Phi D_1$; then, the dynamics of the error system are given by:

$$\begin{aligned}\dot{\delta} &= H\delta + \varphi_1\omega,\\ e &= P\delta + \varphi_2\omega.\end{aligned} \tag{8}$$

Notice that Equation (6) can be solvable if and only if:

$$\mathrm{rank}\begin{bmatrix} N'E \\ \Phi A \\ C \end{bmatrix} = n. \tag{9}$$

With Equation (9), the solution of Equation (6) can be described as:

$$\begin{array}{lll} H = \Gamma_H + \eta_1\Delta_P, & \psi = \Gamma_\psi + \eta_1\Delta_Q, & J = \Gamma_J + \eta_1\Delta_R,\\ P = \Gamma_P + \eta_2\Delta_P, & Q = \Gamma_Q + \eta_2\Delta_Q, & R = \Gamma_R + \eta_1\Delta_R,\\ \varphi_1 = \Gamma_{\varphi_1} + \eta_1\Delta_{\varphi_1}, & \varphi_2 = \Gamma_{\varphi_2} + \eta_2\Delta_{\varphi_2}, \end{array} \tag{10}$$

where η_1 and η_2 are of appropriate dimension. Define the following matrices:

$$\Gamma_P = \Omega^+\begin{bmatrix}I\\0\\0\end{bmatrix},\quad \Delta_P = (I-\Omega\Omega^+)\begin{bmatrix}I\\0\\0\end{bmatrix},\quad \Gamma_H = N'A\Gamma_P,$$

$$\Gamma_Q = \Omega^+\begin{bmatrix}0\\I\\0\end{bmatrix},\quad \Delta_Q = (I-\Omega\Omega^+)\begin{bmatrix}0\\I\\0\end{bmatrix},\quad \Gamma_\psi = N'A\Gamma_Q,$$

$$\Gamma_R = \Omega^+\begin{bmatrix}0\\0\\I\end{bmatrix},\quad \Delta_R = (I-\Omega\Omega^+)\begin{bmatrix}0\\0\\I\end{bmatrix},\quad \Gamma_J = N'A\Gamma_R,$$

$$\begin{array}{ll}\Gamma_{\varphi_1} = -N'D_1 - \Gamma_\psi\Phi D_1, & \Delta_{\varphi_1} = -\Delta_Q\Phi D_1,\\ \Gamma_{\varphi_2} = \Gamma_Q\Phi D_1, & \Delta_{\varphi_2} = \Delta_Q\Phi D_1,\end{array}$$

where $\Omega = \begin{bmatrix} N'E \\ \Phi A \\ C \end{bmatrix}$.

The following theorem gives the sufficient conditions for error system (8) to be stable and $\|e\|_{L_2} < \gamma \|\omega\|_{L_2}$ with (6) and (7).

Theorem 1. For a prescribed level of noise $\gamma > 0$, under $\delta = 0$, the error system (8) with (6) and (7) is asymptotically stable for $w = 0$, and satisfies $\|e\|_{L_2} < \gamma \|w\|_{L_2}$ for $w \neq 0$, if there exists a matrix $X = X^T > 0$ and matrices X_{η_1} and η_2, such that the following linear matrix inequality (LMI) is satisfied:

$$\Sigma = \begin{bmatrix} \sigma_1 & \sigma_2 & \sigma_3 \\ \sigma_2^T & -\gamma^2 I & \sigma_4 \\ \sigma_3^T & \sigma_4^T & -I \end{bmatrix} < 0, \tag{11}$$

where:

$$\sigma_1 = \Gamma_H^T X + X\Gamma_H + \Delta_P^T X_{\eta_1}^T + X_{\eta_1} \Delta_P, \quad \sigma_2 = X\Gamma_{\varphi_1} + X_{\eta_1} \Delta_{\varphi_1},$$
$$\sigma_3 = \Gamma_P + \eta_2 \Delta_P, \quad \sigma_4 = \Gamma_{\varphi_2}^T + \Delta_{\varphi_2}^T \eta_2^T, \quad X_{\eta_1} = X\eta_1.$$

Proof of Theorem 1. According to (10) and (11), one can obtain:

$$\begin{bmatrix} H^T X + XH & X\varphi_1 & P^T \\ \varphi_1^T X & -\gamma^2 I & \varphi_2^T \\ P & \varphi_2 & -I \end{bmatrix} < 0. \tag{12}$$

The Lyapunov function is chosen as $V = \delta^T X \delta$. The derivative of V is obtained as:

$$\dot{V}(t) = \dot{\delta}^T X \delta + \delta^T X \dot{\delta}$$
$$= \delta^T (H^T X + XH)\delta + w^T \varphi_1^T X \delta + \delta^T X \varphi_1 w.$$

With $w = 0$ and (12), $\dot{V} < 0$ is satisfied; hence, the system (8) is asymptotically stable.

$$\dot{V} + e^T e - \gamma^2 w^T w$$
$$= \begin{bmatrix} \delta^T & w^T \end{bmatrix} \begin{bmatrix} H^T X + XH + P^T P & X\varphi_1 + P^T \varphi_2 \\ \varphi_1^T X + \varphi_2^T P & \varphi_2^T \varphi_2 - \gamma^2 I \end{bmatrix} \begin{bmatrix} \delta \\ w \end{bmatrix}$$

By the Schur complement to (12), one obtains:

$$\begin{bmatrix} H^T X + XH + P^T P & X\varphi_1 + P^T \varphi_2 \\ \varphi_1^T X + \varphi_2^T P & \varphi_2^T \varphi_2 - \gamma^2 I \end{bmatrix} < 0.$$

Therefore:

$$\dot{V} < \gamma^2 w^T w - e^T e,$$
$$\int_0^\infty \dot{V}(\tau) d\tau < \int_0^\infty \gamma^2 w^T(\tau) w(\tau) d\tau - \int_0^\infty e^T(\tau) e(\tau) d\tau.$$

Under the zero initial condition, $V(\infty) < \gamma^2 \|w\|^2 - \|e\|^2$. Hence, the error system satisfies $\|e\|_{L_2} < \gamma \|w\|_{L_2}$ for $w \neq 0$.

Inserting the solution of (6) into (12), Theorem 1 is obtained. Then, the theorem is proved. □

From Theorem 1, the prescribed level of noise γ determines the feasibility of (11). According to robust control theory, γ can be selected by the following optimization problems:

$$\min(\gamma)$$
$$\text{s.t.} \quad X = X^T > 0, \tag{13}$$
$$\Sigma < 0.$$

This optimization problem can be solved with the YALMIP toolbox [27].

The proof process of Theorem 1 embodies the following observer design steps:

1. Model the battery system as a descriptor system (2);
2. Determine the matrix Φ by $\Phi E = 0$;
3. Determine the matrix N' by the (9);

4. Choose the prescribed level of noise γ by optimization problems (13);
5. Solve the feasible solution of (11) given by Theorem 1;
6. Calculate the matrices H, J, P, Q, R, φ_1, and φ_2;
7. Convert the virtual output into the actual measurable output by $\bar{y} = y - Du$.

From the above steps, there are some parameters that need to be chosen. γ determines the disturbance rejection level of the observer, which usually cannot be a large value. N' only needs to satisfy (9), and the numerical size of each element in the matrix N' has little effect on the final result. Therefore, compared with the existing method for SOC estimation, the proposed H_∞ observer does not require complex tuning.

4. Results and Discussion

In order to illustrate the superiority of the proposed H_∞ observer, this paper will compare it with the PI observer [7] and SMO [15]. To ensure the fairness of the test, the parameters of battery Figure 2 are shown in Table 1.

Table 1. Parameters of the lithium battery.

C_N	C_C	R_e	R_c	R_t
18,000 F	200 F	0.003 Ω	0.003 Ω	0.001 Ω

The piecewise algebraic relationship between v_{oc} and SOC is $v_{oc} = 1.2s + 3$; bring the battery parameters and v_{oc} functions into the model, and the battery modeling is complete.

It can be verified that the battery system whose parameters are shown in Table 1 satisfies Assumption 1; therefore, we can design an H_∞ observer of the construction (4) by Theorem 1.

Take a non-zero solution of the equation $\Phi E = 0$ as $\Phi = \begin{bmatrix} 0 & 0 & 1 \end{bmatrix}$. Note that $N' = \begin{bmatrix} 1 & 0 & 0 \\ 0 & 1 & 1 \end{bmatrix}$ satisfies (9). Based on the $\gamma_{min} = 0.7124$ from the optimization problem (13), we take $\gamma = 1.1$, use the YALMIP toolbox to solve the (11), and obtain the H_∞ observer as:

$$\dot{z} = \begin{bmatrix} -0.6694 & -0.5393 \\ 0.3236 & -1.3969 \end{bmatrix} z + \begin{bmatrix} -0.0027 & -1.6735 \\ -0.0003 & 0.8091 \end{bmatrix} u + \begin{bmatrix} 1.0971 \\ 1.1272 \end{bmatrix} y, \quad (14)$$

$$\hat{x} = \begin{bmatrix} 0.9095 & -0.0754 \\ -0.7095 & 0.9409 \\ 0.8825 & -0.2646 \end{bmatrix} z - \begin{bmatrix} 0.0004 & 0.2262 \\ 0.0003 & 0.1774 \\ 0.0013 & -2.2062 \end{bmatrix} u + \begin{bmatrix} 0.1508 \\ 0.1182 \\ 0.5292 \end{bmatrix} y.$$

As a comparison, the PI observer applied in the technique proposed in [7] is:

$$\dot{\hat{x}} = \begin{bmatrix} -0.0111 & 0.0111 & 0 \\ -0.8333 & 0.8333 & 0 \\ 0.8222 & 0 & -0.8222 \end{bmatrix} \hat{x} + \begin{bmatrix} -0.0000324 \\ -0.0025 \\ 0.003322 \end{bmatrix} u \quad (15)$$

$$+ \begin{bmatrix} 0.1845 \\ 0.005 \\ 0.2 \end{bmatrix} (y - \hat{y}) + \begin{bmatrix} 0.1 \\ 0.005 \\ 0.2 \end{bmatrix} \alpha,$$

$$\dot{\alpha} = 0.01(y - \hat{y}),$$
$$\hat{y} = \begin{bmatrix} 0 & 0 & 1 \end{bmatrix} \hat{x},$$

where $\hat{x}$ and $\hat{y}$ are the estimate of $\begin{bmatrix} v_{oc}^T & v_c^T & v_t^T \end{bmatrix}^T$ and v_t, respectively. Notice that $u = i$. The SMO proposed in [15] is shown as:

$$\dot{\hat{x}} = \begin{bmatrix} -0.8333 & 0.8333 \\ 0.00926 & -0.00926 \end{bmatrix} \hat{x} + \begin{bmatrix} 0.0025 \\ 0.000027 \end{bmatrix} u - \begin{bmatrix} 1.667 \\ 0.0185 \end{bmatrix} (y - \hat{y}) + \begin{bmatrix} 0.0025 \\ 0.000027 \end{bmatrix} v,$$

$$v = \begin{cases} -\dfrac{661.376}{y - \hat{y}} \left(6.048 \times 10^{-3}(y - \hat{y}) + 6.929 \times 10^{-5}(y - \hat{y})^{1.4} \right) & y - \hat{y} \neq 0 \\ 0 & y - \hat{y} = 0 \end{cases}, \quad (16)$$

$$\hat{y} = \begin{bmatrix} 0.5 & 0.5 \end{bmatrix} \hat{x} + 0.0025u,$$

where $\hat{x}$ and $\hat{y}$ are the estimatse of $\begin{bmatrix} v_c^T & v_{oc}^T \end{bmatrix}^T$ and v_t, respectively. The input u is the current i.

The constant current discharge experiment, to evaluate the performance of the observer, is employed as follows: choose a discharge current of 5 A whose discharge period is 3980 s, and discharge for 180 s. Figure 3 shows the current of the constant current discharge experiment. For fairness, the known initial SOC of the battery model is 0.8, and Figure 4 shows the estimate errors from the different observers in this experiment.

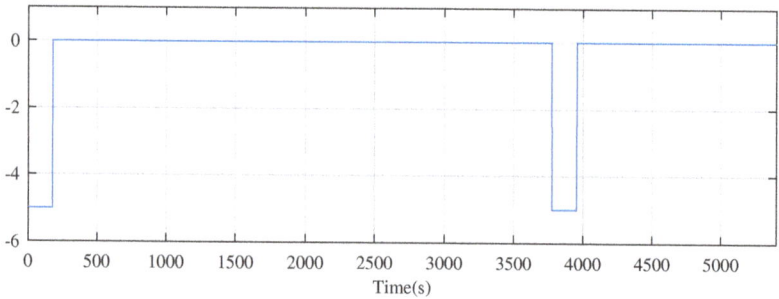

Figure 3. The current of the constant current discharge experiment.

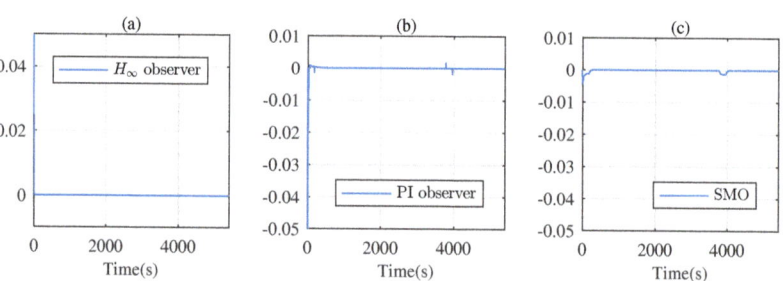

Figure 4. SOC estimate error under the constant current discharge experiment: (a) based on the H_∞ observer; (b) based on the PI observer; (c) based on SMO.

From Figure 4, the H_∞ observer and the PI observer can converge to zero quicker, with respect to the SMO. However, at each instant of discharge, the PI observer and SMO need a short period of adjustment to reach the steady state again, while the H_∞ observer based on the descriptor system overcomes this drawback.

The initial conditions of the SOC are set as 0.3 for observers and 0.8 for the battery. In this case, the simulation result is shown in Figure 5. Due to the inaccurate initial SOC, there is large error of SOC from each observer at the initial moment. However, the estimate error of the H_∞ observer converges to zero within 20 s, while the estimate error of the PI observer and SMO converge to zero within 500 s and 200 s, respectively. So, the H_∞ observer is not sensitive to accurate initial SOC and has a fast convergence speed.

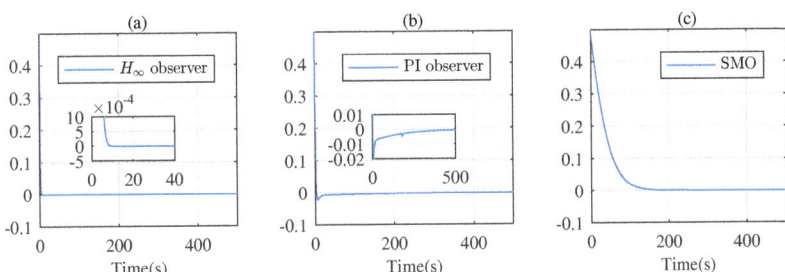

Figure 5. SOC estimate error under the constant current discharge experiment with inaccurate initial SOC: (**a**) based on the H_∞ observer; (**b**) based on the PI observer; (**c**) based on SMO.

The dynamic stress test (DST) is a standard test condition proposed by the Advanced Battery Association of the United States to simulate urban driving condition for electric vehicles. It is commonly applied to test the dynamic performance of SOC observers. The current of DST is shown in Figure 6.

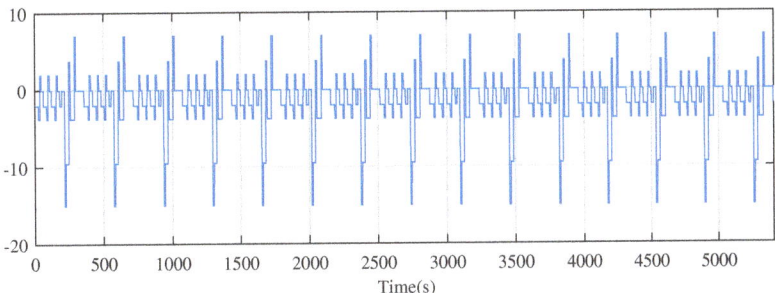

Figure 6. The current of the dynamic stress test.

Under the DST, the estimated SOC is shown in Figure 7. Generally, each observer can track the SOC. However, from Figure 7b, the true SOC is covered by the estimated SOC from the H_∞ observer completely, which means the H_∞ observer exhibits better performance for tracking the real SOC, with respect to the PI observer and SMO.

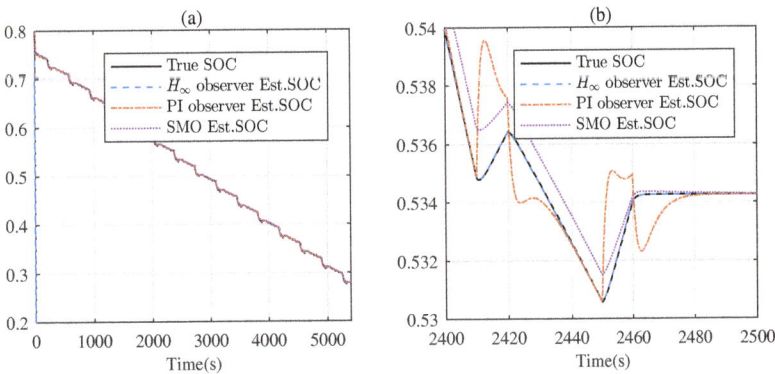

Figure 7. Real SOC and its estimate under the DST: (**a**) full graph; (**b**) zoomed graph.

The estimate error under the DST is plotted in Figure 8. Because of the reservation of the current dynamic performance, there does not exist pulse mode in the estimation error of the H_∞ observer. Figures 7 and 8 illustrate the outstanding dynamic performance.

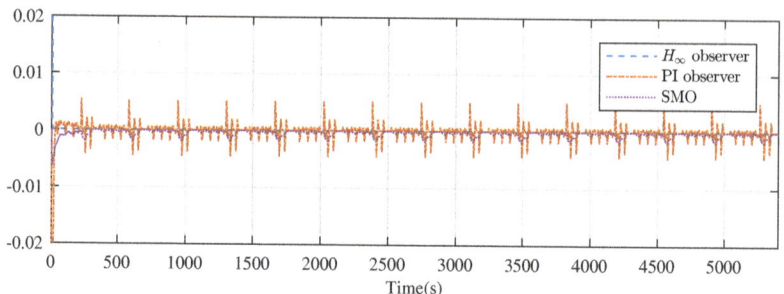

Figure 8. Error of SOC for DST from the H_∞ observer, PI observer, and SMO.

We are also interested in measuring the performance of the proposed observer's deleted structure (4) in the presence of parameter perturbations to see if there are improvements in the robustness, with respect to the PI observer and SMO. So, the case when the capacitance and resistance parameter perturbations are given (Figure 9) is considered. Figure 10 shows the behavior of the true SOC and its estimate when the uncertainty is present.

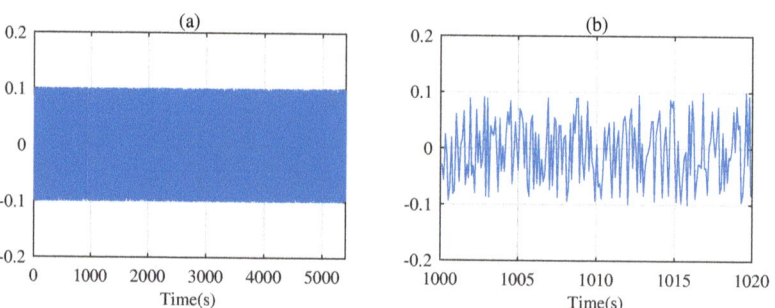

Figure 9. The uncertainty factor in the battery model: (**a**) full graph; (**b**) zoomed graph.

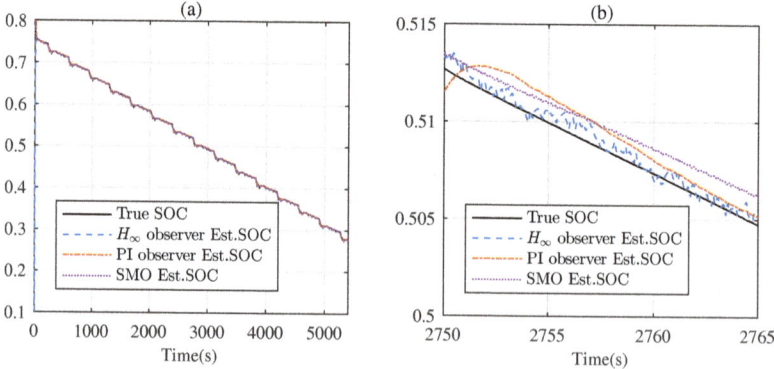

Figure 10. The real SOC and its estimate from the H_∞ observer, PI observer, and SMO under parameter perturbations: (**a**) full graph; (**b**) zoomed graph.

From Figure 10, due to the parameter perturbations, each observer is unable to track the true SOC accurately. However, the estimate error of SOC from the H_∞ observer is less than the PI observer and SMO, which illustrates the robustness of the schema proposed in this paper.

5. Conclusions

To improve the accuracy of SOC observers and decrease the effects of the parameter perturbations, an H_∞ observer, based on descriptor systems, is designed to estimate SOC. Firstly, the battery is modeled as a descriptor system without the extra derivative operation of nonlinear piecewise constraints. For the uncertainty, robustness nonlinear H_∞ theory is employed to solve the observer. Furthermore, the design steps of a type of battery SOC observer are given systematically. The simulation results show that the H_∞ observer based on descriptor systems is more effective, and the observation accuracy is higher with respect to the PI observer and SMO. The method proposed in this paper is not sensitive to battery parameter changes. Therefore, the proposed H_∞ observer based on descriptor systems provides a new and effective online method to estimate SOC in REVs. The extension of our work to nonlinear H_∞ observers is under study.

Author Contributions: Conceptualization S.M., S.L., F.M. and A.P.; methodology and software, S.M. and H.C.; writing—original draft preparation, S.M. and S.L. All authors have read and agreed to the published version of the manuscript.

Funding: This research was funded by the Natural Science Foundation of Hebei Province (F2019501012) and the National Natural Science Foundation of China (12162007).

Institutional Review Board Statement: Not applicable.

Informed Consent Statement: Not applicable.

Conflicts of Interest: The authors declare no conflict of interest.

Abbreviations

The following abbreviations are used in this manuscript:

REVs	Renewable energy vehicles
SOC	State of charge
OCVM	Open-circuit voltage method
CCM	Coulomb counting method
KF	Kalman filter
SMO	Sliding-mode observer
PI	Proportional-integral
OCV	Open-circuit voltage
RC	Resistance–capacitance
LMI	Linear matrix inequality
DST	Dynamic stress test
RMSE	Root mean square error
MAE	Maximum absolute error

References

1. Chen, X.; Zhang, T.; Ye, W.; Wang, Z.; Iu, H.H.C. Blockchain-Based Electric Vehicle Incentive System for Renewable Energy Consumption. *IEEE Trans. Circuits Syst. II Express Briefs* **2021**, *68*, 396–400. [CrossRef]
2. Zhang, S.; Guo, X.; Dou, X.; Zhang, X. A rapid online calculation method for state of health of lithium-ion battery based on coulomb counting method and differential voltage analysis. *J. Power Sources* **2020**, *479*, 228740. [CrossRef]
3. Soylu, E., Soylu, T., Bayir, R. Design and Implementation of SOC Prediction for a Li-Ion Battery Pack in an Electric Car with an Embedded System. *Entropy* **2017**, *19*, 146. [CrossRef]
4. Sethia, G.; Nayak, S.K.; Majhi, S. An Approach to Estimate Lithium-Ion Battery State of Charge Based on Adaptive Lyapunov Super Twisting Observer. *IEEE Trans. Circuits Syst. I Regul. Pap.* **2021**, *68*, 1319–1329. [CrossRef]
5. Shrivastava, P.; Soon, T.K.; Idris, M.Y.I.B.; Mekhilef, S. Overview of Model-Based Online State-of-Charge Estimation Using Kalman Filter Family for Lithium-Ion Batteries. *Renew. Sustain. Energy Rev.* **2019**, *113*, 109233. [CrossRef]
6. Barillas, J.K.; Li, J.; Guenther, C.; Danzer, M.A. A comparative study and validation of state estimation algorithms for Li-ion batteries in battery management systems. *Appl. Energy* **2015**, *155*, 455–462. [CrossRef]
7. Xu, J.; Mi, C.C.; Cao, B.; Deng, J.; Chen, Z.; Li, S. The State of Charge Estimation of Lithium-Ion Batteries Based on a Proportional-Integral Observer. *IEEE Trans. Veh. Technol.* **2014**, *63*, 1614–1621.

8. Prajapati, V.; Hess, H.; William, E.J.; Gupta, V.; Huff, M.; Manic, M.; Rufus, F.; Thakker, A.; Govar, J. A Literature Review of State Of-Charge Estimation Techniques Applicable to Lithium Poly-Carbon Monoflouride (LI/CFx) Battery. In Proceedings of the India International Conference on Power Electronics 2010 (IICPE2010), New Delhi, India, 28–30 January 2011; IEEE: New York, NY, USA, 2011; pp. 1–8.
9. Rahimi-Eichi, H.; Baronti, F.; Chow, M.Y. Online Adaptive Parameter Identification and State-of-Charge Coestimation for Lithium-Polymer Battery Cells. *IEEE Trans. Ind. Electron.* **2013**, *61*, 2053–2061. [CrossRef]
10. Du, J.; Liu, Z.; Wang, Y.; Wen, C. An adaptive sliding mode observer for lithium-ion battery state of charge and state of health estimation in electric vehicles. *Control Eng. Pract.* **2016**, *54*, 81–90. [CrossRef]
11. Li, K.; Zhou, P.; Lu, Y.; Han, X.; Zheng, Y. Battery life estimation based on cloud data for electric vehicles. *J. Power Sources* **2020**, *468*, 228192. [CrossRef]
12. Afshar, S.; Morris, K.; Khajepour, A. State-of-Charge Estimation Using an EKF-Based Adaptive Observer. *IEEE Trans. Control Syst. Technol.* **2019**, *27*, 1907–1923. [CrossRef]
13. Qiao, Z.; Liang, L.; Hu, X.; Xiong, N.; Hu, G. H_∞-Based Nonlinear Observer Design for State of Charge Estimation of Lithium-Ion Battery with Polynomial Parameters. *IEEE Trans. Veh. Technol.* **2017**, *99*, 1–13.
14. Chang, F.; Zheng, Z. An SOC Estimation Method Based on Sliding Mode Observer and the Nernst Equation. In Proceedings of the 2015 IEEE Energy Conversion Congress and Exposition (ECCE), Montreal, QC, Canada, 20–24 September 2015; IEEE: New York, NY, USA, 2015; pp. 6187–6190.
15. Chen, Q.; Jiang, J.; Ruan, H.; Zhang, C. Simply Designed and Universal Sliding Mode Observer for the SOC Estimation of Lithium-ion Batteries. *IET Power Electron.* **2017**, *10*, 697–705. [CrossRef]
16. Shen, P.; Ouyang, M.; Han, X.; Feng, X.; Lu, L.; Li, J. Error Analysis of the Model-Based State-of-Charge Observer for Lithium-Ion Batteries. *IEEE Trans. Veh. Technol.* **2018**, *67*, 8055–8064. [CrossRef]
17. Zhang, F.; Liu, G.; Fang, L.; Wang, H. Estimation of Battery State of Charge With H_∞ Observer: Applied to a Robot for Inspecting Power Transmission Lines. *IEEE Trans. Ind. Electron.* **2012**, *59*, 1086–1095. [CrossRef]
18. Huang, J.; Ma, X.; Che, H.; Han, Z. Further Result on Interval Observer Design for Discrete-Time Switched Systems and Application to Circuit Systems. *IEEE Trans. Circuits Syst. II Express Briefs* **2020**, *67*, 2542–2546. [CrossRef]
19. Meng, F.; Pang, A.; Dong, X.; Han, C.; Sha, X. H_∞ Optimal Performance Design of an Unstable Plant under Bode Integral Constraint. *Complexity* **2018**, *2018*, 4942906. [CrossRef]
20. Meng, F.; Wang, D.; Yang, P.; Xie, G. Application of Sum of Squares Method in Nonlinear H_∞ Control for Satellite Attitude Maneuvers. *Complexity* **2019**, *2019*, 5124108. [CrossRef]
21. Liu, C.Z.; Zhu, Q.; Li, L.; Liu, W.Q.; Wang, L.Y.; Xiong, N.; Wang, X.Y. A State of Charge Estimation Method Based on H_∞ Observer for Switched Systems of Lithium-Ion Nickel-Manganese-Cobalt Batteries. *IEEE Trans. Ind. Electron.* **2017**, *64*, 8128–8137. [CrossRef]
22. Chen, W.; Chen, W.T.; Saif, M.; Li, M.F. Simultaneous Fault Isolation and Estimation of Lithium-Ion Batteries via Synthesized Design of Luenberger and Learning Observers. *IEEE Trans. Control Syst. Technol.* **2013**, *22*, 290–298. [CrossRef]
23. Kim, I.S. Nonlinear State of Charge Estimator for Hybrid Electric Vehicle Battery. *IEEE Trans. Power Electron.* **2008**, *23*, 2027–2034.
24. Zheng, G.; Efimov, D.; Bejarano, F.J.; Perruquetti, W.; Wang, H. Interval observer for a class of uncertain nonlinear singular systems. *Automatica* **2016**, *71*, 159–168. [CrossRef]
25. Berger, T.; Reis, T. Observers and Dynamic Controllers for Linear Differential-Algebraic Systems. *SIAM J. Control Optim.* **2017**, *55*, 3564–3591. [CrossRef]
26. Duan, G.R. Analysis and Design of Descriptor Linear Systems. In *Advances in Mechanics and Mathematics*; Springer New York: New York, NY, USA, 2010; Volume 23.
27. Löfberg, J. YALMIP: A Toolbox for Modeling and Optimization in MATLAB. In Proceedings of the CACSD Conference, Taipei, Taiwan, 2–4 September 2004; IEEE: New York, NY, USA, 2004.

Article

A New Nonlinear Dynamic Speed Controller for a Differential Drive Mobile Robot

Ibrahim A. Hameed [1,*], Luay Hashem Abbud [2], Jaafar Ahmed Abdulsaheb [3], Ahmad Taher Azar [4,5,*], Mohanad Mezher [6], Anwar Ja'afar Mohamad Jawad [7], Wameedh Riyadh Abdul-Adheem [8], Ibraheem Kasim Ibraheem [8,*] and Nashwa Ahmad Kamal [9]

1 Department of ICT and Natural Sciences, Norwegian University of Science and Technology, Larsgårdsve-gen, 2, 6009 Ålesund, Norway
2 Air Conditioning and Refrigeration Techniques Engineering Department, Al-Mustaqbal University College, Hillah 51001, Iraq
3 Department of Electronics and Communication, College of Engineering, Uruk University, Baghdad 10001, Iraq
4 College of Computer and Information Sciences, Prince Sultan University, Riyadh 11586, Saudi Arabia
5 Faculty of Computers and Artificial Intelligence, Benha University, Benha 13518, Egypt
6 Faculty of Pharmacy, The University of Mashreq, Baghdad 10001, Iraq
7 Department of Computer Techniques Engineering, Al-Rafidain University College, Baghdad 46036, Iraq
8 Department of Electrical Engineering, College of Engineering, University of Baghdad, Baghdad 10001, Iraq
9 Faculty of Engineering, Cairo University, Giza 12613, Egypt
* Correspondence: ibib@ntnu.no (I.A.H.); aazar@psu.edu.sa or ahmad.azar@fci.bu.edu.eg or ahmad_t_azar@ieee.org (A.T.A.); ibraheemki@coeng.uobaghdad.edu.iq (I.K.I.)

Abstract: A disturbance/uncertainty estimation and disturbance rejection technique are proposed in this work and verified on a ground two-wheel differential drive mobile robot (DDMR) in the presence of a mismatched disturbance. The offered scheme is the an improved active disturbance rejection control (IADRC) approach-based enhanced dynamic speed controller. To efficiently eliminate the effect produced by the system uncertainties and external torque disturbance on both wheels, the IADRC is adopted, whereby all the torque disturbances and DDMR parameter uncertainties are conglomerated altogether and considered a generalized disturbance. This generalized disturbance is observed and cancelled by a novel nonlinear sliding mode extended state observer (NSMESO) in real-time. Through numerical simulations, various performance indices are measured, with a reduction of 86% and 97% in the *ITAE* index for the right and left wheels, respectively. Finally, these indices validate the efficacy of the proposed dynamic speed controller by almost damping the chattering phenomena and supplying a high insusceptibility in the closed-loop system against torque disturbance.

Keywords: mobile robot; speed controller; active disturbance rejection control; extended state observer; chattering phenomenon; torque disturbance; system uncertainties

1. Introduction

Generally, in most engineering applications, disturbances/uncertainties (D/Us) are widely presented and negatively affect the performance of the control systems [1]. Control engineering strives to minimize D/Us, and feedforward methods may attenuate or reject the effect of disturbances that can be detected through measurement [2]. Nevertheless, exogenous disturbances cannot be calculated or are exceptionally difficult to calculate. The first spontaneous thought to treat this challenge is to build an observer to estimate the disturbance. Then, an activation signal can be established to compensate for the exogenous disturbance effect. The simplicity of this indication can be expanded to also reject uncertainties. The unmodeled effects of uncertainties or dynamics can be estimated as a proportion of the overall disturbance. As a consequence, a new term was introduced for disturbance activity, which is known as "total disturbance", which describes the accumulation of exogenous disturbances, unmodeled dynamics, and uncertain conditions in plants. This class of techniques is denoted as estimation and attenuation of disturbance/uncertainty (EAD/U). Several EAD/U structures have been individually suggested. Han first suggested an extended state observer (ESO) in the 1990s [3]. An ESO is generally viewed as playing a major essential role in the technique termed active disturbance rejection control (ADRC) [4]. ADRC consists of three essential parts: a tracking differentiator (TD), extended state observer (ESO), and nonlinear state error feedback controller (NLSEF).

In precision assembly applications, ADRC has been used as a whole configuration; it has been used to perform high-accuracy control of ball screw feed drives [5]. Likewise, a double-loop ADRC scheme was utilized for an active hydraulic suspension system [6]. Taking into account the fact that ADRC is very useful in the field of robotics, this method is particularly useful for the control of quad helicopters because of its capability to handle nonlinear models with significant unsettling influences with vulnerability [7]. Moreover, many engineering systems with ADRC have proven successful [8–10]. The main objective of this work is to design a controller that provides an active rejection of the bounded mismatched total disturbances, which have a direct effect on the performance of permanent magnet direct current (PMDC) motors of the DDMR. The controller guarantees a minimum orientation error despite disturbances. Exogenous disturbance involves disturbances including friction torques, fluctuations of the load, changes in parameters for the actuators, and external disturbances that occur due to collisions with obstacles.

The contribution of this paper lies in applying an improved version of Han's classical ADRC to motion control of a DDMR, which is a nonlinear, multi-input–multi-output (MIMO) system, as an extension of our four previous published papers [11–14]. The proposed IADRC is constructed by combining three primary units. The first unit is the improved nonlinear tracking differentiator (INTD), which is used to obtain a smooth and accurate differentiation of any nonlinear signal. The INTD also declines signals with frequencies outside a certain frequency band. The second unit in the proposed controller is the improved nonlinear state error feedback (INSEF) controller. This unit is derived by combining the nonlinear gains and the classical PID controller with a new control structure. The last unit is the sliding mode extended state observer (SMESO), which is an expansion of the linear extended state observer (LESO) method; to reduce the chattering in the control signal, the nonlinearity and a sliding mode term are added to the LESO to obtain the proposed SMESO, which performs better than the LESO.

The remainder of this work is structured as follows: Section 2 presents the main results of the IADRC. In Section 3, the convergence of the proposed observers, in addition to stability analysis of the closed-loop system, is investigated. Handling of mismatched disturbances is analyzed within the context of the ADRC in Section 4. Mathematical modeling of the DDMR and PMDC is introduced in Section 5. Section 6 presents the numerical simulations of the proposed IADRC control scheme on DDMR. Finally, the work is concluded in Section 7.

2. The Main Results: Improved Active Disturbance Rejection Control (IADRC)

Classical active disturbance rejection control is a powerful controlling method that was first suggested by J. Han [4]. Classical ADRC can be structured by gathering a linear extended state observer (LESO), a tracking differentiator (TD), and a nonlinear state error feedback (NLSEF); the entire structure is presented in [4,15,16].

The enhanced configuration of the improved active disturbance rejection control (IADRC) is shown in Figure 1. The following subsections discuss each part of the proposed control scheme supported by necessary explanations.

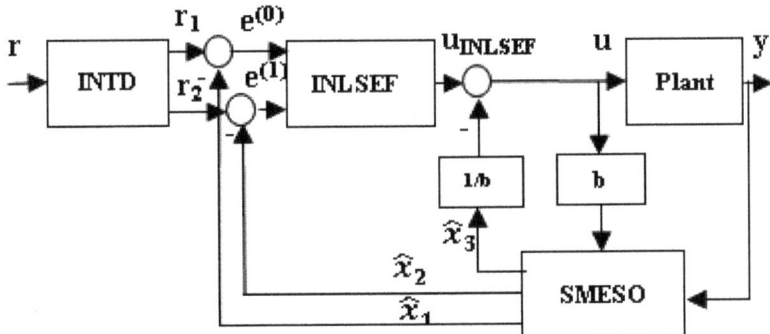

Figure 1. Schematic diagram of the second-order IADRC.

2.1. The Improved Nonlinear TD (INTD)

The INTD is the improved version of the classical tracking differentiator. The improvement is achieved by adopting a smooth sigmoid nonlinear function $\varphi(.) = tanh(\cdot)$, instead of a $sign(\cdot)$ function. The reason behind choosing the sigmoid function $tanh(\cdot)$ is that the $\varphi(.) = tanh(\cdot)$ near the origin provides a slope with a smooth shape, which reduces the chattering phenomenon and speeds up the convergence of the proposed tracking differentiator in a significant way. Moreover, adding nonlinearity to the design of the TD increases the robustness of the proposed TD against noise. Another improvement is introduced by integrating nonlinear and linear parts. This TD presents an enhanced dynamic performance relative to Han's TD. An INTD for second-order systems has been designed using the hyperbolic tangent function [11,17],

$$\begin{cases} \dot{r}_1 = r_2 \\ \dot{r}_2 = -R^2 \varphi(r_1(t) - r(t)) - R r_2 \end{cases} \quad (1)$$

where $\varphi(r_1(t) - r(t)) = \tanh\left(\frac{\beta r_1 - (1-\alpha)r}{\gamma}\right)$, r is the reference signal, and r_1 and r_2 are the tracking reference and its derivative, respectively. The coefficients R, β, γ, and α are tuning coefficients, with $0 < \alpha \langle 1, \beta \rangle 1, \gamma > 0$, and $R > 0$. The configuration with the proposed INTD can effectively eliminate the chattering phenomenon and measurement noise and provide swift and smooth tracking of the desired reference signal. To check the stability of the proposed tracking differentiator, the Lyapunov stability approach is utilized [11].

Definition 1 (simple sigmoid functions) [18]. *a function ($\varphi : \mathbb{R} \to (-1,1)$) is supposed to be a sigmoid. The sigmoid function meets the following conditions.*
1. *The function $\varphi(\cdot)$ is smooth, i.e., $\varphi(x) \in C^\infty$;*
2. *$\varphi(\cdot)$ is an odd function;*
3. *The function $\varphi(\cdot)$ satisfies $\lim_{x \to \pm\infty} |\varphi(x)| = 1$.*

Assumption 1. *The function $\varphi(.)$ in definition (4.1) is an odd function with $\psi(y) = \int_0^y \varphi(u) du \geq 0$, where u is a variable without any special physical meaning.*

The proposed INTD has the following advantages relative to other tracking differentiators:

(i) The proposed tracking differentiator is built using a smooth nonlinear function ($\varphi(\cdot)$) instead of the $sign(\cdot)$ function used in most conventional nonlinear differentiators. This is an essential step toward preventing a chattering phenomenon from the output derivatives;

(ii) A second improvement is accomplished by combining the linear and the nonlinear terms. The benefits of this are clear in suppressing high-frequency components in the signal, such as noise. With this feature, the proposed GTD also achieves better performance than other tracking differentiators;

(iii) The saturation feature of the function $\varphi(\cdot)$ increases the robustness against noisy signals because for large errors, even with a wide range of noise, it is mapped to a small domain set of the function $\varphi(\cdot)$ (see Figure 2, range and domain sets A);

(iv) Increasing the slope of the continuous function $\varphi(\cdot)$ near the origin significantly accelerates the convergence of the proposed tracking differentiator (see Figure 2, range and domain sets B).

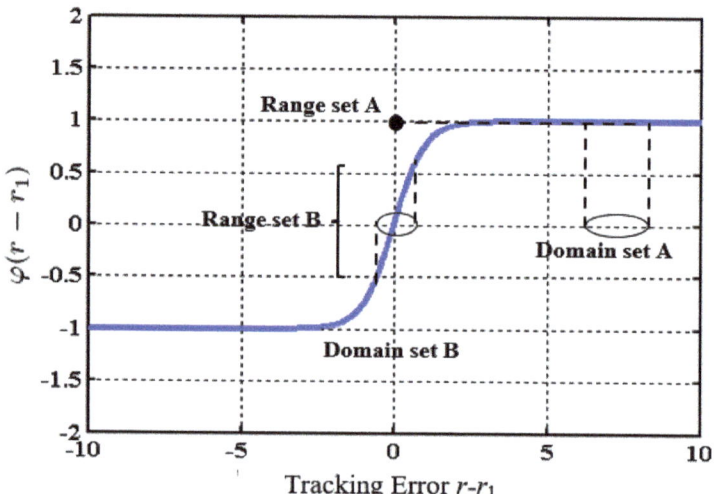

Figure 2. The domain and range sets of the function $\varphi(\cdot)$

The convergence of the proposed INTD is investigated in the next theorem.

Theorem 1. *Consider the dynamic system (1). If the signal $r(t)$ is differentiable and $sup_{t\in[0,\infty)}|\dot{r}(t)| = B < \infty$, then the solution of (1) is convergent in the sense that, $r_1(t)$ is convergent to $r(t)$ as $R \to \infty$.*

Proof. Let, $t = \frac{\tau}{R}$. Then

$$\dot{r}_i(t) = \frac{dr_i(t)}{d\tau}\frac{d\tau}{dt} = R\frac{dr_i(t)}{d\tau}, i \in \{1,2\} \quad (2)$$

Combining (1) and (2) yields

$$\begin{cases} R\frac{dr_1(\frac{\tau}{R})}{d\tau} = r_2(\frac{\tau}{R}) \\ R\frac{dr_2(\frac{\tau}{R})}{d\tau} = -R^2\varphi(r_1(\frac{\tau}{R}) - r(\frac{\tau}{R})) - Rr_2(\frac{\tau}{R}) \end{cases} \quad (3)$$

which leads to

$$\begin{cases} \frac{dr_1(\frac{\tau}{R})}{d\tau} = \frac{1}{R}r_2(\frac{\tau}{R}) \\ \frac{dr_2(\frac{\tau}{R})}{d\tau} = -R\varphi(r_1(\frac{\tau}{R}) - r(\frac{\tau}{R})) - r_2(\frac{\tau}{R}) \end{cases} \quad (4)$$

Assume

$$\begin{cases} z_1(\tau) = r_1(\frac{\tau}{R}) - r(\frac{\tau}{R}), \\ z_2(\tau) = \frac{1}{R}r_2(\frac{\tau}{R}) \end{cases} \quad (5)$$

which results in

$$\begin{cases} \frac{dz_1(\tau)}{d\tau} = \frac{dr_1(\frac{\tau}{R})}{d\tau} - \frac{dr(\frac{\tau}{R})}{d\tau} \\ \frac{dz_2(\tau)}{d\tau} = \frac{1}{R}\frac{dr_2(\frac{\tau}{R})}{d\tau} \end{cases} \quad (6)$$

This, together with (4), yields,

$$\begin{cases} \frac{dz_1(\tau)}{d\tau} = \frac{1}{R} r_2\left(\frac{\tau}{R}\right) - \frac{dr\left(\frac{\tau}{R}\right)}{d\tau}, \\ \frac{dz_2(\tau)}{d\tau} = \frac{1}{R}\left[-R\varphi\left(r_1\left(\frac{\tau}{R}\right) - r\left(\frac{\tau}{R}\right)\right) - r_2\left(\frac{\tau}{R}\right)\right] \end{cases} \quad (7)$$

Then,

$$\begin{cases} \dot{z}_1(\tau) = \frac{1}{R} r_2\left(\frac{\tau}{R}\right) - \frac{dr\left(\frac{\tau}{R}\right)}{d\tau}, \\ \dot{z}_2(\tau) = -\varphi\left(r_1\left(\frac{\tau}{R}\right) - r\left(\frac{\tau}{R}\right)\right) - \frac{1}{R} r_2\left(\frac{\tau}{R}\right) \end{cases} \quad (8)$$

Substituting (5) and (8), we obtain,

$$\begin{cases} \dot{z}_1(\tau) = z_2(\tau) - \frac{dr\left(\frac{\tau}{R}\right)}{d\tau}, \\ \dot{z}_2(\tau) = -\varphi(z_1(\tau)) - z_2(\tau) \end{cases} \quad (9)$$

Select the candidate Lyapunov function ($V(z)$) as

$$V(z) = \int_0^{z_1} \varphi(v)\, dv + \frac{1}{2} z_2^2(\tau) \quad (10)$$

The total derivative of $V(z)$ with respect to τ along the trajectory of the system (9) is given as,

$$\dot{V}(z) = \varphi(z_1)\dot{z}_1 + z_2 \dot{z}_2 \quad (11)$$

This, together with (8), yields,

$$\dot{V}(z) = \varphi(z_1)\left[z_2(\tau) - \frac{dr\left(\frac{\tau}{R}\right)}{d\tau}\right] + z_2[-\varphi(z_1(\tau)) - z_2(\tau)] \quad (12)$$

which is derived from

$$\dot{V}(z) = -\varphi(z_1)\frac{dr\left(\frac{\tau}{R}\right)}{d\tau} - z_2^2 \quad (13)$$

Finally, we obtain

$$\dot{V}(z) \leq |\varphi(z_1)||\dot{r}(t)|\frac{1}{R} \quad (14)$$

According to Assumption 1 and Definition 1,

$$\dot{V}(z) \leq \frac{B}{R} \quad (15)$$

$$\lim_{R \to \infty} \dot{V}(z) < 0 \quad (16)$$

Then, the solution of (9) is globally asymptotically stable (GAS) by invoking LaSalle's invariance principle [19]. It follows that $\lim_{R \to \infty} z_1 = 0$. According to (5), we obtain

$$\lim_{R \to \infty} r_1 = r \quad (17)$$

□

2.2. The Improved Nonlinear State Error Feedback Controller (INSEFC)

Consider the following observable nth-order nonlinear affine-in-control system,

$$\begin{cases} \xi^{(n)} = f\left(\xi, \dot{\xi}, \ldots, \xi^{(n-1)}, t\right) + bu \\ y = \xi \end{cases} \quad (18)$$

where $u(t) \in C(\mathbb{R}, \mathbb{R})$ is the control input, $y(t) \in C(\mathbb{R}, \mathbb{R})$ is the measured output, $b \in \mathbb{R}$ is the input gain, and $f \in C(\mathbb{R}^n \times \mathbb{R}, \mathbb{R})$ is a nonlinear function. It is necessary to design a nonlinear feedback controller ($\Psi : \mathbb{R} \to \mathbb{R}$) such that the control effort ($u(t)$) is at its minimum while achieving the following:

1. The closed-loop system is asymptotically stable in the presence of external disturbances, system uncertainties, and measurement noise;
2. The output ($y(t)$) is forced to track a known reference signal ($r(t)$), i.e., $\lim_{t \to \infty} |r(t) - y(t)| = 0$, satisfying the transient response specifications;
3. The chattering phenomenon in the control signal ($u(t)$) is reduced.

The original version of the nonlinear state error feedback (SEF) functions in the form of *fal*(.) was first proposed by Han [4] and expressed as,

$$fal(e, \alpha, \delta) = \begin{cases} \frac{e}{\delta^{1-\alpha}} & |e| \leq \delta \\ |e|^\alpha sgn(e) & |e| > \delta \end{cases} \qquad (19)$$

where δ is a small number used to express the domain of the linear function near zero [3], and $0 < \alpha < 1$. The $fal(\cdot)$ is a nonsmooth, piecewise, continuous, nonlinear saturation and a monotonously increasing function [20–23]. The curve of the $fal(\cdot)$ function when $\delta = 0.1$ is shown in Figure 3a. The curve of the $fal(\cdot)$ function when $\alpha = 0.25$ is shown in Figure 3b. The $fal(\cdot)$ function is nonsmooth at the inflection point [24], and when the value of δ is too small, it is still easy for the phenomenon of high-frequency chattering to appear. This is true even for large δ values [25].

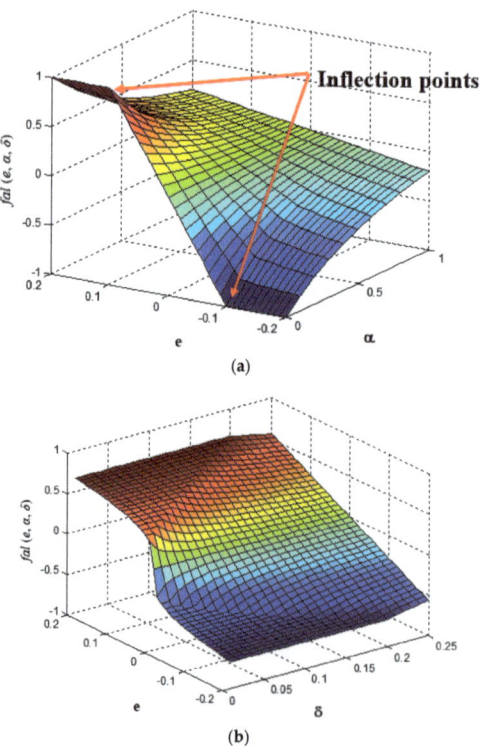

Figure 3. The curve of the $fal(\cdot)$ function: (a) $\delta = 0.1$; (b) $\alpha = 0.25$.

When $\alpha = 0.75$, the $fal(\cdot)$ function is almost linear. In practical terms, the value of α is generally selected as $\delta = 0.01$ [26] and can be further tuned and determined by experiments [27].

The improved nonlinear state error feedback control (INSEFC) law provides more shape flexibility within a wide range of the state error vector. This behavior improves both the performance and the robustness of the controlled system.

The enhanced nonlinear control law uses exponential functions and $sign(.)$, and it is established as follows,

$$u_{INLSEF} = \Psi(e) = k(e)^T f(e) + u_{integrator} \qquad (20)$$

where e is the $n \times 1$ state error vector, which is defined as,

$$e = \begin{bmatrix} e^{(0)} & \ldots e^{(i)} \ldots & e^{(n-1)} \end{bmatrix}^T \qquad (21)$$

where $e^{(i)}$ is the state error derivative of an nth order and expressed as,

$$e^{(i)} = r_{i+1} - \hat{\xi}_{i+1} \qquad (22)$$

$k(e)$ is a function of nonlinear gains and expressed as,

$$k(e) = \begin{pmatrix} k_1(e) \\ \vdots \\ k_i(e) \\ \vdots \\ k_n(e) \end{pmatrix} = \begin{pmatrix} \left(k_{11} + \dfrac{k_{12}}{1+\exp\left(\mu_1(e^{(0)})^2\right)}\right) \\ \vdots \\ \left(k_{i1} + \dfrac{k_{i2}}{1+\exp\left(\mu_i(e^{(i-1)})^2\right)}\right) \\ \vdots \\ \left(k_{n1} + \dfrac{k_{n2}}{1+\exp\left(\mu_n(e^{(n-1)})^2\right)}\right) \end{pmatrix} \qquad (23)$$

where k_{i1}, k_{i2}, and μ_i are positive coefficients, and $i \in \{1,2,\ldots,n\}$,. The advantage of $k(e)_i$ is that it improves the nonlinear controller's ability to detect even small errors. When $e^{(i-1)} = 0$, $k(e)_i = k_{i1} + k_{i2}/2$, while as $e^{(i-1)}$ increases, $k(e)_i \approx k_{i1}$. For values of $e^{(i-1)}$ in between, the value of $k(e)_i$ lies in the sector of $[k_{i1}, k_{i1}+k_{i2}/2]$, as shown in Figure 4.

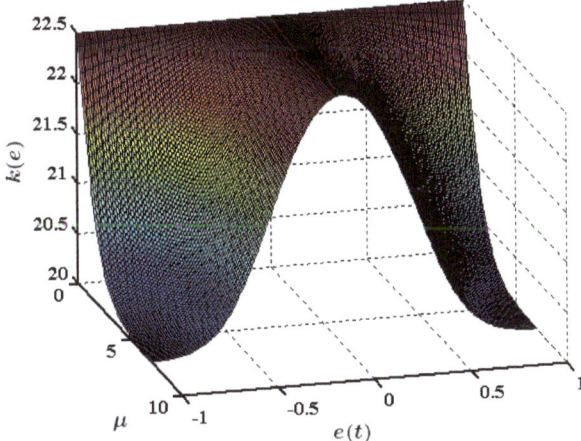

Figure 4. Characteristics of the nonlinear gain function ($k_i(e)$) for $k_{i1} = 20$ and $k_{i2} = 5$.

The function $f(e)$ is expressed as,

$$f(e) = \left[\left|e^{(0)}\right|^{\alpha_1} sign(e) \quad \cdots \quad \left|e^{(i)}\right|^{\alpha_i} sign\left(e^{(i)}\right) \cdots \quad \left|e^{(n-2)}\right|^{\alpha_n} sign\left(e^{(n-1)}\right) \right]^T \quad (24)$$

Equation (24) shows significant features in the nonlinear term $|e|^{\alpha} sign(e)$. For $\alpha_i \ll 1$, the term rapidly switches its state, as shown in Figure 5a. This feature makes the error function ($f(e)$) sensitive to small error values. When α exceeds 1, the nonlinear term becomes less sensitive to small variations in e (see Figure 5b).

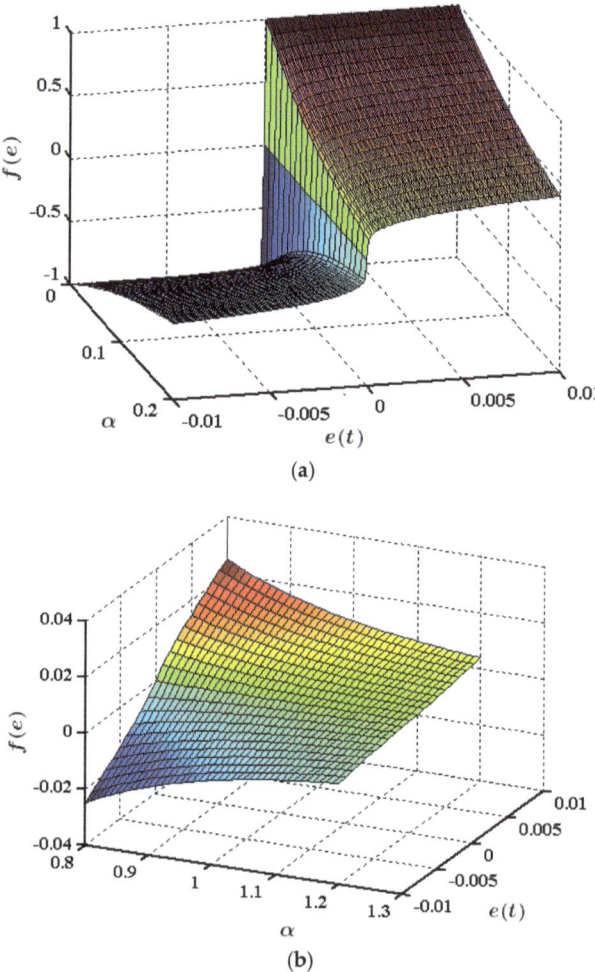

Figure 5. Characteristics of the nonlinear error function ($f(e)$): (a) $0 \leq \alpha \leq 0.2$; (b) $0.8 \leq \alpha \leq 1.2$.

The control signal (u) can be limited using the nonlinear hyperbolic function ($tanh(\cdot)$) in the form,

$$u = \delta \tanh\left(\frac{u_{INLSEF}}{\delta}\right) \quad (25)$$

where u_{INLSEF} is defined in (17) and has the following features:
(i) Any real number $(-\infty, \infty)$ is mapped to a number in the range of $[-\delta, \delta]$;

(ii) The $tanh(\cdot)$ function is symmetric about the origin, and only zero-valued inputs are mapped to zero outputs;
(iii) The control action (u) is limited via mapping but not clipped. Therefore, there are no strong harmonics in the high-frequency range.

Figure 6 shows the control signal (u) against $e(t)$ and $\dot{e}(t)$, considering (25).

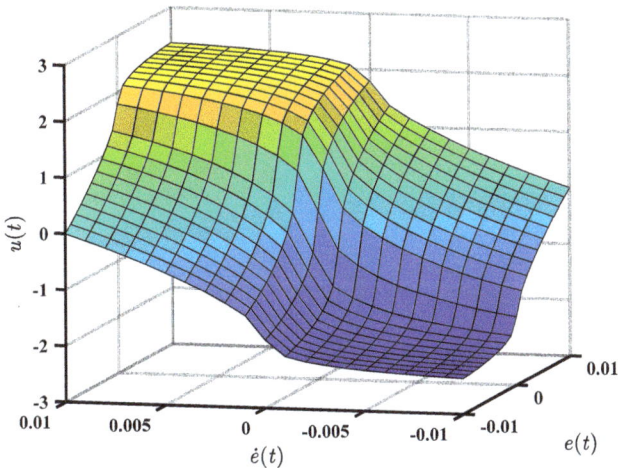

Figure 6. The characteristics of the control signal (u) of (25): $n = 2$, $k_{11} = 20$, $k_{12} = 5$, $k_{21} = 20$, $k_{22} = 5$, $\mu_1 = 2.5$, $\mu_2 = 1.5$, $\alpha_1 = 0.5$, $\alpha_2 = 0.5$, and $\delta = 2.5$.

Theorem 2. *Consider the following observable second-order nonlinear control system ($n = 2$)*

$$\begin{cases} \ddot{\varsigma} = f\left(\varsigma, \dot{\varsigma}\right) + bu, \\ y = \varsigma. \end{cases} \quad (26)$$

as shown in Figure 7a. The PD controller is described as,

$$u = k_p e + k_d \dot{e} \quad (27)$$

where the tracking error is $e = r - y$. Then, the linear control law (u) can be generalized to the form $u = \Psi(e)$ (see Figure 7b) such that Ψ is sector-bounded and satisfies $\Psi(0) = 0$.

Proof. Let $x_1 = x$, and $x_2 = \dot{x}$. Then, the system (26) can be represented as,

$$\begin{cases} \dot{\varsigma}_1 = \varsigma_2, \\ \dot{\varsigma}_2 = f(\varsigma_1, \varsigma_2) + bu, \\ y = \varsigma_1 \end{cases} \quad (28)$$

Consider a convergent TD, which is described as $\lim_{t \to \infty} |r_1 - r| = 0$, $\lim_{t \to \infty} |r_2 - \dot{r}| = 0$. Let a convergent state observer be characterized by $\lim_{t \to \infty} |\tilde{\varsigma}_1 - \varsigma_1| = 0$, and $\lim_{t \to \infty} |\tilde{\varsigma}_2 - \varsigma_2| = 0$. Since the tracking error is $e = y - r$, $\dot{e} = \dot{y} - \dot{r}$; then, the two errors can be defined as $\lim_{t \to \infty} e = \lim_{t \to \infty} (\tilde{\varsigma}_1 - r_1)$ and $\lim_{t \to \infty} \dot{e} = \lim_{t \to \infty} (\tilde{\varsigma}_2 - r_2)$. Finally, as $t \to \infty$, the control law (25) takes the following form: $u = k_p(r_1 - \tilde{\varsigma}_1) + k_d(r_2 - \tilde{\varsigma}_2)$.

This formula can be expanded for an nth-order system to take the following form: $u = K^T e$, where $K = (k_1, k_2, \ldots, k_n)^T$ is the gain vector, $e = \left(e, \dot{e}, \ldots, e^{(n-1)}\right)^T$ is the tracking error vector, and the linear combination can be generalized to a nonlinear combination formula described as $u = \Psi(e)$. □

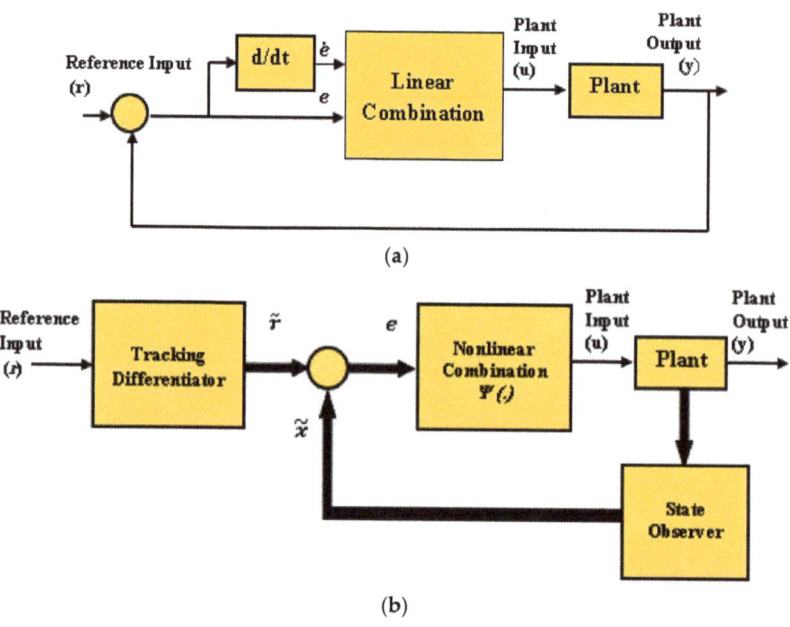

Figure 7. The SISO system in Theorem 1. (**a**) Linear combination control law; (**b**) nonlinear combinational control law.

2.3. Sliding Mode Extended State Observer (SMESO)

In state space form, the suggested SMESO can be expressed as follows,

$$\dot{\hat{\zeta}} = F\hat{X} + B_1 u + B_2 g(y - \hat{\zeta}_1) \tag{29}$$

where $\hat{\zeta} \in R^{(n+1) \times 1}$ is a vector that comprises the observed total disturbance and states of the plant, $\hat{X} \in R^{(n+1) \times 1}, B_1 \in R^{(n+1) \times 1}, B_2 \in R^{(n+1) \times 1}$, and $F \in R^{(n+1) \times (n+1)}$.

$$\zeta = [\zeta_1 \ \zeta_2 \ \cdots \ \zeta_{n+1}]^T, \dot{\hat{\zeta}} = [\dot{\hat{\zeta}}_1 \ \dot{\hat{\zeta}}_2 \ \cdots \ \dot{\hat{\zeta}}_{n+1}]^T$$

$$F = \begin{bmatrix} 0 & 1 & 0 & 0 & \cdots & 0 \\ 0 & 0 & 1 & 0 & \cdots & 0 \\ 0 & 0 & 0 & 1 & \cdots & 0 \\ 0 & \vdots & \vdots & \vdots & \ddots & \vdots \\ 0 & 0 & 0 & 0 & \cdots & 1 \\ 0 & 0 & 0 & 0 & 0 & 0 \end{bmatrix} \tag{30}$$

$$B_1 = [0 \ 0 \ \cdots 1 \ 0]^T, B_2 = [\beta_1 \ \beta_2 \ \cdots \ \beta_{n+1}]^T$$

Now, $g(y - \hat{\zeta}_1) = K_\alpha |y - \hat{\zeta}_1|^\alpha sign(y - \hat{\zeta}_1) + K_\beta |y - \hat{\zeta}_1|^\beta (y - \hat{\zeta}_1)$, where K_α, α, K_β, and β are appropriate design parameters. With $n = 2$, the SMESO can be expressed as,

$$\begin{cases} \dot{\hat{\zeta}}_1 = x_2 + \beta_1 (K_\alpha |y - \hat{\zeta}_1|^\alpha sign(y - \hat{\zeta}_1) + K_\beta |y - \hat{\zeta}_1|^\beta (y - \hat{\zeta}_1)) \\ \dot{\hat{\zeta}}_2 = \hat{\zeta}_3 + bu + \beta_2 (K_\alpha |y - \hat{\zeta}_1|^\alpha sign(y - \hat{\zeta}_1) + K_\beta |y - \hat{\zeta}_1|^\beta (y - \hat{\zeta}_1)) \\ \dot{\hat{\zeta}}_3 = \beta_3 (K_\alpha |y - \hat{\zeta}_1|^\alpha sign(y - \hat{\zeta}_1) + K_\beta |y - \hat{\zeta}_1|^\beta (y - \hat{\zeta}_1)) \end{cases} \quad (31)$$

The SMESO is the nonlinear modified version of the LESO. The proposed SMESO is the third part of the IADRC, which considers the main part that is used to actively estimate what is known as the "total disturbance". Compared with the LESO, SMESO performs better when it comes to reducing chattering in control signals. In [13], the proposed SMESO demonstrated in detail that estimation error converges to zero asymptotically for nonlinear gain functions. With a sliding term, estimation accuracy is increased for the nonlinear extended state observer. As a result, the proposed method achieves excellent performance when it comes to smoothed control signals, requiring less control energy to accomplish the intended result [13].

3. Convergence and Stability Analysis

In this section, the convergence of the proposed SMESO and the stability of the closed-loop system are investigated in detail to validate the proposed design techniques.

3.1. Convergence Analysis of the Proposed SMESO

To prove the convergence of the SMESO, the following assumptions are needed.

Assumption 2. *There exists an upper bound for the time derivative of the generalized disturbance (i.e., at least $\dot{L} \in C^1$ and $sup_{t \in [0,\infty)} |\dot{L}| = M < \infty$, where $\in \mathbb{R}$);*

Assumption 3. *L is a continuously differentiable function;*

Assumption 4. *$V : \mathbb{R}^{n+1} \to \mathbb{R}^+$ and $W : \mathbb{R}^{n+1} \to \mathbb{R}^+$ are continuously differentiable functions with [16],*

$$\lambda_1 \|\eta\|^2 \leq V(\eta) \leq \lambda_2 \|\eta\|^2, \quad W(\eta) = \|\eta\|^2 \quad (32)$$

$$\sum_{i=1}^{n-1} \frac{\partial V(\eta)}{\eta_i} \left(\eta_{i+1} - a_i k \left(\frac{\eta_1}{\omega_0^\rho} \right) \cdot \eta_1 \right) - \frac{\partial V(\eta)}{\partial y_n} a_n k \left(\frac{\eta_1}{\omega_0^n} \right) \eta_1 \leq -W(\eta) \quad (33)$$

Theorem 3. *(SMESO convergence). Given the system of (18) and SMESO of (29), it follows that under assumptions A3 and A5, for any initial conditions,*

(i) $\lim\limits_{t \to \infty} |\zeta_i(t) - \hat{\zeta}_i(t)| = O\left(\frac{1}{\omega_0^{n+2-i}}\right)$

(ii) $\lim\limits_{\substack{t \to \infty \\ \omega_0 \to \infty}} |\zeta_i(t) - \hat{\zeta}_i(t)| = 0$

where ζ_i and $\hat{\zeta}_i$ symbolize the state of (18) and (29), respectively, where $i \in \{1, 2, \ldots, n+1\}$.

Proof. Let $e_i = \xi_i - \hat{\xi}_i$, $1 \in \{1, 2, \ldots, n+1\}$. Correspondingly, let

$$\eta_i = \omega_0^{n-i} e_i \left(\frac{t}{\omega_0} \right), \quad i \in \{1, 2, \ldots, n+1\} \quad (34)$$

Then, the dynamics of the estimation error can be expressed in a time scale as,

$$\begin{cases} \frac{d\eta_1}{dt} = \eta_2 - a_1 k\left(\frac{\eta_1}{\omega_0^{n-1}}\right)\eta_1 \\ \frac{d\eta_2}{dt} = \eta_3 - a_2 k\left(\frac{\eta_1}{\omega_0^{n-1}}\right)\eta_1 \\ \vdots \\ \frac{d\eta_n}{dt} = \eta_n - a_n k\left(\frac{\eta_1}{\omega_0^{n-1}}\right)\eta_1 \\ \frac{d\eta_{n+1}}{dt} = \frac{\Delta_h}{\omega_0^2} - a_{n+1} k\left(\frac{\eta_1}{\omega_0^{n-1}}\right)\eta_1 \end{cases} \quad (35)$$

Let the candidate Lyapunov functions $(V, W : \mathbb{R}^{n+1} \to \mathbb{R}^+)$ denoted by $V(\eta) = \langle P\eta, \eta \rangle = \eta^T P \eta$, where $\eta \in \mathbb{R}^{n+1}$, and P is a positive definite symmetric matrix. Consider (22) of assumption A4 with $\lambda_1 = \lambda_{min}(P)$ and $\lambda_2 = \lambda_{max}(P)$, where $\lambda_{min}(P)$ and $\lambda_{max}(P)$ are the minimum and maximum eigenvalues of P, respectively. $\dot{V}$ with regard to t over η (over the solution of (35))is determined as follows:

$$\dot{V}(\eta)\Big|_{\text{along (35)}} = \sum_{i=1}^{n+1} \frac{\partial V(\eta)}{\partial \eta_i}\dot{\eta}_i(t) \quad (36)$$

Then,

$$\dot{V}(\eta)\Big|_{\text{along (35)}} = \sum_{i=1}^{n-1} \frac{\partial V(\eta)}{\eta_i}\left(\eta_{i+1}(t) - a_i k\left(\frac{\eta_1(t)}{\omega_0^n}\right).\eta_1(t)\right) - \frac{\partial V(\eta)}{\partial \eta_n} a_n k\left(\frac{\eta_1(t)}{\omega_0^n}\right).\eta_1(t) + \frac{\partial V(\eta)}{\partial \eta_{n+1}} \frac{M}{\omega_0^2} \quad (37)$$

Consider (33) of assumption A4; then,

$$\dot{V}(\eta)\Big|_{\text{along (35)}} \leq -W(\eta) + \frac{\partial V(\eta)}{\partial \eta_{n+1}} \frac{M}{\omega_0^2} \quad (38)$$

As $V(\eta) \leq \lambda_{max}(P)\|\eta\|^2$ and $\left|\frac{\partial V(\eta)}{\partial \eta_{n+1}}\right| \leq \|\frac{\partial V(\eta)}{\partial \eta}\|$, then $\left|\frac{\partial V}{\partial \eta_{n+1}}\right| \leq 2\lambda_{max}(P)\|\eta\|$. As $V(\eta) \leq \lambda_{max}(P)\|\eta\|^2 = \lambda_{max}(P)W(\eta)$. Thus, $-W(\eta) \leq -\frac{V(\eta)}{\lambda_{max}(P)}$. Finally, because $\lambda_{min}(P)\|\eta\|^2 \leq V(\eta)$, this leads to $\|\eta\| \leq \sqrt{\frac{V(\eta)}{\lambda_{min}(P)}}$. Accordingly, and given assumption A4, $\dot{V}(\eta)$ becomes,

$$\dot{V}(\eta) \leq -\frac{V(\eta)}{\lambda_{max}(P)} + \frac{M}{\omega_0^2}2\lambda_{max}(P)\frac{\sqrt{V(\eta)}}{\sqrt{\lambda_{min}(P)}}. \text{ Since } \frac{d}{dt}\sqrt{V(\eta)} = \frac{1}{2}\frac{1}{\sqrt{V(\eta)}}\dot{V}(\eta), \text{ then,}$$

$$\frac{d}{dt}\sqrt{V(\eta)} \leq \frac{1}{2}\frac{1}{\sqrt{V(\eta)}}\left(-\frac{V(\eta)}{\lambda_{max}(P)} + \frac{M}{\omega_0^2}2\lambda_{max}(P)\frac{\sqrt{V(\eta)}}{\sqrt{\lambda_{min}(\eta)}}\right) \quad (39)$$

which gives

$$\frac{d}{dt}\sqrt{V(\eta)} \leq -\frac{\sqrt{V(\eta)}}{2\lambda_{max}(P)} + \frac{M}{\omega_0^2}\frac{\lambda_{max}(P)}{\sqrt{\lambda_{min}(P)}} \quad (40)$$

which can be solved as

$$\sqrt{V(\eta)} \leq \frac{2M\lambda_{max}^2(P)}{\omega_0^2\sqrt{\lambda_{min}(P)}}\left(1 - e^{-\frac{t}{2\lambda_{max}(P)}}\right) + \sqrt{V(\eta(0))}e^{-\frac{t}{2\lambda_{max}(P)}}$$

According to assumption A4, we have $\lambda_{min}(P)\|\eta\|^2 \leq V(\eta)$. This leads to $\|\eta\| \leq \sqrt{\frac{V(\eta)}{\lambda_{min}(P)}}$. Then,

$$\|\eta\| \leq \sqrt{\frac{1}{\lambda_{min}(P)}\left(\frac{2M\lambda_{max}^2(P)}{\omega_0^2\sqrt{\lambda_{min}(P)}}\left(1-e^{-\frac{t}{2\lambda_{max}(P)}}\right)+\sqrt{V(\eta(0))}e^{-\frac{t}{2\lambda_{max}(P)}}\right)}$$

which yields

$$\|\eta\| \leq \frac{2M\lambda_{max}^2(P)}{\omega_0^2\lambda_{min}(P)}\left(1-e^{-\frac{t}{2\lambda_{max}(P)}}\right)+\sqrt{\frac{V(\eta(0))}{\lambda_{min}(P)}}e^{-\frac{t}{2\lambda_{max}(P)}} \tag{41}$$

It follows from (34) that,

$$|\xi_i - \hat{\xi}_i| \leq \frac{1}{\omega_0^{n-i}}\|\eta(\omega_0 t)\|$$

It follows from (41) that,

$$|\xi_i - \hat{\xi}_i| \leq \frac{1}{\omega_0^{n-i}}\left(\frac{2M\lambda_{max}^2(P)}{\omega_0^2\lambda_{min}(P)}\left(1-e^{-\frac{\omega_0 t}{2\lambda_{max}(P)}}\right)+\sqrt{\frac{V(\eta(0))}{\lambda_{min}(P)}}e^{-\frac{\omega_0 t}{2\lambda_{max}(P)}}\right)$$

Finally,

$$\lim_{t\to\infty}|\xi_i - \hat{\xi}_i| = \frac{1}{\omega_0^{n+2-i}}\frac{2M\lambda_{max}^2(P)}{\lambda_{min}(P)} = O\left(\frac{1}{\omega_0^{n+2-i}}\right) \tag{42}$$

and

$$\lim_{\substack{t\to\infty \\ \omega_0\to\infty}} |\xi_i - \hat{\xi}_i| \tag{43}$$

□

3.2. Stability Analysis of the Closed-Loop System

In this section, the closed-loop stability is investigated for a general nonlinear SISO uncertain system with an ADRC controller.

Assumption 5. *The states ξ_i ($i = 1, 2, \ldots, n$) and the generalized disturbance $\xi_{n+1} = f$ of an n-dimensional uncertain nonlinear SISO system are estimated by a convergent ESO, which produces the estimated states $\hat{\xi}_i$, $i \in \{1, 2, \ldots, n\}$ of the plant and the estimated generalized disturbance $\hat{\xi}_{n+1}$ as $t \to \infty$, i.e.,*

$$\lim_{t\to\infty}|\xi_i - \hat{\xi}_i| = 0, \ i \in \{1, 2, \ldots, n\}, \tag{44}$$

and

$$\lim_{t\to\infty}|f - \hat{\xi}_{n+1}| = 0 \tag{45}$$

Assumption 6. *A tracking differentiator produces a trajectory (r_i, $i \in \{1, 2, \ldots, n\}$) with minimum set point change. The trajectory converges to a reference trajectory ($r^{(i-1)}$) for $i \in \{1, 2, \ldots, n\}$ with $r^{(n)} = 0$ as $t \to \infty$, i.e.,*

$$\lim_{t\to\infty}|r^{(i-1)} - r_i| = 0, \ i \in \{1, 2, \ldots, n\} \tag{46}$$

Theorem 4. Consider a n-dimensional uncertain nonlinear SISO system expressed as

$$\begin{cases} \dot{\zeta}_i = \zeta_{i+1}, \; i \in \{1,2,\ldots,n-1\} \\ \dot{\zeta}_n = f(\zeta_1,\zeta_2,\ldots,\zeta_\rho,w,t) + u \\ y = \zeta_1 \end{cases} \quad (47)$$

The system (47) is controlled by the linearization control law (LCL) signal (u) expressed by,

$$u = v - \hat{\zeta}_{n+1} \quad (48)$$

where v is expressed by,

$$v = k_1(\tilde{e}_1)\tilde{e}_1 + k_2(\tilde{e}_2)\tilde{e}_2 + \ldots + k_n(\tilde{e}_n)\tilde{e}_n \quad (49)$$

where $\tilde{e}_i = r_i - \hat{\zeta}_i$, $i \in \{1,2,\ldots,n\}$ is the tracking error, and $k_i : \mathbb{R} \to \mathbb{R}^+$, $i \in \{1,2,\ldots,n\}$; assume that assumptions A5 and A6 hold. Then

$$\lim_{t \to \infty} |\tilde{e}_i| = 0, \; i \in \{1,2,\ldots,n\} \quad (50)$$

Proof. The tracking error ($\tilde{e}_i, i \in \{1,2,\ldots,n\}$) of the closed-loop system is the error between the reference trajectory and the corresponding plant estimated states expressed as,

$$\tilde{e}_i = r_i - \hat{\zeta}_i, i \in \{1,2,\ldots,n\}$$

After convergence occurs, the tracking error is described by,

$$\tilde{e}_i = r^{(i-1)} - \zeta_i, i \in \{1,2,\ldots,n\} \quad (51)$$

For the system given in (33), the states (ζ_i) are expressed in terms of the plant output, which is expressed as,

$$\zeta_i = y^{(i-1)}, i \in \{1,2,\ldots,\rho\} \quad (52)$$

Substitute (52) in (51), and the tracking error is expressed by,

$$\tilde{e}_i = r^{(i-1)} - y^{(i-1)}, i \in \{1,2,\ldots,\rho\} \quad (53)$$

Differentiating the tracking error ($e_i, i \in \{1,2,\ldots,n\}$) with regard to time yields

$$\dot{\tilde{e}}_i = r^{(i)} - y^{(i)} = \tilde{e}_{i+1}, i \in \{1,2,\ldots,n\} \quad (54)$$

It follows that the tracking error dynamics $\tilde{e}_i, i \in \{1,2,\ldots,n\}$ are expressed as

$$\begin{cases} \dot{\tilde{e}}_1 = \tilde{e}_2, \\ \dot{\tilde{e}}_2 = \tilde{e}_3, \\ \vdots \\ \dot{\tilde{e}}_n = r^{(n)} - y^{(n)} = r^{(n)} - \dot{\zeta}_n \end{cases} \quad (55)$$

This, together with (47), yields,

$$\begin{cases} \dot{\tilde{e}}_1 = \tilde{e}_2, \\ \dot{\tilde{e}}_2 = \tilde{e}_3, \\ \vdots \\ \dot{\tilde{e}}_n = r^{(n)} - (f + u) \end{cases} \quad (56)$$

From (48) and (56), we obtain,

$$\begin{cases} \dot{\tilde{e}}_1 = \tilde{e}_2, \\ \dot{\tilde{e}}_2 = \tilde{e}_3, \\ \quad \vdots \\ \dot{\tilde{e}}_n = r^{(n)} - v + \hat{\xi}_{n+1} - f \end{cases} \tag{57}$$

It follows from (45) and (57) that,

$$\begin{cases} \dot{\tilde{e}}_1 = \tilde{e}_2, \\ \dot{\tilde{e}}_2 = \tilde{e}_3, \\ \quad \vdots \\ \dot{\tilde{e}}_n = r^{(n)} - v \end{cases} \tag{58}$$

The tracking error dynamics given in (58) associated with the control law (v) designed in (49) produce the following dynamics

$$\begin{cases} \dot{\tilde{e}}_1 = \tilde{e}_2, \\ \dot{\tilde{e}}_2 = \tilde{e}_3, \\ \quad \vdots \\ \dot{\tilde{e}}_n = r^{(n)} - k_1(\tilde{e}_1)\tilde{e}_1 - k_2(\tilde{e}_2)\tilde{e}_2 - \ldots - k_n(\tilde{e}_n)\tilde{e}_n \end{cases} \tag{59}$$

Based on assumption A6, the dynamics given in (59) can be represented in compact form as,

$$\dot{\tilde{e}} = A\tilde{e} \tag{60}$$

where

$$A = \begin{pmatrix} 0 & 1 & 0 & \cdots & 0 & 0 \\ 0 & 0 & 1 & \cdots & 0 & 0 \\ \vdots & \cdots & \cdots & \cdots & \vdots & \vdots \\ 0 & 0 & 0 & \cdots & 1 & 0 \\ 0 & 0 & 0 & \cdots & 0 & 1 \\ -k_1(\tilde{e}_1) & -k_2(\tilde{e}_2) & -k_3(\tilde{e}_3) & \cdots & -k_{n-1}(\tilde{e}_{n-1}) & -k_p(\tilde{e}_n) \end{pmatrix} \tag{61}$$

and $\tilde{e} = (\tilde{e}_1, \tilde{e}_2, \ldots, \tilde{e}_n)^T$

The characteristic polynomial of A is expressed by

$$|\lambda I - A| = \lambda^n + k_n(\tilde{e}_n)\lambda^{n-1} + k_{n-1}(\tilde{e}_{n-1})\lambda^{n-2} + \ldots + k_1(\tilde{e}_1) \tag{62}$$

The design parameters of the proposed controller are selected to ensure that the roots of the characteristic polynomial (43) have a strictly negative real part, which makes (61) asymptotically stable. Hence, $\lim_{t \to \infty} |\tilde{e}_i| = 0$. □

Remark 1. *The error vector is calculated up to the relative degree (n) of the system because the ESO estimate system states up to n, i.e., $e_i = r_i - \hat{\xi}_i$, $i \in \{1, 2, \ldots, n\}$. This implies that the vector $k(e)$ of (23) and the vector $f(e)$ of (24) are of size n.*

Corollary 1. *Consider the nonlinear system and the control signal given in Theorem 2. The control signal (v) is expressed as $v = \sum_{i=1}^{n} k_i(\tilde{e}_i) f_i(\tilde{e}_i)$, where $k_i(\tilde{e}_i) = \left(k_{i1} + \frac{k_{i2}}{1+\exp(\mu_i \tilde{e}_i^2)} \right)$, and $f_i(\tilde{e}_i) = |\tilde{e}_i|^{\alpha_i} \text{sign}(\tilde{e}_i)$ for $i \in \{1, 2, \ldots, n\}$. Moreover, if assumptions A5 and A6 hold, then*

$$\lim_{t\to\infty}|r_i - \hat{\xi}_i| = 0, \ i \in \{1,2,\ldots,\rho\} \text{ for a suitable set of the design parameters } k_{i1}, k_{i2}, \mu_i, \text{ and } \alpha_i \text{ with } i \in \{1,2,\ldots,n\}.$$

Proof. Since

$$k_i(\tilde{e}_i)f_i(\tilde{e}_i) = \left(k_{i1} + \frac{k_{i2}}{1+exp(\mu_i\tilde{e}_i^2)}\right)|\tilde{e}_i|^{\alpha_i}sign(\tilde{e}_i), \ i \in \{1,2,\ldots,n\} \quad (63)$$

Equation (63) can be expressed as,

$$k_i(\tilde{e}_i)f_i(\tilde{e}_i) = \begin{cases} 0 & \tilde{e}_i = 0 \\ \hbar_i(\tilde{e}_i)\tilde{e}_i & \tilde{e}_i \neq 0 \end{cases} \quad (64)$$

where the function $\hbar_i : \mathbb{R}/\{0\} \to \mathbb{R}^+$ is an even nonlinear gain function, and:

$$\hbar_i(\tilde{e}_i) = \left(k_{i1} + \frac{k_{i2}}{1+exp(\mu_i\tilde{e}_i^2)}\right)|\tilde{e}_i|^{\alpha_i-1}, \ i \in \{1,2,\ldots,n\} \quad (65)$$

The expression (65) is time-varying because it is a function of $\tilde{e}_i$. For simplicity, consider that the parameters $k_{i2} = 0$ and $\alpha_i = 1$ and that the expression (65) is reduced to $\hbar_i(\tilde{e}_i) = k_{i1}$. Consider the tracking error dynamics given in (59) with $n = 2$, which provides

$$\begin{cases} \dot{\tilde{e}}_1 = \tilde{e}_2, \\ \dot{\tilde{e}}_n = -k_{11}\tilde{e}_1 - k_{21}\tilde{e}_2 \end{cases} \quad (66)$$

The characteristic equation of (66) is expressed as,

$$|\lambda I - A| = \lambda^2 + k_{21}\lambda + k_{11} \quad (67)$$

The roots of the characteristic equation (67) are $\lambda_{1,2} = -\frac{k_{21}}{2} + \frac{\sqrt{k_{21}^2 - 4k_{11}}}{2}$ for $k_{21}^2 < 4k_{11}$, which leads to a complex conjugate with a negative real part. Then, $\tilde{e}_1 \to 0$ and $\tilde{e}_2 \to 0$ at $t \to \infty$.

In Theorem 4, we assumed that $r^{(n)} = 0$ for the case of $r^{(n)}$ in (59) not satisfying assumption A 6, i.e., $r^{(n)} \neq 0$. Then, for $n = 2$,

$$\begin{cases} \dot{\tilde{e}}_1 = \tilde{e}_2, \\ \dot{\tilde{e}}_2 = -k_{11}\tilde{e}_1 - k_{21}\tilde{e}_2 + r^{(2)}(t) \end{cases} \quad (68)$$

Let $q(t) = r^{(2)}(t)$ after taking the Laplace transform of both sides of (68)

$$s\tilde{E}_1(s) = \tilde{E}_2(s)$$

$$s\tilde{E}_2(s) = -k_{11}\tilde{E}_1(s) - k_{21}\tilde{E}_1(s) + Q(s)$$

Solving for $\tilde{E}_1(s)$ and $\tilde{E}_2(s)$ in terms of $Q(s)$, we obtain

$$\tilde{E}_1(s) = \frac{Q(s)}{s^2 + k_{21}s + k_{11}} \quad (69)$$

$$\tilde{E}_2(s) = \frac{sQ(s)}{s^2 + k_{21}s + k_{11}} \quad (70)$$

It can be noticed from (70) that for nonzero $r^{(2)}(t) = q(t)$, the error $\tilde{e}_1(t)$ tracks $r^{(2)}$, which means that at a steady state, $\tilde{e}_1(t)$ is nonzero, depending on $r^{(2)}(t)$. The error $\tilde{e}_2$ is the derivative of $\tilde{e}_1(t)$. □

4. Mismatched Disturbances

To satisfy the matched condition, the ESO assumes that the plant is expressed in the normal form [28,29]. Thus, it can only be applied to systems that can be directly expressed in the normal form or by changing variables. When a system has zero dynamics, performing such a transformation can be challenging. There are also nonlinear systems with disturbances appearing in a different channel of control input; these systems fail to satisfy the matching condition. Therefore, ADRC is no longer able to manipulate this mismatched disturbance as before. For instance, the following nonlinear model belongs to a class of uncertain nonlinear systems in a lower triangular form with mismatched disturbance [30–35],

$$\begin{cases} \dot{\xi}_i = a_i \xi_{i+1} + \phi_i(\xi_1, \ldots, \xi_i) + w_i, \ i \in \{1, 2, \ldots, n-1\} \\ \dot{\xi}_n = \phi_n(\xi_1, \xi_2, \ldots, \xi_n) + w_n + bu, \\ y = \xi_1 \end{cases} \quad (71)$$

where $\xi = (\xi_1(t), \xi_2(t), \ldots, \xi_n(t))^T \in \mathbb{R}^n$ is the system state, $y(t) \in \mathbb{R}$ is the measured output, $u(t) \in \mathbb{R}$ is the control input, $w_i(t) \in \mathbb{R}$, $i \in \{1, 2, \ldots, n\}$ is the unknown exogenous disturbance, and $b \in \mathbb{R}$ is the control coefficient. The function $\phi_i : \mathbb{R}^i \to \mathbb{R}$, $i \in \{1, 2, \ldots, n\}$.

Theorem 5. *A second-order nonlinear system in a lower triangular form with mismatched disturbances can be described as follows,*

$$\begin{cases} \dot{\xi}_1 = a_1 \xi_2 + \phi_1(\xi_1) + w_1 \\ \dot{\xi}_2 = \phi_2(\xi_1, \xi_2) + w_2 + bu \\ y = \xi_1 \end{cases} \quad (72)$$

where $\xi = (\xi_1(t), \xi_2(t))^T \in \mathbb{R}^2$ is the system state, $y(t) \in \mathbb{R}$ is the measured output, $u(t) \in \mathbb{R}$ is the control input, $w_i(t) \in \mathbb{R}$, $i \in \{1, 2\}$ is the unknown exogenous disturbance, and $b \in \mathbb{R}$ is the control coefficient. The function $\phi_i : \mathbb{R}^i \to \mathbb{R}$, $i \in \{1, 2\}$. If the function ϕ_1 and the exogenous disturbance (w_1) are differentiable with regard to t, the system (72) can be transformed into the following form,

$$\begin{cases} \dot{\tilde{\xi}}_1 = \tilde{\xi}_2 \\ \dot{\tilde{\xi}}_2 = f\left(\tilde{\xi}_1, \tilde{\xi}_2, w_1, \dot{w}_1, w_2\right) + \hat{b}u \\ y = \tilde{\xi}_1 \end{cases} \quad (73)$$

where $f\left(\tilde{\xi}_1, \tilde{\xi}_2, w_1, \dot{w}_1, w_2\right) = a_1 \phi_2\left(\tilde{\xi}_1, \frac{\tilde{\xi}_2 - \phi_1(\tilde{\xi}_1) - w_1}{a_1}\right) + \frac{\partial \phi_1(\tilde{\xi}_1)}{\partial \tilde{\xi}_1} \tilde{\xi}_2 + a_1 w_2 + \dot{w}_1$, and $\hat{b} = a_1 b$.

Proof. Let $\tilde{\xi}_1 = \xi_1$ and $\tilde{\xi}_2 = \dot{\xi}_1$. Then,

$$\dot{\tilde{\xi}}_2 = a_1 \dot{\xi}_2 + \frac{\partial \phi_1(\xi_1)}{\partial \xi_1} \dot{\xi}_1 + \dot{w}_1 \quad (74)$$

By substituting (72) in (74), we obtain,

$$\dot{\tilde{\xi}}_2 = a_1 \phi_2\left(\tilde{\xi}_1, \xi_2\right) + \frac{\partial \phi_1(\tilde{\xi}_1)}{\partial \tilde{\xi}_1} \tilde{\xi}_2 + a_1 w_2 + \dot{w}_1 + a_1 bu \quad (75)$$

Since $\xi_2 = \frac{\tilde{\xi}_2 - \phi_1(\tilde{\xi}_1) - w_1}{a_1}$, (75) can be expressed as,

$$\dot{\tilde{\xi}}_2 = a_1 \phi_2\left(\tilde{\xi}_1, \frac{\tilde{\xi}_2 - \phi_1(\tilde{\xi}_1) - w_1}{a_1}\right) + \frac{\partial \phi_1(\tilde{\xi}_1)}{\partial \tilde{\xi}_1} \tilde{\xi}_2 + a_1 w_2 + \dot{w}_1 + a_1 bu \quad (76)$$

Finally, system (72) can be defined as,

$$\begin{cases} \dot{\tilde{\xi}}_1 = \tilde{\xi}_2, \\ \dot{\tilde{\xi}}_2 = f\left(\tilde{\xi}_1, \tilde{\xi}_2, w_1, \dot{w}_1, w_2\right) + \hat{b}u, \\ y = \tilde{\xi}_1 \end{cases} \qquad (77)$$

where $f\left(\tilde{\xi}_1, \tilde{\xi}_2, w_1, \dot{w}_1, w_2\right) = a_1\phi_2\left(\tilde{\xi}_1, \dfrac{\tilde{\xi}_2 - \phi_1(\tilde{\xi}_1) - w_1}{a_1}\right) + \dfrac{\partial\phi_1(\tilde{\xi}_1)}{\partial\tilde{\xi}_1}\tilde{\xi}_2 + a_1w_2 + \dot{w}_1, \hat{b} = a_1b. \square$

Theorem 5 can be generalized easily for nth-order uncertain nonlinear systems in a lower triangular form with mismatched disturbance $w_i(t)$, $i \in \{1, 2 \ldots, n\}$ as in (71).

5. Mathematical Modelling of The Differential Drive Mobile Robot

The mathematical model of the mobile robot mathematical is an approximation of the physical mobile robot, which consists of the dynamical kinematic and actuator models. To restrain the robot's motor dynamics, an internal loop is also involved. Figure 8 illustrates the mobile robot block diagram with an internal control loop [36].

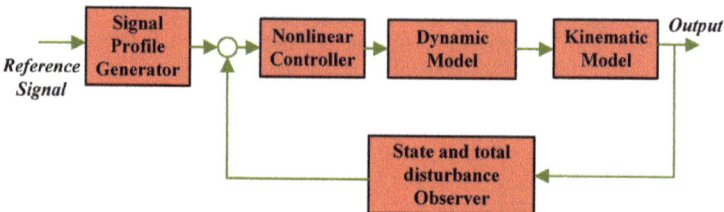

Figure 8. Mobile robot with an internal control loop.

As shown in Figure 8, $w(t)$, $q(t)$, and $p(t)$ represent the reference input velocity, the output of the internal loop (i.e., recent velocity), and the kinematic model output (i.e., robot posture), respectively. The control inputs are the differences between the required and the recent velocities ($e(t) = w(t) - q(t)$), while the control output ($u(t)$) influences the dynamics of the mobile robot as forces or torques. The posture of the mobile robot regarding the origin of the global coordinate system (GCS) is described by the position coordinates (x, y) of its local coordinate system (LCS) origin, with rotation defined by an angle (θ_m) [36].

As shown in Figure 9, the kinematic model can be described by the robot's linear velocity (V_m) and its angular velocity (ω_m). However, it is desirable to describe most control configurations according to the wheel angular velocities (ω_{wr}, ω_{wl}). The general kinematic model of DDMR is defined as [37–42],

$$\begin{cases} \dot{x}' = V_m \cos(\theta_m) \\ \dot{y}' = V_m \sin(\theta_m) \\ \dot{\theta}_m = \omega_m \end{cases} \qquad (78)$$

Linear velocity is computed by averaging the linear velocities of the two wheels in the LCS [37–40],

$$V_m = \dfrac{(V_{wr} + V_{wl})}{2} = r_w \dfrac{(\omega_{wr} + \omega_{wl})}{2} \qquad (79)$$

The DDMR angular velocity is expressed as,

$$\omega_m = \dfrac{(V_{wr} - V_{wl})}{D} = r_w \dfrac{(\omega_{wr} - \omega_{wl})}{D} \qquad (80)$$

where V_m is the longitudinal velocity of the center of mass; ω_m is the angular velocity of DDMR; V_{wr} and V_{wl} are the longitudinal velocities of the left and right wheels, respectively; ω_{wr} and ω_{wl} are the angular tire velocities of the left and right wheels, respectively; and r_w is the nominal radius of the tire.

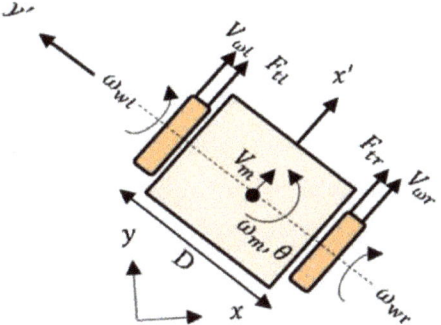

Figure 9. The differential drive mobile robot (DDMR).

In [13], the nonlinear dynamics of the motor wheels were illustrated and presented in detail. The state-space depiction of the overall motor and wheel dynamics is summarized as follows (for the right wheel):

$$J_{eq} n \dot{\omega}_{wr} = -B_{eq} n \omega_{wr} + k_t i_{ar} - \tau'_{lr} \tag{81}$$

$$L_a \frac{di_{ar}}{dt} = -k_b n \omega_{wr} - R_a i_{ar} + v_{ar} \tag{82}$$

$$\tau'_{lr} = \tau_{rext}/n \tag{83}$$

where v_{ar} and v_{al} are the input voltages applied to the right and left motors, respectively; i_{ar} and i_{al} are the armature current of the right and left motors, respectively; τ'_{lr} and τ'_{ll} are the right and left motor-developed torques, respectively; k_t is a torque constant; k_b is a voltage constant; L_a is an electric self-inductance constant; R_a is an electric resistance constant; the total equivalent inertia is denoted as J_{eq}; total equivalent damping is denoted as B_{eq}; n is the ratio of the gearbox; and τ_{rext} and τ_{lext} are the external torque applied at the wheel side for the right land left wheels, respectively. Let $\xi_1 = \omega_{wr}$, $\xi_2 = i_{ar}$, $d = \tau'_{lr}$, and $u = v_{ar}$. Then,

$$\dot{\xi}_1 = -\frac{B_{eq}}{J_{eq}} \xi_1 + \frac{k_t}{J_{eq} n} \xi_2 - \frac{1}{J_{eq} n} d \tag{84}$$

$$\dot{\xi}_2 = -\frac{k_b n}{L_a} \xi_1 - \frac{R_a}{L_a} \xi_2 + \frac{1}{L_a} u \tag{85}$$

Let $b_1 = -\frac{1}{J_{eq} n}$, $b_2 = \frac{1}{L_a}$,

$$f_1(\xi_1, \xi_2) = -\frac{B_{eq}}{J_{eq}} \xi_1 + \frac{k_t}{J_{eq} n} \xi_2 \tag{86}$$

and

$$f_2(\xi_1, \xi_2) = -\frac{k_b n}{L_a} \tag{87}$$

The simplified model with the mismatched uncertainties and external disturbances of the DDMR exactly fits the state-space formulation given in (53). According to Theorem 1, the state-space model with mismatched uncertainties can be transformed into ADRC canonical form with $\hat{b} = \frac{1}{L_a} \frac{k_t}{J_{eq} n}$ for the motor wheel model.

6. Numerical Simulations

The kinematic model of the DDMR with PMDC motors and the proposed IADRC was designed and simulated in the MATLAB®/SIMULINK environment. Numerical simulations of continuous state models were conducted using the MATLAB® ODE45 solver. This Runge–Kutta ODE45 solver produces a fourth-order estimate of error using a fifth-order method. Figure 10 shows the Simulink block diagram of the DDMR and the PMDC motors with IADRC.

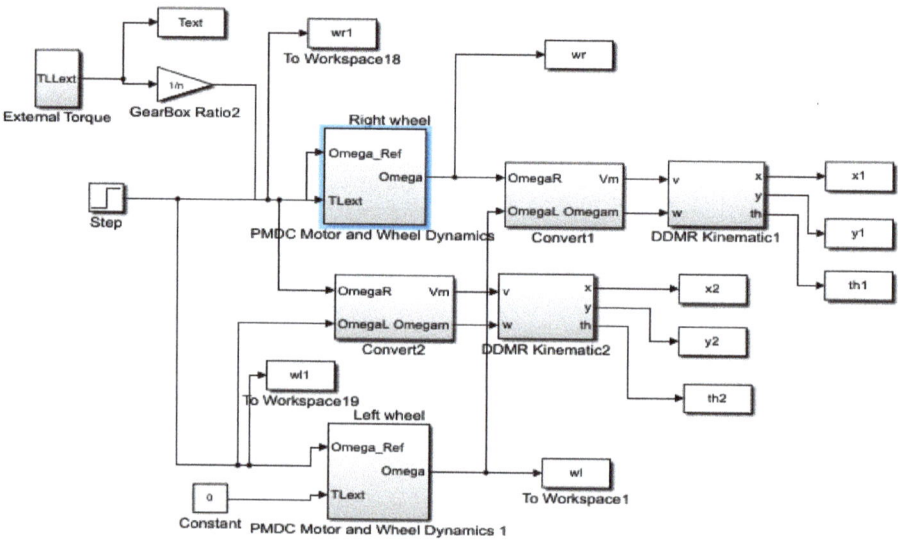

Figure 10. The Simulink® block diagram of the DDMR kinematics and the PMDC motor controlled by the IADRC.

The PMDC motor coefficient values are set to $L_a = 0.82$, $R_a = 0.1557$, $K_t = 1.1882$, $K_b = 1.185$, $B_{eq} = 0.3922$, $J_{eq} = 0.2752$, and $n = 3.0$. The DDMR used in the simulation is assumed to have the following coefficients: $D = 0.40$ and $r_w = 0.075$. The coefficients of the classical ADRC controller are $\delta_1 = 0.4620$, $\delta_2 = 0.24807$, $\alpha_1 = 0.1726$, $\alpha_2 = 0.8730$, $\beta_1 = 30.4$, $\beta_2 = 523.4$, $\beta_3 = 2970.8$, and $R = 100$. The coefficients of the proposed IADRC scheme include the coefficients of the NLSEFC, which are expressed as $k_{11} = 144.6642$, $k_{12} = 8.0475$, $k_{21} = 25.5574$, $k_{22} = 4.8814$, $k_3 = 0.5308$, $\delta = 3.8831$, $\mu_1 = 44.3160$, $\mu_2 = 48.8179$, $\mu_3 = 26.1493$, $\alpha_1 = 0.9675$, $\alpha_2 = 1.4487$, and $\alpha_3 = 3.5032$. The ITD suggested in this paper has a set of coefficients expressed as $\alpha = 0.4968$, $\beta = 2.1555$, $\gamma = 11.9554$, and $R = 16.8199$. $K_\alpha = 0.6265$, $\alpha = 0.8433$, $K_\beta = 0.5878$, $\beta = 0.04078$, $\beta_0 = 30.4$, $\beta_1 = 513.4$, and $\beta_2 = 1570.8$ represent the coefficients of the SMESO used in this work.

The DDMR was tested by applying reference angular velocities for both wheels of 1 rad/s at $t = 0$ and $t = 100$ s. To examine the proposed IADRC performance, an exogenous torque acting as a constant disturbance was applied to the right wheel during the simulation at $t = 30$ and removed after 20 s. Figure 11 shows the applied external disturbance. Figure 12 shows the transient response of the controlled PMDC motor for the right wheel when both the ADRC and the IADRC are applied. The figure shows an enhancement in system response before and during the applied disturbance when the IADRC is adopted; this behavior is evident in Figure 12c,d.

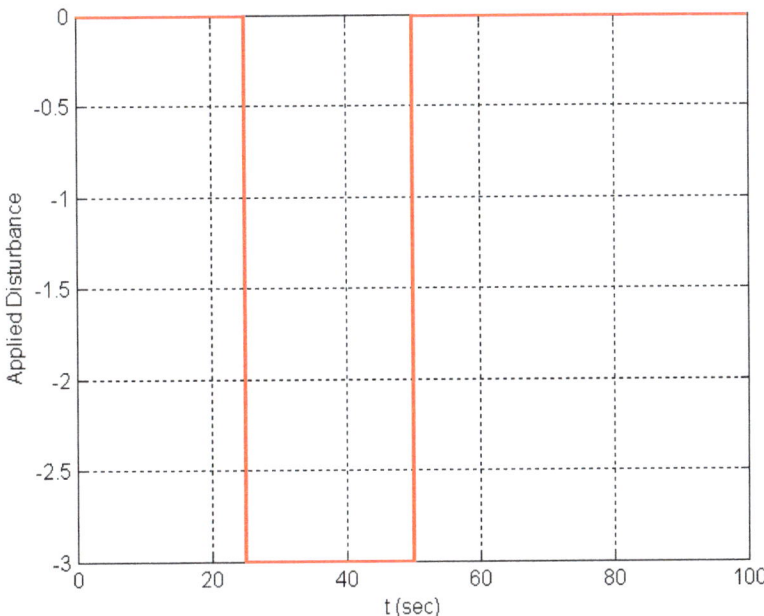

Figure 11. The applied external torque.

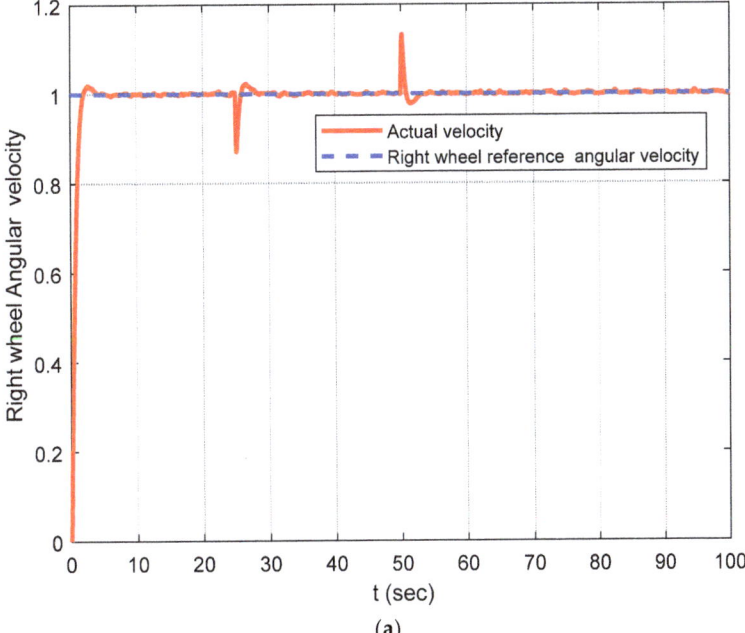

(**a**)

Figure 12. *Cont.*

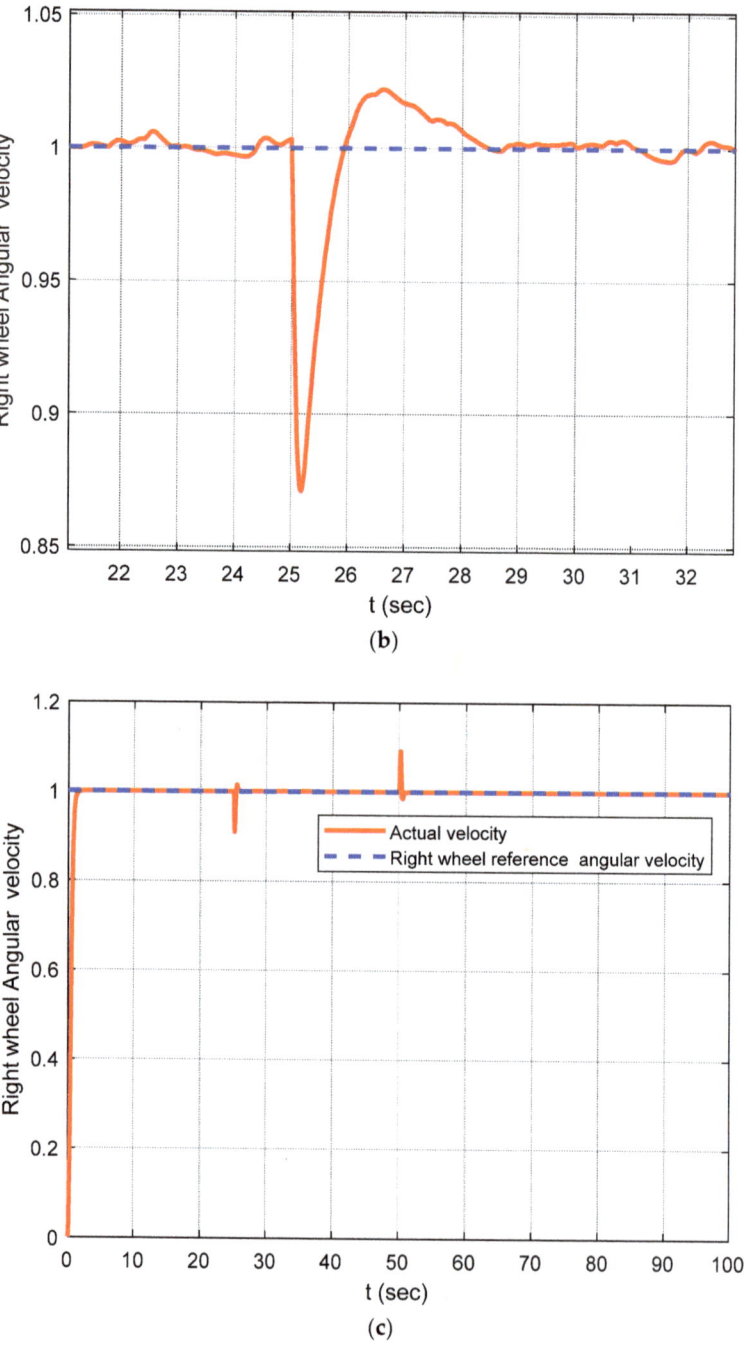

Figure 12. Cont.

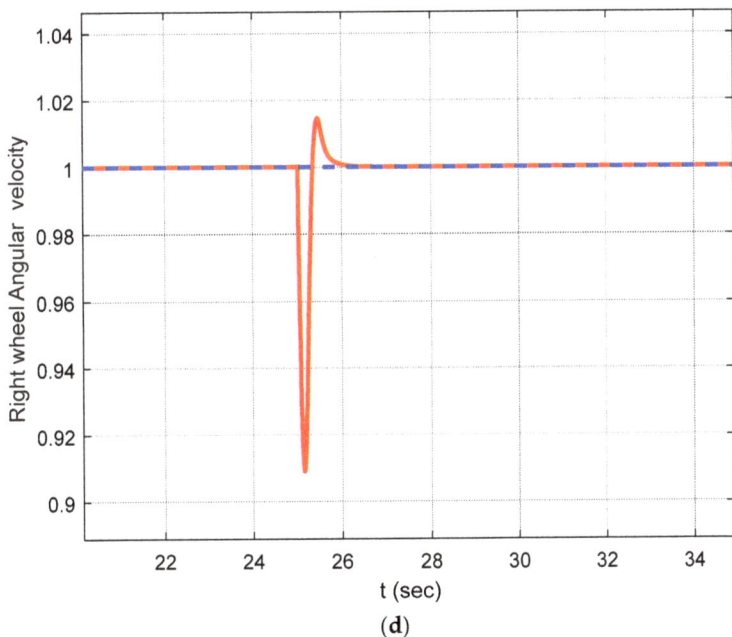

Figure 12. Simulation results: (**a**) the angular velocity of the right wheel using classical ADRC; (**b**) close-up of the response depicted in (**a**); (**c**) the angular velocity of the left wheel using IADRC; (**d**) close-up of the response depicted in (**c**).

The orientation error (e_θ) associated with the tested case is reduced intensely due to the effectiveness of the proposed technique (see Figure 13). Note that $e_\theta = \theta_{ref} - \theta_{actual}$, where θ_{ref} is the orientation of the reference trajectory, and θ_{actual} is the actual orientation. The IADRC produces an error signal with less overshoot (3.4×10^{-3}) than in the ADRC scheme (10.5×10^{-3}). The IADRC also shows a faster convergence for the error signal because of the proposed nonlinearities in the NLSEFC controller, which strongly and quickly damp the error signals.

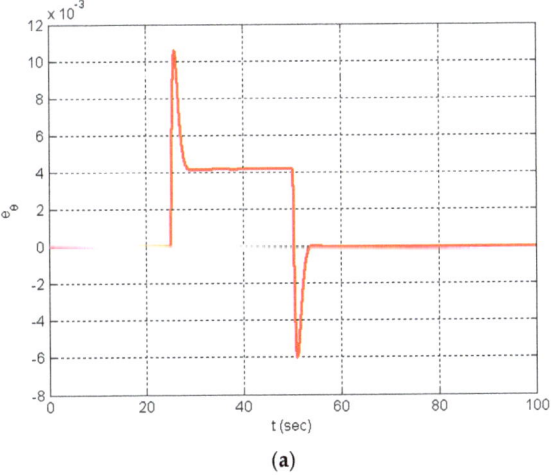

Figure 13. Cont.

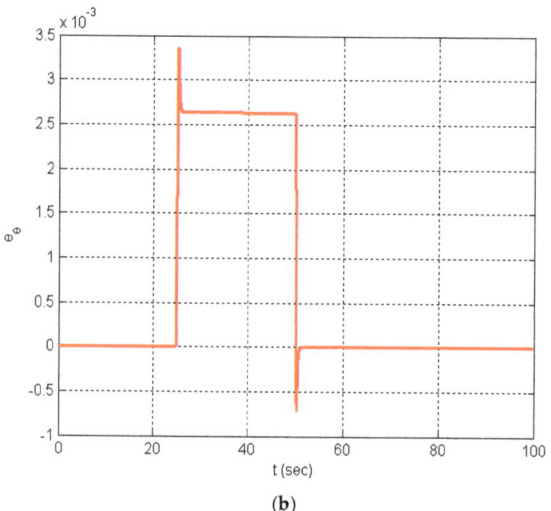

(b)

Figure 13. Simulation results; (**a**) the DDMR orientation error in the case of ADRC; (**b**) the DDMR orientation error in the case of IADRC.

The chattering phenomenon found in the estimated total disturbances produced by the LESO of the conventional ADRC for both wheels (D_r and D_l) are extremely reduced by using the SMESO of the proposed IADRC. The same is true of the control signals that drive the two wheels (u_r and u_l; see Figure 14), where a very smooth control signal is obtained as a result of the slight increase in the overshot (compare Figure 14a,b).

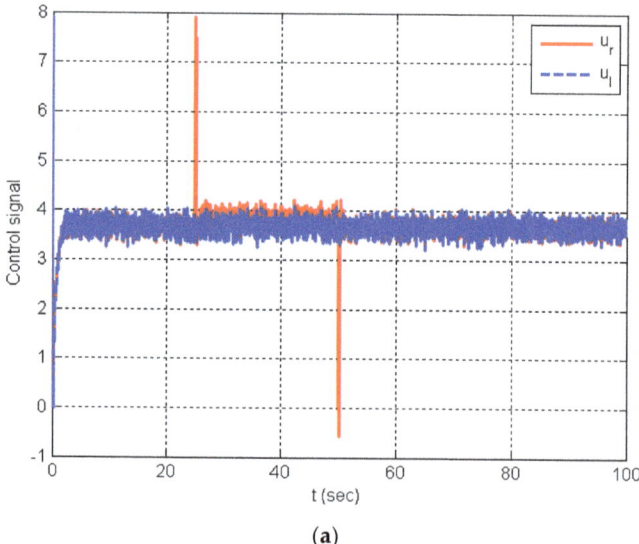

(a)

Figure 14. *Cont.*

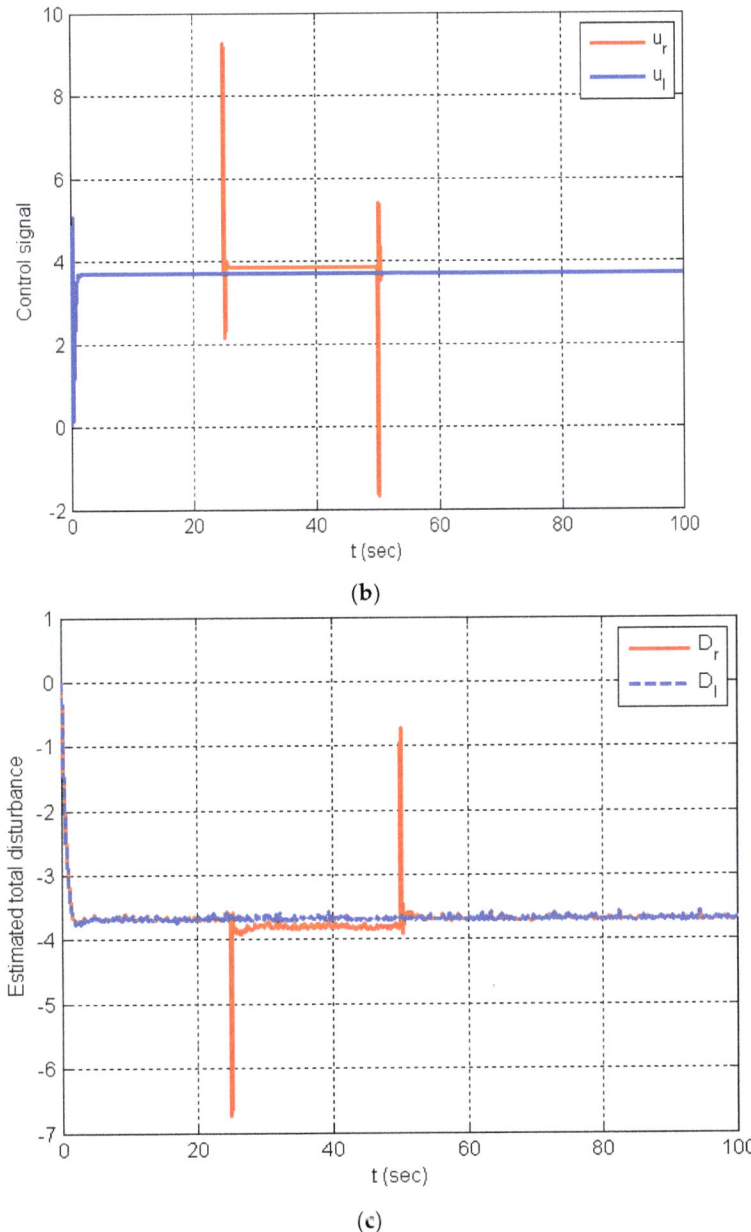

Figure 14. Cont.

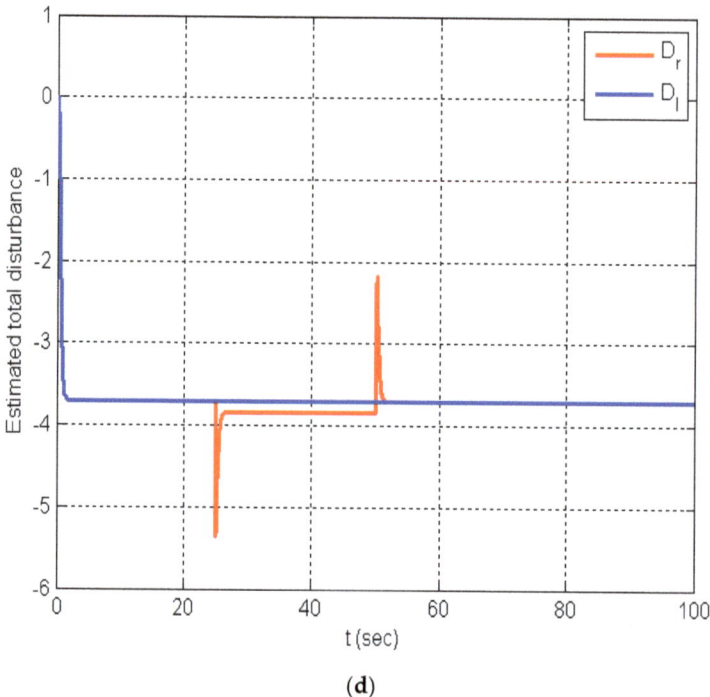

(d)

Figure 14. Simulation results: (**a**) the control signals generated by the ADRC; (**b**) the control signals generated by the IADRC; (**c**) the estimated total disturbance from the LESO; (**d**) the observed total disturbance from the SMESO.

Tables 1 and 2 show the results based on evaluation of several *OPIs*. These indices reflect the performance of the adaptive improved active disturbance rejection control. The results are classified into kinematic and dynamic performance indices.

Table 1. DDMR kinematic performance indices.

Performance Index	Controller	
	ADRC	IADRC
OPI_x	0.0010884970	0.0005257305
OPI_y	0.0016112239	0.0007447036
OPI_θ	0.0000059780	0.0000017459

Table 2. Performance indices of both wheels.

Wheel	Performance Index	Controller	
		ADRC	IADRC
Right	ITAE	13.302889	1.780254
	ISU	1372.090423	1407.300305
Left	ITAE	6.919226	0.146694
	ISU	1343.542226	1372.124019

where $OPI_x = \frac{1}{N}\sum (x_{ref} - x_{actual})^2$,

$OPI_y = \frac{1}{N}\sum (y_{ref} - y_{actual})^2$,

$OPI_\theta = \frac{1}{N}\sum (\theta_{ref} - \theta_{actual})^2$

$ITAE = \sum t|\omega_{ref} - \omega_{actual}|dt$

$ISU = \sum u^2 dt$

Discussion

A new nonlinear error combination ($\Psi(e)$) is proposed, which was used to construct an NLSEFC. When used solely in the feedback loop, it leads to a noticeable improvement in the performance of the closed-loop system in terms of the ISU index for both models. This closeness is due to the common term ($f(e) = |e|^\alpha sign(e)$) included in the structure of the NLSEFC. Furthermore, the nonlinear gain function ($k(e)$), in contrast to the conventional PID controller, produces a variable gain depending on the error value, which, in turn, enhances the transient behavior of the system response. Furthermore, an SMESO is suggested in this paper; the smoothness of the control signal u and the minimum overshoot in the output response are due to using the proposed nonlinear error function ($\cdot$) with the following features: it is a smooth function, and it has high gain near the origin and a small gain with large error values. Finally, a new tracking differentiator, named the INTD, is proposed; it proved superior to the other tracking differentiators by solving the common issues extant in conventional differentiators. One of these issues is the "peaking phenomenon". This phenomenon is reduced by considering the INTD of 4.35 with an optimized set of parameters, i.e., a_1 and a_2. In addition, the proposed INTD eliminates the "phase lag" problem that is extant in most conventional tracking differentiators due to the scaling parameters, i.e., α and β. The input scaling parameter (α) reduces the values of the input signal ($r(t)$) level ($1 - \alpha$), while scaling parameter β amplifies the level of the output signal ($r_1(t)$), thereby accelerating the tracking phase. When these three parts are combined to synthesize the IADRC, the proposed IADRC scheme presented in this paper and applied to DDMR achieves the improvements mentioned above in an easier manner because the nonlinear system is converted into a chain of integrators by the SMESO, which is simply a linearized system controlled by a nonlinear controller. This is reflected in the DDMR in terms of the smooth output response and chatter-free control signal. Moreover, the torque disturbance is canceled by the IADRC scheme and provides very small values for the ITAE and ISU indices, as shown in Tables 1 and 2.

A major improvement in the kinematic indices is achieved for the IADRC against the conventional ADRC, where the OPI_x, OPI_y, and OPI_θ are reduced by 51.7%, 53.78%, and 70.794%, respectively. A significant enhancement in the time-domain response is achieved ($ITAE$ is lowered by 86.6175%) by increasing the ISU, which signifies the power provided to the PMDC motor. In addition, the chattering in the control signal caused by Han's classical ADRC is almost eliminated by the proposed IADRC. Finally, the DDMR orientation error is clearly reduced and swiftly decreases to zero.

7. Conclusions

An improved nonlinear ADRC controller was developed for a DDMR to provide accurate speed tracking in the presence of high external torque disturbance. The proposed IADRC with the SMESO generates an exact estimation of the states and the total disturbance. The proposed IADRC with three parts, namely the SMESO, the NLSEF, and the INTD, provides a committed scheme to enhance the ability of the conventional ADRC to achieve disturbance estimation and attenuation. In conclusion, the simulation results show that the developed IADRC can effectively enhance the performance of the system and improve the accuracy and the speed of the PMDC motor of the DDMR under mismatched uncertainties and torque disturbance. The IADRC eliminates the chattering phenomenon, which is coherent in the conventional ADRC, with minimal increase in the overshoot of the control signal when disturbance occurs. The future directions for our proposed IADRC including extending its applications to include consensus multiagent systems. The first step will be to design a control system for every local agent for consensus disturbance rejection. The second step with involve analysis of the design for network-connected multi-input linear or nonlinear systems using relative state information of the subsystems in the neighborhood. The consensus multiagent system can be configured with in leaderless or leader–follower consensus setups under common assumptions of the network connections.

Author Contributions: Conceptualization, I.A.H., A.T.A., N.A.K. and I.K.I.; Methodology, I.A.H., L.H.A., J.A.A., I.K.I., A.T.A., N.A.K., M.M., A.J.M.J. and W.R.A.-A.; Software, L.H.A., J.A.A., M.M., A.J.M.J. and W.R.A.-A.; Validation, A.T.A., M.M., N.A.K., A.J.M.J. and W.R.A.-A.; Formal analysis, I.A.H., L.H.A., J.A.A., I.K.I., A.T.A., N.A.K., M.M., A.J.M.J. and W.R.A.-A.; Investigation, A.T.A., I.A.H., N.A.K. and L.H.A.; Resources, L.H.A., J.A.A., I.K.I., N.A.K., M.M., A.J.M.J. and W.R.A.-A.; Data curation L.H.A., J.A.A., I.K.I., M.M., A.J.M.J. and W.R.A.-A.; Writing—original draft, I.A.H., L.H.A., J.A.A., I.K.I. and A.T.A.; Writing—review & editing, I.A.H., L.H.A., J.A.A., I.K.I., A.T.A., N.A.K., M.M., A.J.M.J. and W.R.A.-A.; Visualization, A.T.A., N.A.K. and J.A.A.; Supervision, A.T.A. and I.K.I.; Project administration, I.K.I.; Funding acquisition, I.A.H. All authors have read and agreed to the published version of the manuscript.

Funding: This research was funded by the Norwegian University of Science and Technology.

Institutional Review Board Statement: Not applicable.

Data Availability Statement: Not applicable.

Acknowledgments: The authors would like to acknowledge the support of the Norwegian University of Science and Technology for paying the Article Processing Charges (APC) of this publication. The authors would like to thank Prince Sultan University, Riyadh, Saudi Arabia for their support. Special acknowledgement to Automated Systems & Soft Computing Lab (ASSCL), Prince Sultan University, Riyadh, Saudi Arabia. In addition, the authors wish to acknowledge the editor and anonymous reviewers for their insightful comments, which have improved the quality of this publication.

Conflicts of Interest: The authors declare that there is no conflict of interest.

References

1. Åström, K.; Hägglund, T. *Advanced PID Control*; ISA Press: Durham, NC, USA, 2006.
2. Seborg, D.E.; Edgar, T.E. *Process Dynamics and Control*, 2nd ed.; Wiley: Cambridge, MA, USA, 2004.
3. Han, J. Extended state observer for a class of uncertain plants. *Control Decis.* **1995**, *10*, 85–88. (In Chinese)
4. Han, J. From PID to active disturbance rejection control. *IEEE Trans. Ind. Electron.* **2009**, *56*, 900–906. [CrossRef]
5. Zhang, C.; Chen, Y. Tracking Control of Ball Screw Drives Using ADRC and Equivalent-Error-Model-Based Feedforward Control. *IEEE Trans. Ind. Electron.* **2016**, *63*, 7682–7692. [CrossRef]
6. Shi, M.; Liu, X.; Shi, Y. Research n Enhanced ADRC algorithm for hydraulic active suspension. In Proceedings of the International Conference on Transportation, Mechanical, and Electrical Engineering (TMEE), Changchun, China, 16–18 December 2011. [CrossRef]
7. Chenlu, W.; Zengqiang, C.; Qinglin, S.; Qing, Z. Design of PID and ADRC based quadrotor helicopter control system. In Proceedings of the Control and Decision Conference (CCDC), Yinchuan, China, 28–30 May 2016. [CrossRef]
8. Wu, Y.; Zheng, Q. ADRC or adaptive controller—A simulation study on artificial blood pump. *Comput. Biol. Med.* **2015**, *66*, 135–143. [CrossRef]
9. Rahman, M.M.; Chowdhury, A.H. Comparative study of ADRC and PID based Load Frequency Control. In Proceedings of the International Conference on Electrical Engineering and Information Communication Technology (ICEEICT), Savar, Bangladesh, 21–23 May 2015. [CrossRef]
10. Ahmed, A.; Asad Ullah, H.; Haider, I.; Tahir, U.; Attique, H. Analysis of Middleware and ADRC based Techniques for Networked Control. In Proceedings of the 16th International Conference on Sciences and Techniques of Automatic Control & Computer Engineering, Monastir, Tunisia, 21–23 December 2015.
11. Ibraheem, I.K.; Abdul-Adheem, W.R. On the Improved Nonlinear Tracking Differentiator based Nonlinear PID Controller Design. *Int. J. Adv. Comput. Sci. Appl.* **2016**, *7*, 234–241.
12. Abdul-Adheem, W.R.; Ibraheem, I.K. From PID to Nonlinear State Error Feedback Controller. *Int. J. Adv. Comput. Sci. Appl.* **2017**, *8*, 2017.
13. Abdul-Adheem, W.R.; Ibraheem, I.K. Improved Sliding Mode Nonlinear Extended State Observer based Active Disturbance Rejection Control for Uncertain Systems with Unknown Total Disturbance. *Int. J. Adv. Comput. Sci. Appl.* **2016**, *7*, 80–93.

4. Ibraheem, I.K.; Abdul-Adheem, W.R. An Improved Active Disturbance Rejection Control for a Differential Drive Mobile Robot with Mismatched Disturbances and Uncertainties. *arXiv* **2018**, arXiv:1805.12170.
5. Huang, Y.; Xue, W.C. Active disturbance rejection control: Methodology and theoretical analysis. *ISA Trans.* **2014**, *53*, 963–976. [CrossRef]
6. Guo, B.Z.; Zhao, Z.L. *Active Disturbance Rejection Control for Nonlinear Systems: An Introduction*; John Wiley & Sons: Singapore/Hoboken, NJ, USA, 2016.
7. Azar, A.T.; Abed, A.M.; Abdulmajeed, F.A.; Hameed, I.A.; Kamal, N.A.; Jawad, A.J.M.; Abbas, A.H.; Rashed, Z.A.; Hashim, Z.S.; Sahib, M.A.; et al. A New Nonlinear Controller for the Maximum Power Point Tracking of Photovoltaic Systems in Micro Grid Applications Based on Modified Anti-Disturbance Compensation. *Sustainability* **2022**, *14*, 10511. [CrossRef]
8. Menon, A.R.; Mehrotra, K.; Mohan, C.; Ranka, S. Characterization of a Class of Sigmoid Functions with Applications to Neural Networks. Electrical Engineering and Computer Science Technical Reports. 1994. Available online: http://surface.syr.edu/eecs_techreports/152 (accessed on 1 May 2022).
9. Khalil, H.K. *Nonlinear Systems*; Prentice-Hall: Hoboken, NJ, USA, 1996.
10. Wu, S.; Dong, B.; Ding, G.; Wang, G.; Liu, G.; Li, Y. Backstepping sliding mode force/position control for constrained reconfigurable manipulator based on extended state observer. In Proceedings of the 12th World Congress on Intelligent Control and Automation (WCICA), Guilin, China, 12–15 June 2016; pp. 477–482.
11. Yang, H.; Yu, Y.; Yuan, Y.; Fan, X. Back-stepping control of two-link flexible manipulator based on an extended state observer. *Adv. Sp. Res.* **2015**, *56*, 2312–2322. [CrossRef]
12. Xia, Y.; Yang, H.; You, X.; Li, H. Adaptive control for attitude synchronisation of spacecraft formation via extended state observer. *IET Control Theory Appl.* **2014**, *8*, 2171–2185.
13. Lin, Y.P.; Lin, C.L.; Suebsaiprom, P.; Hsieh, S.L. Estimating evasive acceleration for ballistic targets using an extended state observer. *IEEE Trans. Aerosp. Electron. Syst.* **2016**, *52*, 337–349. [CrossRef]
14. Duan, H.; Tian, Y.; Wang, G. Trajectory Tracking Control of Ball and Plate System Based on Auto-Disturbance Rejection Controller. In Proceedings of the 7th Asian Control Conference, Hong Kong, China, 27–29 August 2009; pp. 471–476.
15. Dejun, L.; Changjin, C.; Zhenxiong, Z. Permanent magnet synchronous motor control system based on auto disturbances rejection controller. In Proceedings of the International Conference on Mechatronic Science, Electric Engineering and Computer (MEC), Jilin, China, 8–11 August 2011.
16. Liu, B.; Jin, Y.; Chen, C.; Yang, H. Speed Control Based on ESO for the Pitching Axis of Satellite Cameras. *Math. Probl. Eng.* **2016**, *2016*, 2138190. [CrossRef]
17. Li, J.; Xia, Y.; Qi, X.; Gao, Z. On the Necessity, Scheme, and Basis of the Linear-Nonlinear Switching in Active Disturbance Rejection Control. *IEEE Trans. Ind. Electron.* **2017**, *64*, 1425–1435. [CrossRef]
18. Chen, Z.; Xu, D. Output Regulation and Active Disturbance Rejection Control: Unified Formulation and Comparison. *Asian J. Control* **2015**, *18*, 1668–1678. [CrossRef]
19. Xue, W.; Huang, Y. On performance analysis of ADRC for a class of MIMO lower-triangular nonlinear uncertain systems. *ISA Trans.* **2014**, *53*, 955–962. [CrossRef]
20. Yang, J.; Ding, Z. Global output regulation for a class of lower triangular nonlinear systems: A feedback domination approach. *Automatica* **2017**, *76*, 65–69. [CrossRef]
21. Guo, B.-Z.; Wu, Z.-H. Output tracking for a class of nonlinear systems with mismatched uncertainties by active disturbance rejection control. *Syst. Control Lett.* **2017**, *100*, 21–31. [CrossRef]
22. Qian, C.; Li, S.; Frye, M.T.; Du, H. Global finite-time stabilisation using bounded feedback for a class of non-linear systems. *IET Control Theory Appl.* **2012**, *6*, 2326–2336.
23. Zhu, Q. Stabilization of stochastic nonlinear delay systems with exogenous disturbances and the event-triggered feedback control. *IEEE Trans. Autom. Control* **2019**, *64*, 3764–3771. [CrossRef]
24. Ding, K.; Zhu, Q. Extended dissipative anti-disturbance control for delayed switched singular semi-Markovian jump systems with multi-disturbance via disturbance observer. *Automatica* **2021**, *128*, 109556. [CrossRef]
25. Yang, X.; Wang, H.; Zhu, Q. Event-triggered predictive control of nonlinear stochastic systems with output delay. *Automatica* **2020**, *140*, 110230. [CrossRef]
26. Cerkala, J.; Jadlovska, A. Nonholonomic Mobile Robot with Differential Chassis Mathematical Modelling And Implementation In Simulink with Friction in Dynamics. *Acta Electrotech. Et Inform.* **2015**, *15*, 3–8. [CrossRef]
27. Salem, F.A. Dynamic and Kinematic Models and Control for Differential Drive Mobile Robots. *Int. J. Curr. Eng. Technol.* **2013**, *3*, 253–263.
28. Dhaouadi, R.; Hatab, A.A. Dynamic Modelling of Differential-Drive Mobile Robots using Lagrange and Newton-Euler Methodologies: A Unified Framework. *Adv. Robot Autom.* **2013**, *2*, 107.
29. Sousa, R.L.S.; Forte, M.D.D.N.; Nogueira, F.G.; Torrico, B.C. Trajectory Tracking Control of a Nonholonomic Mobile Robot with Differential Drive. In Proceedings of the Biennial Congress of Argentina (ARGENCON), Buenos Aires, Argentina, 15–17 June 2016. [CrossRef]
30. Ajeil, F.; Ibraheem, I.K.; Azar, A.T.; Humaidi, A.J. Autonomous Navigation and Obstacle Avoidance of an Omnidirectional Mobile Robot Using Swarm Optimization and Sensors Deployment. *Int. J. Adv. Robot. Syst.* **2020**, *17*, 1729881420929498. [CrossRef]

41. Ammar, H.H.; Azar, A.T. Robust Path Tracking of Mobile Robot Using Fractional Order PID Controller. In Proceedings of the International Conference on Advanced Machine Learning Technologies and Applications (AMLTA2019), Cairo, Egypt, 28–30 March 2019; Hassanien, A., Azar, A., Gaber, T., Bhatnagar, R.F., Tolba, M., Eds.; Advances in Intelligent Systems and Computing. Springer: Cham, Switzerland, 2020; Volume 921, pp. 370–381.
42. Zidani, G.; Drid, S.; Chrifi-Alaoui, L.; Benmakhlouf, A.; Chaouch, S. Backstepping Controller for a Wheeled Mobile Robot. In Proceedings of the 4th International Conference on Systems and Control, Sousse, Tunisia, 28–30 April 2015.

Disclaimer/Publisher's Note: The statements, opinions and data contained in all publications are solely those of the individual author(s) and contributor(s) and not of MDPI and/or the editor(s). MDPI and/or the editor(s) disclaim responsibility for any injury to people or property resulting from any ideas, methods, instructions or products referred to in the content.

Article

A Cartesian-Based Trajectory Optimization with Jerk Constraints for a Robot

Zhiwei Fan [1,2,3,*], Kai Jia [1,2,4], Lei Zhang [1,2,4], Fengshan Zou [1,2,4], Zhenjun Du [4], Mingmin Liu [4], Yuting Cao [1,2,3] and Qiang Zhang [5]

1 State Key Laboratory of Robotics, Shenyang Institute of Automation, Chinese Academy of Sciences, Shenyang 110016, China
2 Institutes for Robotics and Intelligent Manufacturing, Chinese Academy of Sciences, Shenyang 110169, China
3 University of Chinese Academy of Sciences, Beijing 100049, China
4 SIASUN Robot & Automation Co., Ltd., Shenyang 110169, China
5 School of Automation, Jiangsu University of Science and Technology, No. 666 Changhui Road, Zhenjiang 212100, China
* Correspondence: fanzhiwei@sia.cn

Abstract: To address the time-optimal trajectory planning (TOTP) problem with joint jerk constraints in a Cartesian coordinate system, we propose a time-optimal path-parameterization (TOPP) algorithm based on nonlinear optimization. The key insight of our approach is the presentation of a comprehensive and effective iterative optimization framework for solving the optimal control problem (OCP) formulation of the TOTP problem in the $(s, \dot{s})$-phase plane. In particular, we identify two major difficulties: establishing TOPP in Cartesian space satisfying third-order constraints in joint space, and finding an efficient computational solution to TOPP, which includes nonlinear constraints. Experimental results demonstrate that the proposed method is an effective solution for time-optimal trajectory planning with joint jerk limits, and can be applied to a wide range of robotic systems.

Keywords: time-optimal trajectory planning; iterative optimization; jerk limits; time-optimal path parameterization; phase plane

1. Introduction

Presently, industrial robotics has a wide range of applications, including welding, palletizing, grinding and polishing, assembly, and painting [1–3]. After decades of research, the problem of time-optimal trajectory planning (TOTP) of robots along specified paths has been extensively studied to optimize operation time and improve the efficiency of automated industrial robot operations [4]. TOTP is based on interpolation and introduces the concepts of constraint and optimization to maximize the performance of the robot and ensure the shortest time, while making the trajectory smooth and the operation run smoothly [5]. Time-optimal path parameterization (TOPP) is a fast method for determining critical conditions for navigating a pre-defined smooth path in a robot system's configuration space while respecting physical constraints [6]. Although finding the time-optimal parameterization of a path subject to second-order constraints is a well-studied problem in robotics, TOPP subject to third-order constraints (such as jerk and torque rate) has received relatively little attention and remains largely open. Moreover, joint space trajectory planning cannot visualize the end position of the robotic arm, and Cartesian space trajectory planning is often used in many specific industrial scenarios such as welding, cutting, or machining that require operation on a predetermined path. Therefore, a TOTP algorithm that satisfies the joint third-order constraints in Cartesian space is urgently needed.

1.1. Related Works

Over the years, many academics have worked on the issue of TOTP for industrial robots. This problem can be roughly divided into three main families of methods: Numerical Integration (NI), Convex Optimization (CO), and Dynamic Programming (DP).

The NI-based strategy was initiated by Bobrow et al. [7] and further developed by other researchers. Kunz et al. [8] provided a circular-blends route differentiability approach to ensure that the trajectory precisely follows the specified path of differentiable joint space. Pham [9] provided a comprehensive solution to the problem of dynamic singularities. Pham et al. [10] proposed TOPP3, a novel TOPP algorithm that addresses third-order constraints, as well as the problem of singularities that may hinder the integration of motion profiles and the smooth connection of optimal profiles.Shen et al. [11,12] proposed various new characteristics of the NI method for TOTP along the defined path, and provided explicit mathematical confirmation of these traits. Lu et al. [13] proposed a time-optimal motion planning method for sculpted surface robot machining that takes joint space and tool tip motion constraints into account. They solved the time-optimal tool motion planning in robot machining using an efficient numerical integration method based on the Pontryagin maximum principle. Methods based on NI explicitly calculate the optimal control at each position along the path, instead of performing an implicit search such as the CO-based method, which makes them very fast. However, finding the switch points between the acceleration and deceleration phases is necessary, and the main reason for their failure.

The CO-based strategy has been expanded upon by numerous researchers after being introduced by Verscheure et al. [14]. Xiao et al. [15] used the cubic polynomial fitting method to construct the maximum pseudo-speed curve that meets the torque and speed limits. Debrouwere et al. [16] proposed an effective sequential convex programming (SCP) method to solve the corresponding nonconvex optimal control problems as a difference of convex (DC) function. Pham et al. [6] presented a TOPP approach based on reachability analysis (TOPP-RA), which iteratively computes the reachable and controllable sets at discrete points along the path by solving linear programming problems (LP). Nagy et al. [17] considered kinematics and dynamics constraints and generated the time-optimal velocity distribution for the LP control problem using the sequential optimization method. Ma et al. [18] converted a nonconvex jerk limit into a linear acceleration constraint and indirectly introduced it into CO for TOTP. This method preserves CO's convexity and does not increase the number of optimization variables, resulting in a quick calculation speed. CO-based methods are easy to implement and quite robust, and can consider multiple optimization objectives beyond just time. However, the optimization problem they solve is very large. The number of variables and constraint inequalities scales with the discretization grid size, resulting in implementation that is an order of magnitude slower than the NI-based method.

The DP-based approach was first developed by [19] and has since been expanded and improved upon by numerous researchers. Kaserer et al. [20] proposed a DP-based method for solving the optimal path-tracking problem, which uses interpolation in the phase plane. This approach considers joint speed, acceleration, torque, and mechanical power, as well as joint jerk and torque rate limitations. Kaserer et al. [21] extended this method to solve the time-optimal path-tracking problem for cooperative grasping tasks involving two robots, while also accounting for robot speed, acceleration, jerk, and torque constraints. Barnett et al. [22] introduced the bisection algorithm (BA), a novel technique that extends DP approaches to tackle more complex problems with a larger number of constraints. These approaches, which break down the larger problem into smaller subproblems, become increasingly advantageous as the number of constraints grows, compared to direct transcription methods. Methods based on DP are simple to implement and do not suffer from local minima problems, and they traverse all states at each path point (rather than requiring convex space or convex function assumptions such as the CO-based method). However, the state space to be searched is huge, resulting in implementation being one (or even more) orders of magnitude slower than the CO-based method. Additionally, the DP method cannot truly achieve the global optimal point due to the issue of grid precision.

There are several alternative approaches to the TOTP problem beyond the three groups mentioned above [23–28]. Nevertheless, these approaches also neglect the third-order constraint and do not perform planning in Cartesian space. Table 1 summarizes and compares the similarities and differences of the above three methods in the following five aspects: the requirement for calculating switching points, the ability to consider multiple optimization objectives, the ability to achieve the optimal point (rather than approximately achieving the optimal point), the planning space, and the highest constraint order.

Table 1. A brief overview of related methods.

Methods		Calculate Switch Points	Optimization Objectives (Simple, Multiple)	Achieve Optimal Point	Planning Space	Constraint Order (Second-Order, Third-Order)
NI-based	[7,11,12]	Need	Simple	Yes	Joint/Cartesian	Second-order
	[8,9]	Need	Simple	Yes	Joint	Second-order
	[10,13]	Need	Simple	Yes	Joint	Third-order
CO-based	[6,17]	Not need	Simple	Yes	Joint	Third-order
	[14,15]	Not need	Multiple	Yes	Joint/Cartesian	Second-order
	[16]	Not need	Multiple	Yes	Joint/Cartesian	Third-order (Limit)
	[18]	Need	Simple	Yes	Joint	Third-order (Limit)
DP-based	[19,20]	Not need	Multiple	No	Joint	Third-order
	[21]	Not need	Multiple	No	Joint/Cartesian	Third-order
	[22]	Not need	Multiple	No	Joint	Second-order
	Ours	Not need	Multiple	Yes	Joint/Cartesian	Third-order

1.2. Motivations and Contributions

Motivated by previous approaches, this paper proposes a TOTP algorithm that considers joint third-order limits in a Cartesian coordinate system, maximizing the robot operation efficiency while maintaining smoothness and minimizing time. To achieve this, kinematic feasibility is ensured by introducing joint velocity, acceleration, and jerk constraints on the path parameters s, which are then relocated to the Cartesian space using a constraint transfer method based on Lie theory (We use the Lie group SE(3) to represent the motion of the robot end-effector in Cartesian space. The detailed description of using Lie theory for robot forward and inverse kinematic analysis and the Jacobian matrix derivation process is presented in the Appendix A), reducing the number of decision variables. After establishing the optimal control problem (OCP) formulation of the TOTP problem in the $(s, \dot{s})$ phase plane, the TOPP-RA algorithm is extended to the Cartesian space to obtain an initial solution, and a constraint relaxation approach is used to simplify nonconvex state-update constraints. The method is validated through simulation experiments on a ROS-based platform and real-world experiments on an actual robot, demonstrating effectiveness, generality, and robustness. This paper makes several contributions to the field of optimal trajectory generation:

- A comprehensive and effective framework for iterative optimization is presented to establish the OCP formulation of the TOTP problem, which is described by the path parameter s;
- Given an efficient computational solution for computing the nonlinear TOPP in Cartesian space while satisfying third-order constraints in joint space;

- Experiments have demonstrated that the proposed method can effectively generate smoother trajectories that satisfy jerk constraints on a wide range of robot systems.

The remainder of this paper is organized as follows. Section 2 outlines the key features of the OCP model used for the TOPP algorithm. In Section 3, we present the Cartesian-based TOPP-RA method and describe the proposed TOPP algorithm based on iterative optimization. Section 4 reports extensive experimental results. Finally, in Section 5, we provide concluding remarks.

2. Problem Statement

In this section, we establish the TOPP problem as an OCP in Cartesian space, which includes joint third-order constraints and an objective function in the $(s, \dot{s})$ phase plane. The details of these constraints will be formulated in the following subsections.

2.1. General Description

In a n-dof robot system, the state profiles in configuration space are denoted by $\mathbf{x}(t) = [\mathbf{q}(t); \dot{\mathbf{q}}(t); \ddot{\mathbf{q}}(t)]$, where $\mathbf{q} \in \mathbb{R}^n$ represents the configuration of the system. The control inputs $\mathbf{u}(t)$ represent the third derivative of the joint angles, $\dddot{\mathbf{q}}(t)$, in configuration space. The following is a standard OCP that can be used to describe the time-optimal speed planning problem [29]:

$$\min J(\mathbf{x}(t), \mathbf{u}(t))$$
$$\text{s.t. } \dot{\mathbf{x}}(t) = f_{Status-update}(\mathbf{x}(t), \mathbf{u}(t)),$$
$$\mathbf{x}_{min} \leq \mathbf{x}(t) \leq \mathbf{x}_{max}, \mathbf{u}_{min} \leq \mathbf{u}(t) \leq \mathbf{u}_{max}, t \in [0, T]; \quad (1)$$
$$\mathbf{x}(0) = \mathbf{x}_{init}, \mathbf{u}(0) = \mathbf{u}_{init}, \mathbf{x}(T) = \mathbf{x}_{goal}, \mathbf{u}(T) = \mathbf{u}_{goal}.$$

$f_{Status-update} = 0$ forms the status-update process. $[\mathbf{x}_{min}, \mathbf{x}_{max}, \mathbf{u}_{min}, \mathbf{u}_{max}]$ describes the allowable regions of state and control profiles. $[\mathbf{x}_{init}, \mathbf{u}_{init}, \mathbf{x}_{goal}, \mathbf{u}_{goal}]$ denotes the start and end conditions of the state and control profiles. T represents the total time, which is unidentified now.

To translate the above model to a TOPP problem in the $(s, \dot{s})$ phase plane, we propose a function $\mathbf{p}(s)_{s \in [0, s_{end}]}$ that represents a geometric path in the Cartesian space, and is piece-wise C^2-continuous. We introduce a time parameterization that itself represents the parameter of the path, as a piece-wise C^2, increasing scalar function $s : [0, T] \rightarrow [0, s_{end}]$. The trajectory is then recovered as $\mathbf{p}(s(t))_{t \in [0,T]}$ [30]. In the rest of this section, we introduce how to complete the transformation of the TOPP problem through $s : [0, T] \rightarrow [0, s_{end}]$.

2.2. Objective Function

To minimize the total time of robot movement, the objective function $J(\mathbf{x}(t), \mathbf{u}(t))$ is defined as

$$J = T = \int_{t=0}^{T} 1 dt \quad (2)$$

Replace the previous equation with $ds/ds = 1$ and change the integral limits from $[0, T]$ (time) to $[0, s_{end}]$ (s) [31]. Formula (2) is updated as follows:

$$J = \int_{t=0}^{T} 1 dt = \int_{t=0}^{T} \frac{ds}{ds} dt = \int_{s=0}^{s_{end}} \frac{dt}{ds} ds = \int_{s=0}^{s_{end}} \frac{1}{\dot{s}} ds \quad (3)$$

Therefore, to minimize the time, $\dot{s}^{-1}(s)$ should be as small as possible. In other words, $\dot{s}(s)$ must be as large as possible while still satisfying the various constraints mentioned later. This means that the state trajectory must follow the boundary of the phase diagram plotted by $(s, \dot{s})$, which is naturally aligned with TOPP-RA method.

2.3. Constraints

In a TOPP problem, there are generally three types of constraints: status-update constraints, constraints on the states/control profiles, and two-point boundary constraints [32].

2.3.1. Status-Update Constraints

The state-update/kinematic constraints of a robot describe the kinematic feasibility of the robot's motion. Using forward and inverse kinematics, the configuration **q** in the joint space can be converted to the corresponding Cartesian space representation **p** (see the Appendix A for transformation method). As a result, the state and control profiles can be expressed in terms of the geometric path **p**(s), which can then be further transformed to a form represented by path parameters s, as shown in Equation (4).

$$\frac{d}{ds}\begin{bmatrix}\dot{s}\\\ddot{s}\end{bmatrix} = \begin{bmatrix}\frac{\ddot{s}}{\dot{s}}\\\frac{\dddot{s}}{\dot{s}}\end{bmatrix}, s \in [0, s_{end}] \quad (4)$$

The status-update function can be rewritten by performing a second-order Taylor series expansion at s_i.

$$\dot{s} = \dot{s}_i + \frac{\ddot{s}_i}{\dot{s}_i}\Delta(s) + \frac{d^2\dot{s}}{ds^2}\bigg|_{s=\zeta}(\Delta(s))^2$$
$$\ddot{s} = \ddot{s}_i + \frac{\dddot{s}_i}{\dot{s}_i}\Delta(s) + \frac{d^2\ddot{s}}{ds^2}\bigg|_{s=\eta}(\Delta(s))^2 \quad (5)$$

where $s \in [s_i, s_{i+1}], \zeta, \eta \in [s_i, s]$ and $\Delta(s) = s - s_i$. Let us define the first-order status-update discretization function as follows:

$$\dot{s} = \dot{s}_i + \frac{\ddot{s}_i}{\dot{s}_i}\Delta(s)$$
$$\ddot{s} = \ddot{s}_i + \frac{\dddot{s}_i}{\dot{s}_i}\Delta(s) \quad (6)$$

The error of the first-order status-update discretization function, denoted by e_{first}^{state}, is as follows:

$$e_{first}^{state} = O(\Delta^2(s)) \quad (7)$$

Similarly, by performing a third-order Tylor series expansion at s_i, the second-order status-update discretization function and its error can be, respectively, rewritten as:

$$\dot{s} = \dot{s}_i + \frac{\ddot{s}_i}{\dot{s}_i}\Delta(s) + (\frac{\dddot{s}_i}{\dot{s}_i^2} - \frac{\ddot{s}_i^2}{\dot{s}_i^3})\Delta^2(s)$$
$$\ddot{s} = \ddot{s}_i + \frac{\dddot{s}_i}{\dot{s}_i}\Delta(s) - \frac{\ddot{s}_i\dddot{s}_i}{\dot{s}_i^3}\Delta^2(s) \quad (8)$$

$$e_{second}^{state} = O(\Delta^3(s)) \quad (9)$$

2.3.2. States/Control Profiles Constraints

The state/control constraints of a robot refer to the physical constraints that the robot must adhere to during its motion process. Typically, these constraints involve the robot's state variables, such as position, velocity, acceleration, joint angles, and so on. The constraints on the robot's states and control profiles can be formulated as $\mathbf{x}_{min} \leq \mathbf{x}(t) \leq \mathbf{x}_{max}$ and $\mathbf{u}_{min} \leq \mathbf{u}(t) \leq \mathbf{u}_{max}$, respectively, where $t \in [0, T]$. These constraints essentially limit the speed, acceleration, and jerk of the robot's joints [33], as illustrated in the following equations.

$$\begin{bmatrix} \dot{q}_{min} \\ \ddot{q}_{min} \\ \dddot{q}_{min} \end{bmatrix} \leq \begin{bmatrix} \dot{q}(t) \\ \ddot{q}(t) \\ \dddot{q}(t) \end{bmatrix} \leq \begin{bmatrix} \dot{q}_{max} \\ \ddot{q}_{max} \\ \dddot{q}_{max} \end{bmatrix}, t \in [0, T] \tag{10}$$

The derivatives of the joints are projected into Cartesian space through the Jacobian matrix, as shown in Formula (11), which yields the derivatives of the path parameter s.

$$\begin{aligned} \mathbf{J}\dot{q} &= \mathbf{p}'\dot{s} \\ \mathbf{J}\ddot{q} + \dot{\mathbf{J}}\dot{q} &= \mathbf{p}''\dot{s}^2 + \mathbf{q}'\ddot{s} \\ \mathbf{J}\dddot{q} + 2\dot{\mathbf{J}}\ddot{q} + \ddot{\mathbf{J}}\dot{q} &= \mathbf{p}'''\dot{s}^3 + 3\mathbf{p}''\dot{s}\ddot{s} + \mathbf{p}'\dddot{s} \end{aligned} \tag{11}$$

where $\square'$ is defined as the differentiation of $\square$ with respect to the path parameter s. Henceforth, we shall refer to $s, \dot{s}, \ddot{s}$, and $\dddot{s}$ as the position, velocity, acceleration, and jerk, respectively. By substituting Equation (11) into Equation (10), the inequality constraints on the states/control profiles can be expressed as follows:

$$\begin{bmatrix} \dot{q}_{min} \\ \ddot{q}_{min} \\ \dddot{q}_{min} \end{bmatrix} \leq \begin{bmatrix} \mathbf{a}(s)\dot{s} \\ \mathbf{b}(s)\dot{s}^2 + \mathbf{c}(s)\ddot{s} \\ \mathbf{d}(s)\dot{s}^3 + \mathbf{e}(s)\dot{s}\ddot{s} + \mathbf{f}(s)\dddot{s} \end{bmatrix} \leq \begin{bmatrix} \dot{q}_{max} \\ \ddot{q}_{max} \\ \dddot{q}_{max} \end{bmatrix}, s \in [0, s_{end}], \text{where} \tag{12}$$

$$\begin{aligned} \mathbf{a}(s) &:= \mathbf{J}^{-1}(s)\mathbf{p}'(s), \\ \mathbf{b}(s) &:= \mathbf{J}^{-1}(s)(\mathbf{p}''(s) - \mathbf{J}'(s)\mathbf{J}^{-1}(s)\mathbf{p}'(s)), \\ \mathbf{c}(s) &:= \mathbf{J}^{-1}(s)\mathbf{p}'(s), \\ \mathbf{d}(s) &:= \mathbf{J}^{-1}[\mathbf{p}'''(s) - 2\mathbf{J}'(s)\mathbf{J}^{-1}(s)(\mathbf{p}''(s) - \mathbf{J}'(s)\mathbf{J}^{-1}(s)\mathbf{p}'(s)) - \mathbf{J}''(s)\mathbf{J}^{-1}(s)\mathbf{p}'(s)], \\ \mathbf{e}(s) &:= 3\mathbf{J}^{-1}(s)(\mathbf{p}''(s) - \mathbf{J}'(s)\mathbf{J}^{-1}(s)\mathbf{p}'(s)), \\ \mathbf{f}(s) &:= \mathbf{J}^{-1}(s)\mathbf{p}'(s). \end{aligned} \tag{13}$$

The formulas for calculating each order derivative of the Jacobian matrix $(\mathbf{J}', \mathbf{J}'')$ will be presented in the Appendix A.

2.3.3. Boundary Constraints

Boundary constraints refer to the limitations imposed on the state and control variables of a robot during the initial and final stages of its operation. The constraints $\mathbf{x}(0) = \mathbf{x}_{init}$, $\mathbf{u}(0) = \mathbf{u}_{init}$, $\mathbf{x}(T) = \mathbf{x}_{goal}$, and $\mathbf{u}(T) = \mathbf{u}_{goal}$ define the boundary conditions. These boundary conditions ensure that the state and control profiles at the start moment $s = 0(t = 0)$ and the end moment $s = s_{end}(t = T)$ represent the necessary facts at those moments, respectively.

$$\begin{aligned}{} [\dot{s}(0), \ddot{s}(0), \dddot{s}(0)] &= [\dot{s}_0, \ddot{s}_0, \dddot{s}_0], \\ [\dot{s}(s_{end}), \ddot{s}(s_{end}), \dddot{s}(s_{end})] &= [\dot{s}_{s_{end}}, \ddot{s}_{s_{end}}, \dddot{s}_{s_{end}}]. \end{aligned} \tag{14}$$

In particular, more degrees of freedom are allowed in setting the control profile $\dddot{s}$ at $s = 0(t = 0)$ to ensure the normal operation of the motor.

As a summary of this section, the following OCP is established to represent the TOPP problem based on Cartesian space:

$$\begin{aligned} &\min \ (3) \\ &\text{s.t. Status-update constraints (4);} \\ &\quad \text{States/Control profiles constraints (12) and (13);} \\ &\quad \text{Two-point boundary constraints (14).} \end{aligned} \tag{15}$$

In general, when moving from the initial state to the target state along a predetermined path, speed planning aims to resolve any potential conflicts that may arise between kinematics-based constraints and environmental constraints. However, due to the nonlinear relationship between the state and control variables, an appropriate initial solution is required to solve the OCP (15). There is a problem with the state/control constraints (12) in OCP (15) because the jerk of the robot is not taken into account when solving the initial solution, which can easily lead to leaving out the free space required for kinematic feasibility. Therefore, directly solving OCP (15) may not always be effective. An alternative option we propose is to build an iterative framework in which the kinematic feasibility is adaptively adjusted when it is found to be inappropriate. The details on how to find an effective computational solution to TOPP with nonlinear constraints are described in Section 3.

3. TOPP by Iterative Optimization (TOPP-IO)

This section introduces our proposed Cartesian-based TOPP-IO method. First, we present the initial guess and control group generated by the Cartesian-based TOPP-RA method, followed by an explanation of the principle of the TOPP-IO method.

3.1. Cartesian-Based TOPP-RA Method

Combining with [6], we expanded the TOPP-RA method from joint space to Cartesian space, which we call the Cartesian-based TOPP-RA method. The geometric path in Cartesian space, denoted by $\mathbf{p}(s)$, is divided into N segments with $N+1$ grid points, where $(s_i, \dot{s}_i, \ddot{s}_i, \dddot{s}_i)$ represents the i-th stage state and control profiles, with $i \in [0, 1, \ldots, N]$. The constraints of joint acceleration can be formulated as follows, by taking into account (12) and (13):

$$\mathbf{B}\dot{s}^2 + \mathbf{C}\ddot{s} \leq \ddot{\mathbf{Q}} \qquad (16)$$

where $\mathbf{B} = \begin{bmatrix} \mathbf{b}(s) \\ -\mathbf{b}(s) \end{bmatrix}, \mathbf{C} = \begin{bmatrix} \mathbf{c}(s) \\ -\mathbf{c}(s) \end{bmatrix}$ and $\ddot{\mathbf{Q}} = \begin{bmatrix} \ddot{\mathbf{q}}_{max} \\ -\ddot{\mathbf{q}}_{min} \end{bmatrix}$. The velocity constraints of the joints are expressed as a range of i-stage state variables, $\mathfrak{X}_i = [(\dot{s}_i^2)^{lower}, (\dot{s}_i^2)^{upper}]$, which reflects the allowable velocity of the joints.

$$(\dot{s}_i^2)^{lower} = \max_j \left\{ \frac{\dot{q}_{min,j}}{a_j(s_i)} \mid a_j(s_i) > 0 \quad \text{or} \quad \frac{\dot{q}_{max,j}}{a_j(s_i)} \mid a_j(s_i) < 0 \right\},$$
$$(\dot{s}_i^2)^{upper} = \min_j \left\{ \frac{\dot{q}_{max,j}}{a_j(s_i)} \mid a_j(s_i) > 0 \quad \text{or} \quad \frac{\dot{q}_{min,j}}{a_j(s_i)} \mid a_j(s_i) < 0 \right\}. \qquad (17)$$

where j is the j-th element of $\mathbf{a}(s_i)$, $\dot{\mathbf{q}}_{min}$ and $\dot{\mathbf{q}}_{max}$. The state-update function for constant acceleration over $[s_i, s_{i+1}]$ is given by:

$$\dot{s}_{i+1}^2 = \dot{s}_i^2 + 2\Delta_i \ddot{s}_i \qquad (18)$$

where $\Delta = s_{i+1} - s_i$.

3.1.1. Backward Pass

In considering the segment $[s_i, s_{i+1}]$ and assuming that the $i+1$-th feasible range, $\mathfrak{S}_{i+1}$, is known, the i-th feasible range, $\mathfrak{S}_i = [(\dot{s}_i^2)^-, (\dot{s}_i^2)^+]$, can be calculated using the following formula:

$$\begin{aligned} (\dot{s}_i^2)^- &:= \min \dot{s}_i^2, \\ (\dot{s}_i^2)^+ &:= \max \dot{s}_i^2, \\ \text{s.t.} \quad & \dot{s}_i^2 \in \mathfrak{X}_i, \\ & \dot{s}_i^2 + 2\Delta_i \ddot{s}_i \in \mathfrak{S}_{i+1}, \\ & \mathbf{B}\dot{s}_i^2 + \mathbf{C}\ddot{s}_i \leq \ddot{\mathbf{Q}}. \end{aligned} \qquad (19)$$

Obviously, Formula (19) indicates that for any $\dot{s}_i^2 \in \mathfrak{S}_i$, there always exists a state $\dot{s}_{i+1}^2 \in \mathfrak{S}_{i+1}$ that corresponds to it. In other words, we can always move from the feasible range $\mathfrak{S}_i$ to $\mathfrak{S}_{i+1}$ using the state-update function. By applying Formula (19) recursively, we can obtain a set of transitive feasible ranges, $[\mathfrak{S}_0, \mathfrak{S}_1, \ldots, \mathfrak{S}_n]$. Any state that belongs to the transitive feasible ranges can be transferred to the ending state when the last feasible range set is determined.

3.1.2. Forward Pass

By transferring $\dot{s}_i^2 \in \mathfrak{S}_i$ from step i to $\dot{s}_{i+1}^2 \in \mathfrak{S}_{i+1}$ of step $i+1$, we can recursively reach the final state $\mathfrak{S}_n$. Furthermore, literature [6] has demonstrated that the transition process occurs on a convex polygon. Therefore, selecting control variables that can reach the upper limit of the next $\mathfrak{S}$ will result in the shortest task time. This selection exhibits locally greedy behavior while globally optimizing performance. Once the transitive feasible ranges have been derived from the backward pass, the method for transferring $(\dot{s}_i^2)^*$ to $(\dot{s}_{i+1}^2)^*$ using a greedy algorithm is as follows:

$$(\dot{s}_{i+1}^2)^* := \max(\dot{s}_i^2)^* + 2\Delta_i \ddot{s}_i,$$
$$\text{s.t.} \quad (\dot{s}_i^2)^* + 2\Delta_i \ddot{s}_i \in \mathfrak{S}_{i+1}, \tag{20}$$
$$\mathbf{B}(\dot{s}_i^2)^* + \mathbf{C}\ddot{s}_i \leq \ddot{\mathbf{Q}}.$$

where $(\dot{s}_i^2)^*$ denotes the optimal solution at the i-th grid point. By setting deterministic values of $(\dot{s}_0^2)^* \in \mathfrak{S}_0$ and $\mathfrak{S}_n = \{(\dot{s}_n^2)^*\}$, the solution of Cartesian-based TOPP-RA, $[(\dot{s}_0^2)^*, (\dot{s}_1^2)^*, \ldots, (\dot{s}_n^2)^*]$, is obtained by recursively applying Formula (20).

3.2. Principle of the Proposed TOPP-IO Method

The general principle of the TOPP iterative optimization method is illustrated by the pseudo-codes in Algorithm 1. Given a path $\mathcal{P}$ in Cartesian space, Algorithm 1 first generates an initial conjecture using the ToppraGuess() function to numerically solve (15) without joint jerk limits. This initial conjecture includes the path discretization, all the parameters required to solve (15), and the initial values of all the decision variables. Then, using the full content of this initial conjecture, Algorithm 1 establishes an iterative OCP where an intermediate optimal solution is obtained from each iteration. After the first three lines of initialization, the while loop is applied to iteratively solve the TOPP$_{OCP}$. Similar to (15), the only difference is that we add (4) as a soft constraint to the objective function. Specifically, this iterative OCP solves the following optimization problems.

$$\min (3) + \omega_{soft} \cdot f_{soft}(s)$$
$$\text{s.t. States/Control profiles constraints (12) and (13);} \tag{21}$$
$$\text{Two-point boundary constraints (14).}$$

where $\omega_{sift} > 0$ is a parameter used to weight the softening of the state updating, and $f_{soft}(s)$ is denoted as

$$f_{soft}(s) = \int_{s=0}^{s_{end}} \left\| \begin{bmatrix} \dot{s} \\ \ddot{s} \end{bmatrix} - f_{Status-update}(s) \right\|^2 ds \tag{22}$$

In each iteration of the while loop, the function SolveIteratively(OCP$_{TOPP}$, $\mathcal{G}$) is used to solve (21) using the initial conjecture $\mathcal{G}$. The function StateUpdateInfeasibility($\mathcal{G}$) evaluates the infeasibility degree of the status update determined by $f_{soft}(s)$ as given in (22). When $f_{soft}(s)$ becomes small enough, i.e., close to 0^+, the function GetTrajectoryInformation($\mathcal{G}$) is called to extract information about the optimal trajectory from the solution $\mathcal{G}$.

Algorithm 1: An Iterative Optimal Method for TOPP

Input: Geometric path in Cartesian space $\mathcal{P}$
Output: Optimal trajectory information info_{opti}

1 $\mathcal{G} = \text{ToppraGuess}(\mathcal{P})$;
2 $\omega_{soft} \leftarrow \omega_{soft0}, iter \leftarrow 0, \text{info}_{opti} \leftarrow \varnothing$;
3 **while** $iter < iter_{max}$ **do**
4 $\quad \text{OCP}_{TOPP} \leftarrow \text{BuildIterativeOCP}(\mathcal{G})$;
5 $\quad \mathcal{G} \leftarrow \text{SolveIteratively}(\text{OCP}_{TOPP}, \mathcal{G})$;
6 $\quad f_{soft}(s) \leftarrow \text{StateUpdateInfeasibility}(\mathcal{G})$;
7 $\quad$ **if** $f_{soft}(s) < \epsilon_{soft}$ **then**
8 $\quad\quad \text{info}_{opti} \leftarrow \text{GetTrajectoryInformation}(\mathcal{G})$;
9 $\quad\quad$ **return**;
10 $\quad$ **else**
11 $\quad\quad \omega_{soft} \leftarrow \omega_{soft} \cdot \alpha, iter \leftarrow iter + 1$;
12 $\quad$ **end**
13 **end**
14 **return**;

3.3. Properties Discussion of Algorithm 1

This subsection describes the relevant properties of the proposed TOPP-IO method in Algorithm 1.

First, the iterative process progressively increases the feasibility and optimality of the phase state. It is assumed that the initial solution obtained by the Cartesian frame TOPP-RA does not satisfy the jerk constraint and, hence, is not status-update feasible. In such cases, restoring status-update feasibility becomes the primary goal of minimizing the objective function of OCP_{TOPP}. Therefore, the optimal solution differs from the initial guess by reducing the status-update infeasibility. Although the status-update infeasibility may not be eliminated, the resulting $(s, \dot{s})$ phase diagram is closer to being feasible, providing opportunities for further improvement in succeeding iterations.

Second, optimality is achieved when Algorithm 1 exits from line 9. As the iteration continues, the status-update infeasibility approaches 0^+ and incrementing ω_{soft} expedites the procedure. When the degree of status-update infeasibility is small, the total time (3) in the objective function of (21) dominates. Thus, the objective function of (21) is minimized, closing in on minimizing the original objective function (3) to an accuracy level of ϵ_{soft}.

Third, the OCP_{TOPP} is always feasible, which is a crucial cornerstone of the entire iterative framework. With strict restrictions on CPU runtime and a willingness to accept suboptimal solutions, a feasible solution can be obtained at any point by interrupting the iterative optimization process. With very slow motion always feasible, the solution procedure for each (21) is consistently in the feasible region of the solution space when the initial solution is set to 0. Thus, as long as the obtained $(s, \dot{s})$ phase diagram's near-future period is status-update feasible, the resulting phase states can be transferred to the next iterative OCP_{TOPP} for further enhancements.

4. Simulation and Real-World Experiment Results

In this section, simulation experiments will be used to demonstrate the feasibility, performance, and generality of the proposed method, as well as an industrial robot real machine-verification experiment will be performed, which gives practical significance to the TOPP-IO algorithm. The proposed method is executed on Ubuntu using an Intel i7-7700HQ @ 2.80 GHz CPU and 16-GB RAM, and all optimization problems are solved using CasADi (CasAdi is an open-source software framework for nonlinear optimization and optimal control. It provides a flexible and efficient interface for constructing and solving various optimization problems, including trajectory optimization) [34]. We use the 6-DOF Firefox robot from SIASUN in both the simulation and the real world, in addition to the Pioneer P3-DX robot used in the simulation. The implementation of TOPP-IO was done in C++, and the required communication between systems for these experiments was established. Figure 1 illustrates the architecture of the implementation.

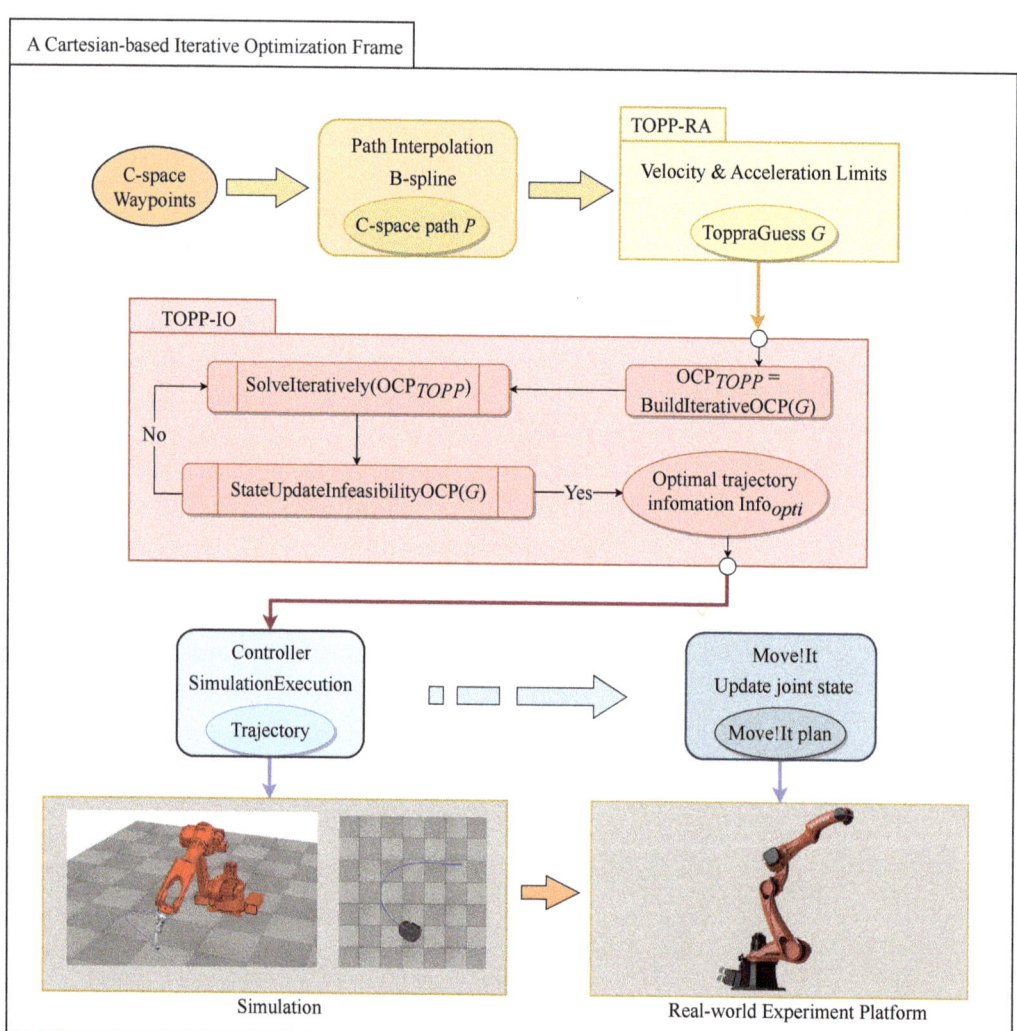

Figure 1. Architecture of implementation.

4.1. Experiment Settings

The joint and wheel velocity, acceleration, and jerk limits are presented in each experiment, respectively, which are critical factors for the safe and efficient operation of robotic systems. To assess the robustness and adaptability of our proposed algorithm, TOPP-IO, we conducted a series of experiments with varying jerk limits. Specifically, we evaluated the performance of TOPP-IO under four different jerk limits: $0.1\times$, $1\times$, $10\times$, and $100\times$ the default value. The basic parameters for the iterative optimization are carefully selected to ensure the convergence and efficiency of the optimization process, which are listed in Table 2.

Table 2. Hyperparameter setting for iterative optimization.

Hyperparameter	Description	Value
$iter_{max}$	Maximum iteration number	5
ω_{soft0}	ω_{soft} initial value	10^5
α	Multiplier to enlarge ω_{soft}	10
ϵ_{soft}	Softened constraints tolerance	16

4.2. Comparison with TOPP-RA Method

This method is built and tested on the random geometric route depicted in Figure 2, subject to joint velocity, acceleration, and jerk limitations which are presented in Table 3. The simulation results are compared with those obtained from the CO algorithm (TOPP-RA) presented in [6] to demonstrate the effectiveness of the proposed strategy in controlling the acceleration surge caused by ignoring the jerk constraints.

Table 3. Velocity, acceleration, and jerk limits of joints.

Limits	Joint1	Joint2	Joint3	Joint4	Joint5	Joint6
Vel. (rad/s)	2	2	2	4	4	4
Acc. (rad/s^2)	5	6	6	12	12	12
Jerk (rad/s^3)	16	16	18	20	28	28

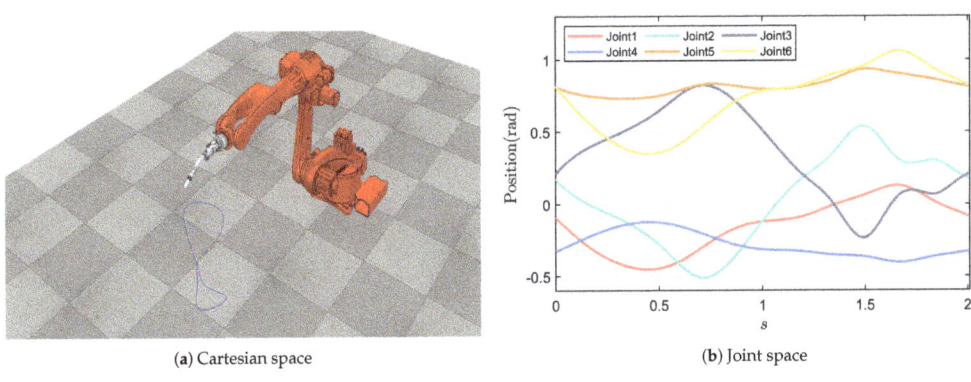

(a) Cartesian space   (b) Joint space

Figure 2. The geometric path on which this approach is implemented and tested.

The results of the two approaches, TOPP-RA and TOPP-IO, in the $(s, \dot{s})$ and $(s, \ddot{s})$ phase planes are presented in Figures 3 and 4, respectively. It can be observed from Figure 4 that TOPP-RA allows for steep slopes of acceleration due to the lack of restriction on jerk, leading to an abrupt shift in acceleration between neighboring path points. This sudden change in acceleration can be seen in the velocity curve of Figure 3, where there is no smooth transition between the acceleration and deceleration portions. Such abrupt changes in acceleration can result in jerky and unstable motion, which is not desirable in many real-world applications. To address this issue, TOPP-IO imposes explicit joint jerk limitations, leading to smoother acceleration profiles between neighboring path points. Figure 4 shows that the TOPP-IO method successfully restricts the acceleration mutation, preventing any abrupt changes in acceleration. Furthermore, the velocity curve of TOPP-IO in Figure 3 exhibits smoother transitions between the portions representing acceleration and deceleration, guaranteeing that the nearby segments will not violate the imposed restrictions.

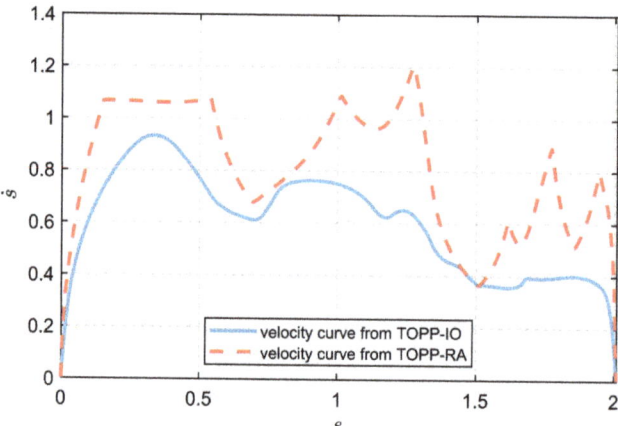

Figure 3. Comparison of the TOPP-RA resultant velocity curve without jerk limitations (red dashed line) and the one obtained from the proposed method with jerk limits (blue solid line).

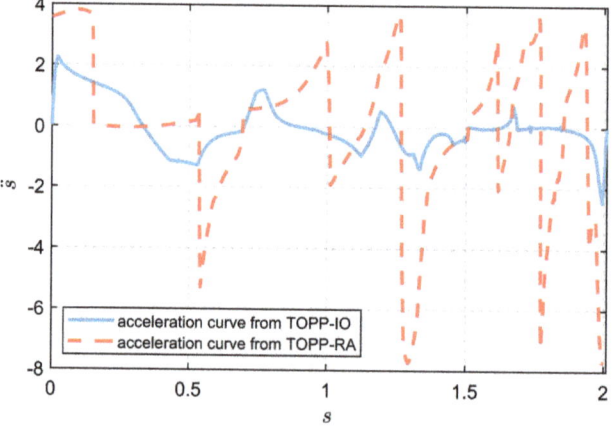

Figure 4. Comparison of the TOPP-RA resultant acceleration curve without jerk limitations (red dashed line) and the one obtained from the proposed method with jerk limits (blue solid line).

To further evaluate the performance of the two approaches, we compare their execution times in Table 4 and display the corresponding speed, acceleration, and jerk curves in Figure 5 for various jerk limits (100×, 10×, 1×, and 0.1×). In the TOPP-RA method, it is evident that the acceleration profiles are bang-bang, satisfying all joint second-order constraints.

With all third-order kinematic constraints, the jerk profiles are bang-bang in the TOPP-IO method, leading to smoother transitions between the portions representing acceleration and deceleration. Without joint jerk limits, the maximum acceleration is about 1638.4 rad/s^3. As the jerk limit is decreased from "none" to 100× and 10× jerk limits, the execution time only slightly increases from 2.81067 s to 2.89393 s and 2.90326 s, respectively, and the smoothing effect of the speed profile is not immediately noticeable. The speed profile becomes smoother as the jerk limits approach 1× jerk limits. Notably, even when the jerk limit is set to 0.1× jerk limits, TOPP-IO can still produce a valid solution, albeit with an increased execution time.

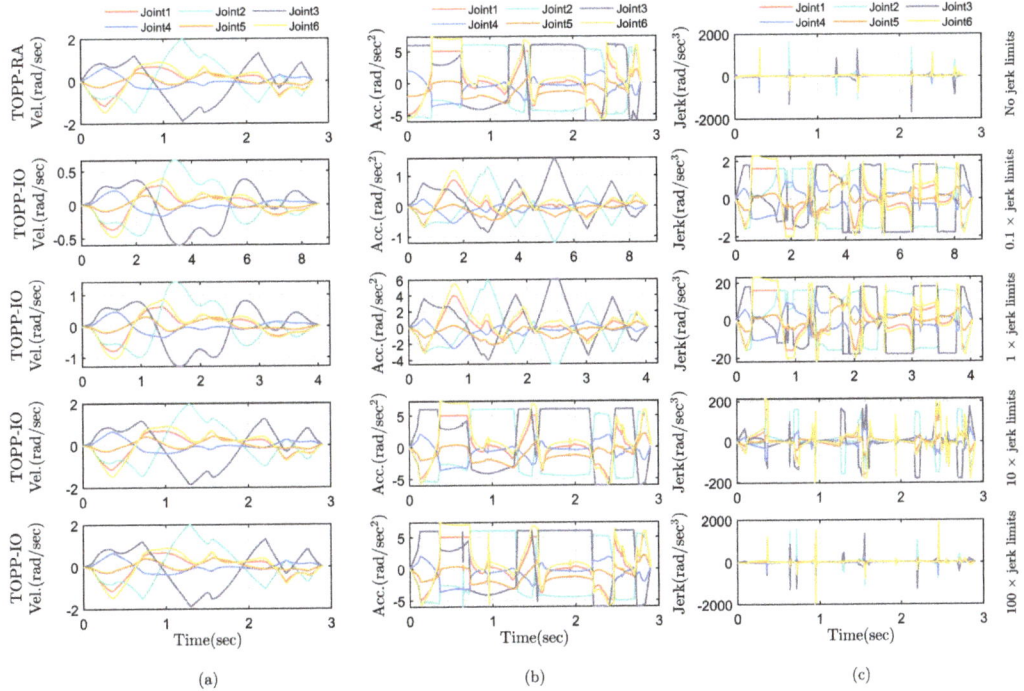

Figure 5. Velocity, acceleration, and jerk profiles for various methods and jerk restrictions. (a) Speed, (b) Acceleration, and (c) Jerk.

Table 4. Execution time of different trajectory planning algorithms and jerk restrictions.

Method	TOPP-RA	TOPP-IO			
Jerk Limits (rad/s³)	-	100×	10×	1×	0.1×
t_e (s)	2.81067	2.89393	2.90326	4.02941	8.63447

4.3. Application on Mobile Robot

Our method applies not only to manipulators but also to a wide range of robots. To demonstrate its flexibility, we computed a ground trajectory for the Pioneer P3-DX, a diff-drive mobile robot. The wheel velocity, acceleration, and jerk limitations are presented in Table 5. Screenshots of the operational phase as well as the wheel speed curve in comparison to TOPP-RA are shown in Figure 6.

Table 5. Velocity, acceleration, and jerk limits of wheels.

Limits	Wheel1	Wheel2
Vel. (rad/s)	2	2
Acc. (rad/s²)	4	4
Jerk (rad/s³)	8	8

The restrictions on wheel jerk and route jerk constraints have similar effects on controlling acceleration mutation. In this study, jerk restrictions were defined as wheel jerk constraints that effectively limit acceleration mutation in the route. The robot trajectories obtained from TOPP-RA and the proposed method are presented in Figure 7. Table 6 indicates that the maximum absolute values of the robot's acceleration and jerk curves obtained from the proposed method are reduced by 60.28% and 69.82%, respectively, compared to those from TOPP-RA.

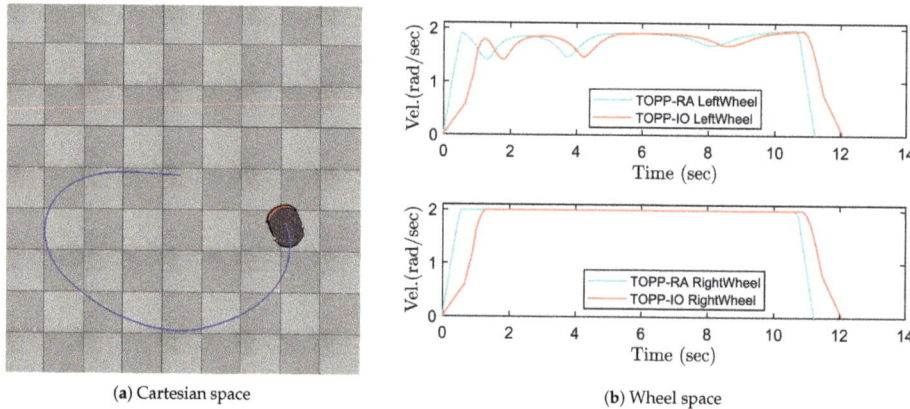

(a) Cartesian space

(b) Wheel space

Figure 6. The ground path on which this approach is implemented and tested.

Table 6. Comparing the maximum absolute value of the robot's acceleration and jerk curves between the two approaches.

	Acceleration (m/s^2)	Jerk (m/s^3)
TOPP-RA	3.98086	26.4858
TOPP-IO	1.58118	7.9937
Degree of decline	60.28%	69.82%

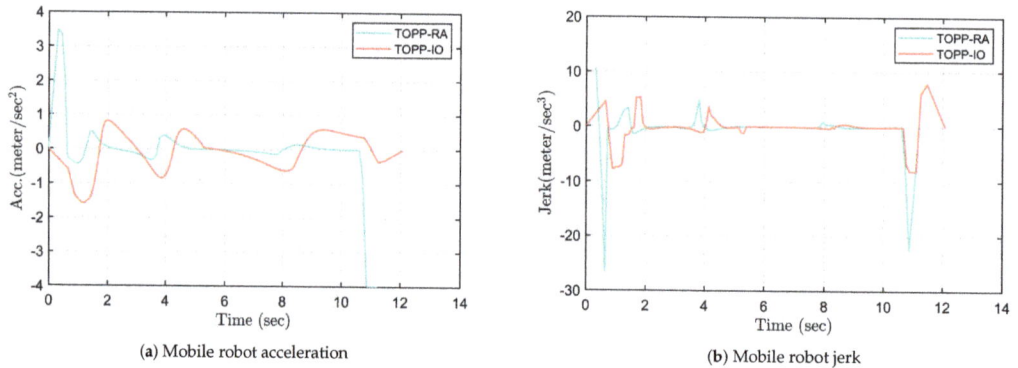

(a) Mobile robot acceleration

(b) Mobile robot jerk

Figure 7. Comparing the resulting mobile robot acceleration (a) and jerk (b) from TOPP-RA and the ones obtained from the proposed approach.

4.4. Real-World Experiments

In real-world experiments, we applied our method to the welding industry where the objective is to complete tasks as quickly, safely, and efficiently as possible. TOPP-IO succeeded in executing the assignment in a timely, safe, and stable manner. The running state of the Firefox robot in the actual world is shown in Figure 8 and is consistent with the simulation results.

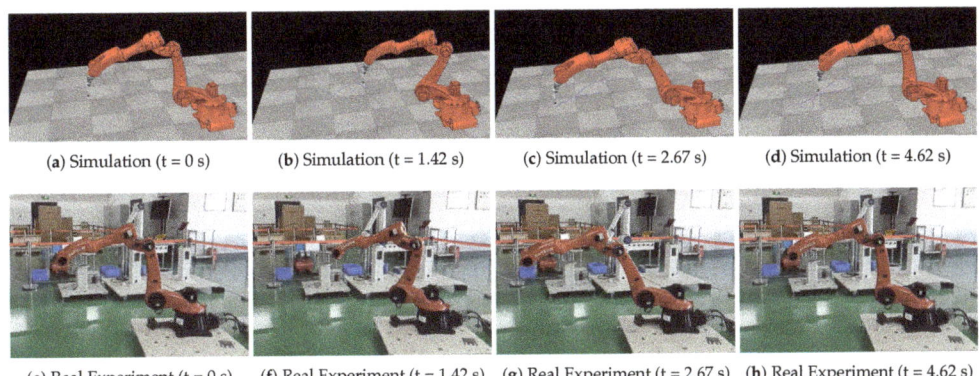

Figure 8. Real-world experiments (e–h) in accordance with the simulation (a–d).

We performed both quantitative and qualitative analyses of our method's performance during the actual operation process. Specifically, we analyzed the position error of each joint and examined the speed-tracking situation using joint1 as an example. Figure 9a shows the joint position error of the TOPP-RA method during actual operation, while Figure 9b displays the joint position error of the TOPP-IO method under the same path. In addition, Table 7 compares the performance of our TOPP-IO method with that of the TOPP-RA method. The results show that the average and maximum position errors of all joints in TOPP-IO have been reduced to different degrees during operation. The absolute values of the average position error and maximum position error have been reduced by about 29% and 27%, respectively, compared to the TOPP-RA method.

Table 7. The absolute values of the average and maximum joint position error on different trajectory planning algorithms.

		Joint1	Joint2	Joint3	Joint4	Joint5	Joint6
Average position error	TOPP-RA (rad)	0.0160	0.0296	0.0293	0.0075	0.0066	0.0192
	TOPP-IO (rad)	0.0112	0.0205	0.0206	0.0053	0.0047	0.0135
	Degree of decline	30.13%	30.67%	29.81%	29.60%	29.65%	29.46%
Maximum position error	TOPP-RA (rad)	0.0518	0.0911	0.0849	0.0270	0.0183	0.0621
	TOPP-IO (rad)	0.0339	0.0624	0.0557	0.0196	0.0127	0.0444
	Degree of decline	34.63%	31.54%	34.36%	27.62%	30.36%	28.47%

To examine the speed-tracking situation, we used joint1 as an example. Figure 10a shows the speed-tracking of the TOPP-RA method during actual operation, while Figure 10b presents the speed curve of our TOPP-IO method, which considers the third-order constraint. Only the second-order constraint of joint space is considered, which leads to snap-point (represented by the gray circle), or the sudden change in joint acceleration, resulting in the inability to track the given speed on the actual physical robot. As shown in the zoomed-in section of Figure 10a, the snap-point causes fluctuations in the speed curve. In contrast, our TOPP-IO method eliminates the snap-point and enables smooth tracking of joint speed. Our method ensures a smooth trajectory and efficient, steady completion of the task while maintaining high speed.

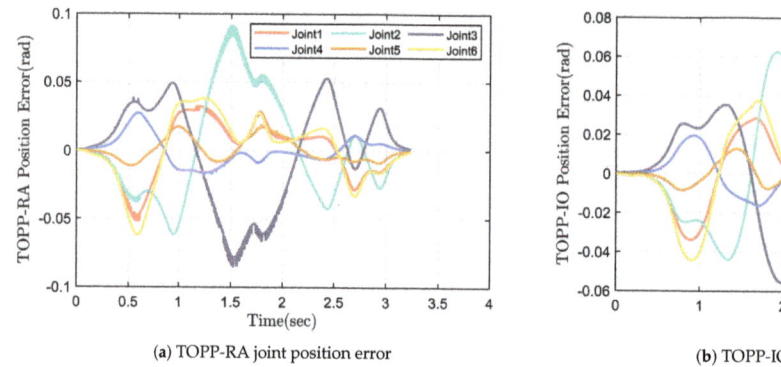

(a) TOPP-RA joint position error

(b) TOPP-IO joint position error

Figure 9. Comparing the resulting joint position error from TOPP-RA (**a**) and the ones obtained from proposed approach (**b**).

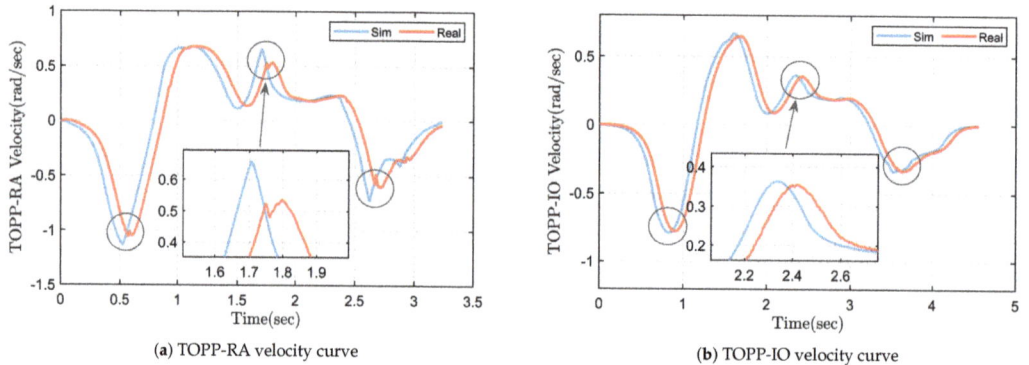

(a) TOPP-RA velocity curve

(b) TOPP-IO velocity curve

Figure 10. Comparing the resulting joint1 velocity curve from TOPP-RA (**a**) and the ones obtained from proposed approach (**b**). (The gray circle represented the snap-point).

5. Conclusions

In this paper, we develop a comprehensive and efficient iterative optimization framework for solving the TOTP problem with joint third-order constraints. The main contributions and results of this paper are as follows:

- The framework is constructed from the bottom up in the Cartesian coordinate system and can be applied to both manipulator and mobile robots;
- Our study has identified two main challenges in the framework: how to consistently represent the TOTP problem in the Cartesian space using the $(s, \dot{s})$ phase plane, while imposing third-order kinematic constraints on each joint, and how to devise an efficient computational solution strategy that uses a constraint relaxation approach to simplify nonconvex constraints without violating them;
- We demonstrated the effectiveness of our proposed framework through both simulation and physical experiments. Compared to the TOPP-RA method, our approach effectively reduced the maximum absolute values of the robot's jerk and the average absolute values of the position error over 60% and 29%, respectively. These are critical factors in ensuring smooth robotic velocity tracking and reducing impact during operation.

Our framework has a few limitations. First, we assume that the path of the end-effector in Cartesian space is predetermined. We use B-spline interpolation to generate continuous, smooth end-effector poses from the given path points. Second, our approach accepts suboptimal solutions when a feasible solution can be obtained at any point by interrupting the iterative optimization process.

Our future work can be divided into two main areas:

- First, we aim to extend our framework to handle both path planning and speed planning simultaneously, which will enable our method to generate feasible solutions more efficiently;
- Second, we plan to explore the potential of the constraint relaxation approaches and achieve real-time performance. Moreover, handling dynamic environments is a challenging and interesting area for future research.

Author Contributions: Conceptualization, Z.F. and Y.C.; methodology, Z.F.; software, Z.F.; validation, Z.F., F.Z., Q.Z. and K.J.; formal analysis, L.Z. and Z.D.; investigation, Z.F.; resources, Z.D.; data curation, M.L.; writing—original draft preparation, Z.F.; writing—review and editing, Z.F. and M.L.; visualization, Z.F. and Y.C.; supervision, K.J.; project administration, L.Z.; funding acquisition, F.Z. and Q.Z. All authors have read and agreed to the published version of the manuscript.

Funding: This research was funded by the National Key R&D Program of China (No. 2020YFB1710905) and the National Natural Science Foundation of China (No. 61903162).

Institutional Review Board Statement: Not applicable.

Informed Consent Statement: Not applicable.

Data Availability Statement: Not applicable.

Conflicts of Interest: The authors declare no conflict of interest.

Abbreviations

The following abbreviations are used in this manuscript:

TOTP	time-optimal trajectory planning
TOPP	time-optimal path parameterization
NI	Numerical Integration
CO	Convex Optimization
DP	Dynamic Programming
DC	difference of convex
SCP	sequential convex programming
TOPP-RA	TOPP approach based on reachability analysis
TOPP-IO	TOPP approach based on iterative optimization
LP	linear programming
BA	bisection algorithm
NLP	nonlinear programming

Appendix A. From Configuration Space to Cartesian Space

In this appendix, we introduce how to determine the position and attitude of the end-effector and their derivatives using the robot joint variables and their derivatives of each order, and then convert them into path parameter s for representation. This process involves forward kinematics, inverse kinematics and the derivatives of Jacobian matrix, which will be described in detail in the following subsections.

Appendix A.1. Forward and Inverse Kinematics

ψ is defined as the screw coordinate of the spiral axis relative to the spatial coordinate system. Its Lie algebra representation is as follows:

$$se(3) = \left\{ \psi = \begin{bmatrix} \rho \\ \phi \end{bmatrix} \in \mathbb{R}^6, \rho \in \mathbb{R}^3, \phi \in so(3), \psi^\wedge = \begin{bmatrix} \phi^\wedge & \rho \\ 0^T & 1 \end{bmatrix} \in \mathbb{R}^{4\times 4} \right\} \quad (A1)$$

The Lie group corresponding to ψ can be expressed as

$$SE(3) = \left\{ \Psi = \begin{bmatrix} R & \varrho \\ 0^T & 1 \end{bmatrix} \in \mathbb{R}^{4\times 4}, R \in SO(3), \varrho \in \mathbb{R}^3 \right\} \quad (A2)$$

where R and ϱ separately denote the directions and positions of the rigid-body. The mapping between the Lie algebra ψ and the corresponding Lie group Ψ is given by:

$$\Psi = \exp(\psi^\wedge) = \begin{bmatrix} \exp(\phi^\wedge) & \frac{(I-\exp(\phi^\wedge))\phi^\wedge \rho + \phi\phi^T \rho}{\|\phi\|^2} \\ 0^T & 1 \end{bmatrix} \quad (A3)$$

Consider a robot system with n degrees of freedom, whose configuration is represented by $\mathbf{q} = [q_1, q_2, \cdots, q_n]^T \in \mathbb{R}^n$. The forward kinematics model of the robot is determined as follows:

$$\Psi_{end}(\mathbf{q}) = \prod_{i=1}^{n} \exp(\psi_i^\wedge q_i) \Psi_S = \begin{bmatrix} R_{end} & \varrho_{end} \\ 0^T & 1 \end{bmatrix} \quad (A4)$$

where Ψ_S represents the pose matrix of the end coordinate system relative to the space coordinate system when the robot is in the initial position; (q_i, ψ_i) denote the joint position and twist, respectively, of the i-th joint; R_{end} and ϱ_{end} separately denote the orientations and positions of the end-effector.

For a given robot model, it is possible to convert the end pose $\mathbf{p}$ expressed in Cartesian space to the corresponding joint values $\mathbf{q}$ in joint space using inverse kinematics. Similarly, the joint values $\mathbf{q}$ can be converted to the end pose $\mathbf{p}$ through forward kinematics. Thus, the reciprocal transformation between $\mathbf{p}$ and $\mathbf{q}$ can be achieved through these two transformations.

Appendix A.2. Explicit Expressions of High-Order Jacobian Derivatives

Jacobian matrix in Formula (13) is as follows:

$$J = J_p \cdot J_b = \begin{bmatrix} A^{-1}(r) & O \\ O & R_{sb}(r) \end{bmatrix} \cdot [\beta_1, \beta_2, \cdots, \beta_n] \quad (A5)$$

where $J_b(\mathbf{q}) \in \mathbb{R}^{6\times n}$ represents the geometric Jacobian matrix; $r \in \mathbb{R}^3$ represents the exponential coordinate of the axis angle, which is reflected in the last three lines of the $\mathbf{p}(s)$. In $J_p \in \mathbb{R}^{6\times 6}$, $A(r) \in \mathbb{R}^{3\times 3}$ and $R_{sb}(r) \in \mathbb{R}^{3\times 3}$ can be expressed as:

$$A(r) = I - \frac{1-\cos\|r\|}{\|r\|^2} r^\wedge + \frac{\|r\|-\sin\|r\|}{\|r\|^3}(r^\wedge)^2, \quad (A6)$$
$$R_{sb}(r) = \exp(r^\wedge).$$

Since r can be expressed by $\mathbf{p}(s)$, the first-order and second-order path parameter derivatives of J_p are expressed as follows:

$$J_p'(\mathbf{p}(s)) = \begin{bmatrix} -A^{-1}(\mathbf{p}(s)) \frac{dA(\mathbf{p}(s))}{ds} A^{-1}(\mathbf{p}(s)) & 0 \\ 0^T & \frac{dR_{sb}}{ds} \end{bmatrix},$$

$$J_p''(\mathbf{p}(s)) = \begin{bmatrix} 2A^{-1}\frac{dA}{ds}A^{-1}\frac{dA}{ds}A^{-1} - A^{-1}\frac{d^2A}{ds^2}A^{-1} & 0 \\ 0^T & \frac{d^2R_{sb}}{ds^2} \end{bmatrix}. \quad (A7)$$

(1) First-order path parameter derivative $\mathbf{J}'$: The first-order derivative of Jacobian matrix $\mathbf{J}$ with respect to the path parameter s is as following:

$$\mathbf{J} = \mathbf{J}'_p \cdot \mathbf{J}_b + \mathbf{J}_p \cdot \mathbf{J}'_b \tag{A8}$$

The matrices $\mathbf{J}_p$ and $\mathbf{J}'_p$ depend on the path parameter s, while $\mathbf{J}_b$ is a function of the joint values $\mathbf{q}$, which can be obtained by forward and inverse kinematics. Each column of $\mathbf{J}_b(\mathbf{q})$ can be represented as an adjoint matrix, given by:

$$\begin{aligned} \beta_i &= Ad_{\Psi_i}^{-1}(\psi_i) = (\Psi_i \psi_i^\wedge \Psi_i^{-1})^\vee \in \mathbb{R}^6, \text{where} \\ \Psi_i &= \prod_{j=1}^{n} \exp(\psi_j^\wedge q_j) \Psi_s. \end{aligned} \tag{A9}$$

According to the chain rule of differentiation, the first-order derivative of the geometric Jacobian matrix with respect to the path parameter s, denoted as $\mathbf{J}'_b$, and its i-th column, denoted as β'_i, can be expressed as:

$$\begin{aligned} \mathbf{J}'_b &= [\beta'_1, \ \beta'_2, \ \ldots, \ \beta'_n], \\ \beta'_i &= \sum_{j=1}^{n} \frac{\partial \beta_i}{\partial q_j} q'_j \\ &= \left[\frac{\partial \beta_i}{\partial q_1}, \ \frac{\partial \beta_i}{\partial q_2}, \ \ldots, \ \frac{\partial \beta_i}{\partial q_n} \right] \mathbf{q}' \\ &= \left[\frac{\partial \beta_i}{\partial q_1}, \ \frac{\partial \beta_i}{\partial q_2}, \ \ldots, \ \frac{\partial \beta_i}{\partial q_n} \right] \mathbf{J}^{-1} \mathbf{p}'. \end{aligned} \tag{A10}$$

In combination with the literature [35], $\frac{\partial \beta_i}{\partial q_j}$ can be calculated using (β_i, β_j) as follows:

$$\frac{\partial \beta_i}{\partial q_j} = \begin{cases} 0, & j < i \\ ad_{\beta_j}(\beta_i) = (\beta_j^\wedge \beta_i^\wedge - \beta_i^\wedge \beta_j^\wedge)^\vee, & j \geq i \end{cases} \tag{A11}$$

(2) Second-order path parameter derivative $\mathbf{J}''$: The second-order derivative of the Jacobian matrix $\mathbf{J}$ with respect to the path parameter s is given by:

$$\mathbf{J} = \mathbf{J}''_p \cdot \mathbf{J}_b + \mathbf{J}'_p \cdot \mathbf{J}'_b + \mathbf{J}_p \cdot \mathbf{J}''_b \tag{A12}$$

According to the chain rule, the second-order path parameter derivative, $\mathbf{J}''_b$, and its i-th column, β''_i, can be denoted as:

$$\begin{aligned} \mathbf{J}''_b &= [\beta''_1, \ \beta''_2, \ \ldots, \ \beta''_n], \\ \beta''_i &= \frac{\partial}{\partial s}\left(\sum_{j=1}^{n} \frac{\partial \beta_i}{\partial q_j} q'_j \right) \\ &= \sum_{j=1}^{n} \frac{\partial \beta_i}{\partial q_j} q''_j + \frac{\partial}{\partial s}\left(\sum_{j=1}^{n} \frac{\partial \beta_i}{\partial q_j} \right) q'_j \\ &= \left[\frac{\partial \beta_i}{\partial q_1}, \ \frac{\partial \beta_i}{\partial q_2}, \ \ldots, \ \frac{\partial \beta_i}{\partial q_n} \right] \mathbf{q}'' \\ &\quad + \left[\frac{\partial}{\partial s}\left(\frac{\partial \beta_i}{\partial q_1}\right), \ \frac{\partial}{\partial s}\left(\frac{\partial \beta_i}{\partial q_2}\right), \ \ldots, \ \frac{\partial}{\partial s}\left(\frac{\partial \beta_i}{\partial q_n}\right) \right] \mathbf{J}^{-1} \mathbf{p}'. \end{aligned} \tag{A13}$$

where

$$\mathbf{q}'' = \mathbf{J}^{-1}(\mathbf{p}'' - \mathbf{J}'\mathbf{J}^{-1}\mathbf{p}') \tag{A14}$$

and

$$\frac{\partial}{\partial s}\left(\frac{\partial \beta_i}{\partial q_j}\right) = \begin{cases} 0, & j < i \\ ad_{\beta_j}(\beta_i') + ad_{\beta_j'}(\beta_i), & j \geq i \end{cases} \quad \text{(A15)}$$

Overall, this appendix establishes the forward kinematics of a robot using Lie theory and symbolically derives the first-order and second-order derivatives of the Jacobian matrix. Higher-order Jacobian matrices could be derived similarly. Furthermore, the inverse kinematics for a specific robot model can be easily obtained.

References

1. Mikolajczyk, T. Manufacturing Using Robot. *Adv. Mater. Res.* **2012**, *463*, 1643–1646. In *Proceedings of the Advanced Materials Research II*; Trans Tech Publications Ltd.: Baech, Switzerland, 2012. [CrossRef]
2. Oztemel, E.; Gursev, S. Literature review of Industry 4.0 and related technologies. *J. Intell. Manuf.* **2020**, *31*, 127–182. [CrossRef]
3. Chiurazzi, M.; Alcaide, J.O.; Diodato, A.; Menciassi, A.; Ciuti, G. Spherical Wrist Manipulator Local Planner for Redundant Tasks in Collaborative Environments. *Sensors* **2023**, *23*, 677. [CrossRef] [PubMed]
4. Gasparetto, A.; Boscariol, P.; Lanzutti, A.; Vidoni, R. Trajectory Planning in Robotics. *Math. Comput. Sci.* **2012**, *6*, 269–279. [CrossRef]
5. Zhang, T.; Zhang, M.; Zou, Y. Time-optimal and Smooth Trajectory Planning for Robot Manipulators. *Int. J. Control. Autom. Syst.* **2021**, *19*, 521–531. [CrossRef]
6. Pham, H.; Pham, Q.-C. A New Approach to Time-Optimal Path Parameterization Based on Reachability Analysis. *IEEE Trans. Robot.* **2018**, *34*, 645–659. [CrossRef]
7. Bobrow, J.E.; Dubowsky, S.; Gibson, J.S. Time-Optimal Control of Robotic Manipulators Along Specified Paths. *Int. J. Robot. Res.* **1985**, *4*, 3–17. [CrossRef]
8. Kunz, T.; Stilman, M. Time-optimal trajectory generation for path following with bounded acceleration and velocity. In *Robotics: Science and Systems VIII*; The MIT Press: Cambridge, MA, USA; London, UK, 2012; pp. 1–8.
9. Pham, Q.C. A General, Fast, and Robust Implementation of the Time-Optimal Path Parameterization Algorithm. *IEEE Trans. Robot.* **2014**, *30*, 1533–1540. [CrossRef]
10. Pham, H.; Pham, Q.C. On the structure of the time-optimal path parameterization problem with third-order constraints. In Proceedings of the 2017 IEEE International Conference on Robotics and Automation (ICRA), Singapore, 29 May–3 June 2017; pp. 679–686. [CrossRef]
11. Shen, P.; Zhang, X.; Fang, Y. Essential Properties of Numerical Integration for Time-Optimal Path-Constrained Trajectory Planning. *IEEE Robot. Autom. Lett.* **2017**, *2*, 888–895. [CrossRef]
12. Shen, P.; Zhang, X.; Fang, Y. Complete and Time-Optimal Path-Constrained Trajectory Planning With Torque and Velocity Constraints: Theory and Applications. *IEEE/ASME Trans. Mechatronics* **2018**, *23*, 735–746. [CrossRef]
13. Lu, L.; Zhang, J.; Fuh, J.Y.H.; Han, J.; Wang, H. Time-optimal tool motion planning with tool-tip kinematic constraints for robotic machining of sculptured surfaces. *Robot.-Comput.-Integr. Manuf.* **2020**, *65*, 101969. [CrossRef]
14. Verscheure, D.; Demeulenäre, B.; Swevers, J.; De Schutter, J.; Diehl, M. Practical time-optimal trajectory planning for robots: A convex optimization approach. *IEEE Trans. Autom. Control.* **2008**, *53*, 1–10.
15. Xiao, Y.; Dong, W.; Du, Z. A time-optimal trajectory planning approach based on calculation cost consideration. In Proceedings of the 2012 IEEE International Conference on Mechatronics and Automation, Chengdu, China, 5–8 August 2012; pp. 1845–1850. [CrossRef]
16. Debrouwere, F.; Van Loock, W.; Pipeleers, G.; Dinh, Q.T.; Diehl, M.; De Schutter, J.; Swevers, J. Time-Optimal Path Following for Robots With Convex-Concave Constraints Using Sequential Convex Programming. *IEEE Trans. Robot.* **2013**, *29*, 1485–1495. [CrossRef]
17. Nagy, Á.; Vajk, I. Sequential Time-Optimal Path-Tracking Algorithm for Robots. *IEEE Trans. Robot.* **2019**, *35*, 1253–1259. [CrossRef]
18. Ma, J.-w.; Gao, S.; Yan, H.-t.; Lv, Q.; Hu, G.-q. A new approach to time-optimal trajectory planning with torque and jerk limits for robot. *Robot. Auton. Syst.* **2021**, *140*, 103744. [CrossRef]
19. Shin, K.; McKay, N. A dynamic programming approach to trajectory planning of robotic manipulators. *IEEE Trans. Autom. Control.* **1986**, *31*, 491–500. [CrossRef]
20. Kaserer, D.; Gattringer, H.; Müller, A. Nearly Optimal Path Following with Jerk and Torque Rate Limits Using Dynamic Programming. *IEEE Trans. Robot.* **2019**, *35*, 521–528. [CrossRef]
21. Kaserer, D.; Gattringer, H.; Müller, A. Time Optimal Motion Planning and Admittance Control for Cooperative Grasping. *IEEE Robot. Autom. Lett.* **2020**, *5*, 2216–2223. [CrossRef]
22. Barnett, E.; Gosselin, C. A Bisection Algorithm for Time-Optimal Trajectory Planning Along Fully Specified Paths. *IEEE Trans. Robot.* **2021**, *37*, 131–145. [CrossRef]
23. Faulwasser, T.; Findeisen, R. Nonlinear Model Predictive Control for Constrained Output Path Following. *IEEE Trans. Autom. Control.* **2016**, *61*, 1026–1039. [CrossRef]
24. Consolini, L.; Locatelli, M.; Minari, A.; Piazzi, A. An optimal complexity algorithm for minimum-time velocity planning. *Syst. Control. Lett.* **2017**, *103*, 50–57. [CrossRef]

25. Steinhauser, A.; Swevers, J. An Efficient Iterative Learning Approach to Time-Optimal Path Tracking for Industrial Robots. *IEEE Trans. Ind. Inform.* **2018**, *14*, 5200–5207. [CrossRef]
26. Consolini, L.; Locatelli, M.; Minari, A. A Sequential Algorithm for Jerk Limited Speed Planning. *IEEE Trans. Autom. Sci. Eng.* **2022**, *19*, 3192–3209. [CrossRef]
27. Petrone, V.; Ferrentino, E.; Chiacchio, P. Time-Optimal Trajectory Planning With Interaction With the Environment. *IEEE Robot. Autom. Lett.* **2022**, *7*, 10399–10405. [CrossRef]
28. Yang, Y.; Xu, H.z.; Li, S.h.; Zhang, L.l.; Yao, X.m. Time-optimal trajectory optimization of serial robotic manipulator with kinematic and dynamic limits based on improved particle swarm optimization. *Int. J. Adv. Manuf. Technol.* **2022**, *120*, 1253–1264. [CrossRef]
29. Singh, S.; Leu, M.C. Optimal Trajectory Generation for Robotic Manipulators Using Dynamic Programming. *J. Dyn. Syst. Meas. Control.* **1987**, *109*, 88–96. [CrossRef]
30. Slotine, J.J.E.; Yang, H.S. Improving the Efficiency of Time-Optimal Path-Following Algorithms. In Proceedings of the 1988 American Control Conference, Atlanta, GA, USA, 15–17 June 1988; pp. 2129–2134. [CrossRef]
31. Consolini, L.; Locatelli, M.; Minari, A.; Nagy, Á.; Vajk, I. Optimal Time-Complexity Speed Planning for Robot Manipulators. *IEEE Trans. Robot.* **2019**, *35*, 790–797. [CrossRef]
32. Li, B.; Ouyang, Y.; Li, L.; Zhang, Y. Autonomous Driving on Curvy Roads Without Reliance on Frenet Frame: A Cartesian-Based Trajectory Planning Method. *IEEE Trans. Intell. Transp. Syst.* **2022**, *23*, 15729–15741. [CrossRef]
33. Guarino Lo Bianco, C.; Faroni, M.; Beschi, M.; Visioli, A. A Predictive Technique for the Real-Time Trajectory Scaling Under High-Order Constraints. *IEEE/ASME Trans. Mechatronics* **2022**, *27*, 315–326. [CrossRef]
34. Andersson, J.A.E.; Gillis, J.; Horn, G.; Rawlings, J.B.; Diehl, M. CasADi—A software framework for nonlinear optimization and optimal control. *Math. Program. Comput.* **2019**, *11*, 1–36. [CrossRef]
35. Fu, Z.; Spyrakos-Papastavridis, E.; Lin, Y.-H.; Dai, J.S. Analytical Expressions of Serial Manipulator Jacobians and their High-Order Derivatives based on Lie Theory. In Proceedings of the 2020 IEEE International Conference on Robotics and Automation (ICRA), Paris, France, 31 May–31 August 2020; pp. 7095–7100. [CrossRef]

Disclaimer/Publisher's Note: The statements, opinions and data contained in all publications are solely those of the individual author(s) and contributor(s) and not of MDPI and/or the editor(s). MDPI and/or the editor(s) disclaim responsibility for any injury to people or property resulting from any ideas, methods, instructions or products referred to in the content.

Article

Event-Triggered Tracking Control for Adaptive Anti-Disturbance Problem in Systems with Multiple Constraints and Unknown Disturbances

Hong Shen [1], Qin Wang [2] and Yang Yi [2,*]

[1] College of Business, Yangzhou University, Yangzhou 225127, China
[2] College of Information Engineering, Yangzhou University, Yangzhou 225127, China
* Correspondence: yiyang@yzu.edu.cn

Abstract: Aimed at the objective of anti-disturbance and reducing data transmission, this article discusses a novel dynamic neural network (DNN) modeling-based anti-disturbance control for a system under the framework of an event trigger. In order to describe dynamical characteristics of irregular disturbances, exogenous DNN disturbance models with different excitation functions are firstly introduced. A novel disturbance observer-based adaptive regulation (DOBAR) method is then proposed, which can capture the dynamics of unknown disturbance. By integrating the augmented triggering condition and the convex optimization method, an effective anti-disturbance controller is then found to guarantee the system stability and the convergence of the output. Meanwhile, both the augmented state and the system output are constrained within given regions. Moreover, the Zeno phenomenon existing in event-triggered mechanisms is also successfully avoided. Simulation results for the A4D aircraft models are shown to verify the availability of the algorithm.

Keywords: dynamic neural networks (DNNs); event-triggered control; anti-disturbance control; adaptive control; saturation constraint; output constraint

1. Introduction

As is well-known, many real-world controlled systems are often subjected to unknown external disturbances [1–5]. Currently, there are various recognized anti-disturbance control algorithms that can be used to eliminate the effects caused by unknown disturbances, such as adaptive theory, robust control and sliding mode control [6–8]. However, the motivation of these methods is to suppress disturbances in the form of feedback rather than feed-forward compensation, which usually makes the reaction time linger and reduces the accuracy [1,2,9]. In order to overcome these limitations, an active feed-forward method of rejecting disturbances based on the disturbance estimation technique is proposed. This method is usually called a disturbance-observer-based control (DOBC) and can proactively offset those unknown disturbances [1,2,4,10–17]. Due to its fast reaction and good compatibility, the DOBC method has been successfully applied to many classical controlled systems, such as permanent magnet synchronous motor (PMSM) systems [11], vehicle control systems [12], Markov jump systems [13], multi-agent systems [15], non-Gaussian distribution systems [16] and so on. However, in order to better estimate disturbances, the DOBC method usually needs to acquire information on the frequency and amplitude of unknown disturbances [1,2]. As a result, most of the DOBC results can only cope with linear or regular disturbances, including constant and harmonic disturbances (see [1,2,14–16] for details). When being affected by those irregular nonlinear disturbances—for example, variable amplitude or frequency disturbances—how to realize the dynamic estimation is a major motivation. In short, exploring more in-depth disturbance observation strategies is one of the most important research objectives.

In either practical systems or theoretical analysis, the problem of control constraints is inevitable. As a typical input constraint phenomenon, actuator saturations frequently occur in almost all control devices and can have a great negative impact on the system performance [18]. Based on this, many researchers began to study effective saturation control algorithms [19–27]. In [19], multiple auxiliary matrices and convex hull partitioning methods were discussed to enlarge the ellipsoidal region of stability. By using bilinear matrix inequalities (BLMIs) or linear matrix inequalities (LMIs) schemes, the polytopic technique was explored to drag the saturation constraint into a designed convex set [20–23]. In order to obtain less conservative results, the sector bounding approach also became popular for describing the saturation function [22]. Moreover, when coupling with other nonlinear characteristics or typical controlled systems, corresponding anti-windup strategies and performance analysis were also discussed in [22–27]. Parallel to the input constraint, both the output and state-constrained controls are also attractive topics driven by both practical and theoretical requirements [28,29]. Among the existing results, the symmetric barrier Lyapunov function (BLF), asymmetric BLF and error transformation proved to be effective in dealing with output constraints [28–31]. However, the aforementioned discussions are only limited to the single-input single-output (SISO) systems or triangular multiple-input multiple-output (MIMO) systems. It is urgent to explore new control methods to guarantee the state or output constraints of general MIMO nonlinear systems. Further, when multiple constraints and unknown disturbances are coupled, how to design an effective anti-disturbance constrained controller is another motivation of the work.

Generally, most controlled systems adopt a time-triggered mechanism (also called periodic sampling mechanism), which is rather convenient for theoretical analysis and conventional engineering applications. However, when the system performance has reached the designed requirements in networked environments, data transmission and calculation do not stop immediately, which will inevitably cause a waste of bandwidths and computing resources to a certain extent [32]. Due to this consideration, the idea of event triggering is proposed by equipping event-triggered schedulers at sensor nodes [33,34]. In the event-triggered control (ETC) framework, control tasks are carried out only after the well-designed triggering criteria are violated, which can availably decrease resource utilization while achieving a satisfactory system performance [35]. Some exciting results regarding ETC systems have successfully addressed traditional problems of robust control, output feedback control, sliding mode control, adaptive control, and so on [34–40]. In practical applications, Ref. [41] proposed an effective decentralized event-triggered algorithm to guarantee the dynamical performance of power systems. Based on the event-triggered theory, the effective attitude tracking control was discussed for the surface vessels [42].

On the basis of the analysis above, this paper explores a novel event-based anti-disturbance constraint control problem for general MIMO systems subject to unknown disturbances and multiple constraints. The proposed scheme has the following characteristics. Firstly, a DNN disturbance model was employed to identify those indescribable irregular disturbances, which further enriches the varieties of disturbances when compared with most existing anti-disturbance results [1,2,11,13–15]. By designing the adaptive law for adjustable parameters of DNNs, an active disturbance-observer-based adaptive control (DOBAC) algorithm was designed to successfully realize the dynamical estimation and rejection of unknown disturbances. Secondly, in order to avoid the waste of resources and achieve favorable dynamical tracking, an event-triggered mechanism with the designed augmented triggering condition was introduced into the controlled system. Further, a composite event-triggered anti-disturbance controller can be smoothly implemented after decoupling the saturated input with the disturbances. Thirdly, unlike many previous non-convex results [20,22], the improved convex optimization algorithm was constructed to simultaneously satisfy the multi-objective control requirements, including the stability of the augmented system, dynamical tracking performance, state constraint, output constraint and non-Zeno phenomenon. It also represents a major expansion with respect to those single-constraint control or dynamical tracking problems. By introducing two kinds of different disturbances, the simulation examples of the A4D model are presented to reflect the significance of the algorithm.

2. Problem Description

Considering the MIMO system with external disturbances and an input constraint as

$$\begin{cases} \dot{x}(t) = Ax(t) + B\mathbf{sat}(u(t) + g(t)) \\ z(t) = Cx(t) \end{cases} \quad (1)$$

where $u(t) \in \mathbb{R}^m$, $z(t) \in \mathbb{R}^p$, $x(t) \in \mathbb{R}^n$ and $g(t) \in \mathbb{R}^m$ are, respectively, the control input, the system output, the state vector and the unknown disturbance. $A \in \mathbb{R}^{n \times n}$, $B \in \mathbb{R}^{n \times m}$ and $C \in \mathbb{R}^{p \times n}$ are the coefficient matrices. $\mathbf{sat}(*)$ stands for the saturation constraint, which is expanded as $\mathbf{sat}(*) = [sat_1(*), \ldots, sat_m(*)]^T$, where $sat_i(*) = sign(*)\min(*, 1)$ stands for the signum function.

To better estimate unknown disturbances, $g(t)$ is described by an external model with adjustable parameters as

$$\begin{cases} \dot{\sigma}(t) = W\sigma(t) + M^*\Phi(\sigma(t)) \\ g(t) = V\sigma(t) \end{cases} \quad (2)$$

where $\sigma(t) \in \mathbb{R}^{n_1}$ represents the middle state of the DNN model, and W and V are corresponding coefficient matrices. In addition, $M^* \in \mathbb{R}^{n_1 \times n_1}$ represents the optimal model parameter matrix, and $\Phi(*)$ can be seen as the activation function of DNNs with $\Phi(*) = [\phi_1(*), \ldots, \phi_{n_1}(*)]^T$. Due to the powerful identification capacity of DNNs (see [43,44]), DNN models ought to be useful identifiers to depict different types of disturbances by selecting different activation functions.

For the purpose of achieving a favorable dynamic tracking performance, an augmented state is defined as

$$\bar{x}(t) = \left[x^T(t), \int_0^t e^T(\tau)d\tau\right]^T \quad (3)$$

where the error is defined by $e(t) := z(t) - z_d$ with z_d standing for the expected system output, and z_d is a nonzero vector. According to (1) as well as (3), the extended system can be expressed by

$$\begin{cases} \dot{\bar{x}}(t) = \bar{A}\bar{x}(t) + \bar{B}\mathbf{sat}(u(t) + g(t)) + \bar{G}z_d \\ z(t) = \bar{C}\bar{x}(t) \end{cases} \quad (4)$$

with

$$\bar{A} = \begin{bmatrix} A & 0 \\ C & 0 \end{bmatrix}, \bar{B} = \begin{bmatrix} B \\ 0 \end{bmatrix}, \bar{G} = \begin{bmatrix} 0 \\ -I \end{bmatrix}, \bar{C} = \begin{bmatrix} C^T \\ 0 \end{bmatrix}^T$$

Moreover, the polyhedron boundary skill is employed to identify the function with saturation. By selecting a matrix P_1, the ellipsoid is constructed as

$$\Lambda(P_1, 1) = \left\{\bar{x}(t) \in \mathbb{R}^{n+p} : \bar{x}^T P_1 \bar{x} \leq 1\right\} \quad (5)$$

Based on this, a polyhedron is structured as

$$L(H) = \left\{\bar{x}(t) \in \mathbb{R}^{n+p} : \left|H^l \bar{x}\right| \leq 1, l \in Q_m\right\} \quad (6)$$

where $Q_m = \{1, 2, \cdots, m\}$, H^l stands for the lth row of H. Further, the lemma is imported.

Lemma 1 ([18–20]). *Let $K, H \in \mathbb{R}^{m \times (n+p)}$. For every $\zeta \in \mathbb{R}^{n+p}$, if $\zeta \in L(H)$, then*

$$\mathbf{sat}(K\zeta) = co\{D_i K\zeta + D_i^- H\zeta, i \in Q\} \quad (7)$$

where $co(*)$ stands for the convex hull representation, and $Q = \{1, \cdots, 2^m\}$. In addition, D_i is a diagonal matrix, in which each element is 0 or 1, and it satisfies $D_i + D_i^- = I$.

3. Event-Triggered PI Controller Design

For reducing the waste of resources in networked environments, an event-trigger-based proportional-integral (PI) controller is designed in this part.

First, the novel augmented event triggering condition is defined as

$$t_{k+1} = \inf\left\{t > t_k : (\bar{x}(t) - \bar{x}(t_k))^T \Psi (\bar{x}(t) - \bar{x}(t_k)) > \delta^2 \bar{x}^T(t) \Psi \bar{x}(t)\right\} \tag{8}$$

where t_k represents the moment at which the event is triggered in kth, and $\bar{x}(t)$ and $\bar{x}(t_k)$ are the augmented states at the current sampling time and the latest triggered time. The scalar δ satisfies $0 \leq \delta < 1$, and $\Psi > 0$ represents a designed positive definite matrix.

Define

$$e_k(t) = \begin{bmatrix} e_{1k} \\ e_{2k} \end{bmatrix}, \quad t \in [t_k, t_{k+1}) \tag{9}$$

with

$$\begin{cases} e_{1k}(t) = x(t) - x(t_k) \\ e_{2k}(t) = \int_0^t e(\tau)d\tau - \int_0^{t_k} e(\tau)d\tau = \int_{t_k}^t e(\tau)d\tau \end{cases} \tag{10}$$

Then, we can derive from (9) and (10) that

$$\dot{e}_k(t) = \begin{bmatrix} \dot{x}(t) \\ e(t) \end{bmatrix} = \begin{bmatrix} \dot{x}(t) \\ Cx(t) - z_d \end{bmatrix} \tag{11}$$

In the event-triggered mechanism, the event trigger monitors whether the events occur. Once the triggering condition $e_k^T(t)\Psi e_k(t) \leq \delta^2 \bar{x}^T(t)\Psi \bar{x}(t) + \delta_1 e^{-\varsigma t}$ is not met, a new event will occur. The event detector then sends the updated data $\bar{x}(t)$ to the control port. Otherwise, the current updated data will be put away.

Based on this, the event-triggered PI state feedback controller is expressed by the form

$$u(t) = -\hat{g}(t) + K\bar{x}(t_k), \quad K = [K_P, K_I], \quad t \in [t_k, t_{k+1}) \tag{12}$$

where K_P and K_I stand for the control gains to be sought.

4. Event-Triggered DOBAC Algorithm Design

For the sake of estimating unknown disturbance $g(t)$ accurately, an adaptive observer with adjustable weight is built. The specific expression of the adaptive DO is described as

$$\begin{cases} \dot{r}(t) = \hat{M}(t)\Phi(\hat{\sigma}(t)) - L(-\tilde{A}\bar{x}(t) - \tilde{G}z_d - \tilde{B}u(t)) + (W + L\tilde{B}V)(-L\bar{x}(t) + r(t)) \\ \hat{\sigma}(t) = -L\bar{x}(t) + r(t) \\ \hat{g}(t) = V\hat{\sigma}(t) \end{cases} \tag{13}$$

where L is the gain to be devised later, $r(t)$ represents the instrumental variable and $\hat{M}(t)$ is the adjustable dynamical weight, and its adaptive law is defined as

$$\dot{\hat{M}}(t) = -\|\hat{\sigma}(t)\|\hat{M}(t) + \gamma P_2 \hat{\sigma}(t) \Phi^T(\hat{\sigma}(t)) \tag{14}$$

where $\gamma > 0$ is a given parameter and $P_2 > 0$ will be solved in the next section.

The following theorem gives the boundedness proof of the adjustable parameter $\hat{M}(t)$.

Theorem 1. *If the adaptive parameter $\hat{M}(t)$ is updated by (14) and the initial condition satisfies $\hat{M}(0) \in \Theta_{\hat{M}}$, then $\hat{M}(t) \in \Theta_{\hat{M}}$ will be guaranteed for all $t \geq 0$, where*

$$\Theta_{\hat{M}} = \{\hat{M}(t) \mid \|\hat{M}\|_F \leq \gamma\sqrt{n_1}\|P_2\|\}$$

is a known compact set.

Proof. Design the function as

$$\Gamma(t) = \frac{1}{2}tr\{\hat{M}^T(t)\gamma^{-1}\hat{M}(t)\}. \tag{15}$$

According to the above Formula (14), we have

$$\dot{\Gamma} = -\gamma^{-1}\|\hat{\sigma}(t)\|\|\hat{M}(t)\|_F^2 + \left\|\hat{\sigma}^T(t)P_2\hat{M}(t)\Phi(\hat{\sigma}(t))\right\| \tag{16}$$

The excitation function is chosen as

$$\Phi(\sigma(t)) = \left[1/\left(e^{-\kappa\sigma_1}+1\right),\cdots,1/\left(e^{-\kappa\sigma_{n_1-1}}+1\right),1\right]^T$$

where κ is a positive constant. The boundary condition $\|\Phi(\hat{\sigma}(t))\| \leq \sqrt{n_1}$ can easily be achieved. Further, (16) is rewritten as

$$\dot{\Gamma} = \|\hat{\sigma}(t)\|\|\hat{M}\|_F\left(\gamma^{-1}\|\hat{M}\|_F - \sqrt{n_1}\|P_2\|\right) \tag{17}$$

which certifies that $\dot{\Gamma}(t) \leq 0$ once the inequality $\|\hat{M}(t)\|_F > \gamma\sqrt{n_1}\|P_2\|$ holds. Hence, if the initial condition satisfies $\hat{M}(0) \in \Theta_{\hat{M}}$, then holds $\hat{M}(t) \in \Theta_{\hat{M}}$ holds. □

The following discussion is concerned with the decoupling problem of a nonlinear saturated input under the event-triggered framework. According to Lemma 1, by choosing $H = [H_1, -V]$ to satisfy $\eta(t) \in L(H), \forall t \in [t_k, t_{k+1})$, one has

$$\text{sat}(u(t) + g(t)) = \sum_{i=1}^{2^m} \chi_i(D_iK + D_i^- H_1)\tilde{x}(t_k) - Ve_\sigma(t) \tag{18}$$

where the scalars χ_i meet the condition $0 \leq \chi_i \leq 1$ and $\sum_{i=1}^{2^m}\chi_i = 1$. $e_\sigma(t) = \hat{\sigma}(t) - \sigma(t)$, $\eta(t) = \left[\tilde{x}^T(t_k), e_\sigma^T(t)\right]^T$.
Introducing the input (18) to the system (4) results in the form

$$\dot{\tilde{x}}(t) = \left(\tilde{A} + \sum_{i=1}^{2^m}\chi_i\tilde{B}(D_iK + D_i^- H_1)\right)\tilde{x}(t) - \sum_{i=1}^{2^m}\chi_i\tilde{B}(D_iK + D_i^- H_1)e_k(t) - \tilde{B}Ve_\sigma(t) + \tilde{G}y_d \tag{19}$$

Defining $\tilde{M}(t) = M^* - \hat{M}(t)$ and applying (2), (13) and (17), we arrive at

$$\dot{e}_\sigma(t) = (W + L\tilde{B}V)e_\sigma(t) - \sum_{i=1}^{2^m}\chi_iL\tilde{B}D_i^-(H_1 - K)\tilde{x}(t) - \tilde{M}(t)\Phi(\hat{\sigma}(t))$$

$$+ \sum_{i=1}^{2^m}\chi_iL\tilde{B}D_i^-(H_1 - K)e_k(t) + M^*(\Phi(\hat{\sigma}(t)) - \Phi(\sigma(t))) \tag{20}$$

Further, by integrating the system (19) with the error dynamic system (20), we can obtain

$$\dot{\xi}(t) = \tilde{A}\xi(t) + \tilde{G}y_d + \tilde{I}(M^*(\Phi(\hat{\sigma}(t)) - \Phi(\sigma(t))) - \tilde{M}\Phi(\hat{\sigma}(t))) \tag{21}$$

where

$$\xi(t) = \begin{bmatrix} \bar{x}(t) \\ e_\sigma(t) \\ e_k(t) \end{bmatrix}, \quad \bar{G} = \begin{bmatrix} \tilde{G} \\ 0 \\ \tilde{G} \end{bmatrix}, \quad \bar{I} = \begin{bmatrix} 0 \\ I \\ 0 \end{bmatrix}, \quad \bar{A} = \begin{bmatrix} \tilde{A} + \prod_{11} & -\bar{B}V & -\prod_{11} \\ -\prod_{21} & W + L\bar{B}V & \prod_{21} \\ \tilde{A} + \prod_{11} & -\bar{B}V & -\prod_{11} \end{bmatrix}$$

$$\prod_{11} = \sum_{i=1}^{2^m} \chi_i \bar{B}(D_i K + D_i^- H_1), \quad \prod_{21} = \sum_{i=1}^{2^m} \chi_i L\bar{B}D_i^-(H_1 - K)$$

In the next section, by importing the convex optimization method, the desirable gains K and L will be given to meet the multi-objective control requirements of the augmented system.

5. Analysis and Proof of Multi-Objective Tracking Control Performance

For the sake of ensuring the performance of the closed-loop system, some related assumptions are necessary.

Assumption 1. *The selected basis function $\Phi(*)$ is assumed to satisfy the following Lipschitz condition:*

$$(\Phi(\sigma) - \Phi(\hat{\sigma}))^T (\Phi(\sigma) - \Phi(\hat{\sigma})) \le e_\sigma^T(t) U_\sigma^T U_\sigma e_\sigma(t) \tag{22}$$

where U_σ is a known positive definite matrix.

Assumption 2. *The optimal parameter M^* is usually an unknown bounded matrix, so there exists a positive definite matrix $\bar{M}$ satisfying the inequality $M^{*T}M^* \le \bar{M}$.*

Assumption 3. *The unknown disturbance $g(t)$ is supposed to satisfy the condition $g^T(t)g(t) \le \theta_g$, where θ_g is a constant. Further, because of $g^T(t)g(t) = \sigma^T(t)V^T V\sigma(t) \le \theta_g$, another inequality condition follows: $\sigma^T(t)\sigma(t) \le \frac{\theta_g}{\lambda_{\min}(V^T V)}$.*

In this section, the following four theorems will give the relevant proofs of dynamic performances of the closed-loop system (19), including the stability, dynamical tracking, output constraint and non-Zeno phenomenon.

Theorem 2. *For given parameters $\mu_i > 0, i = 1, 2, \delta > 0$ and $\delta_1 > 0$, if there exist the matrices $\Psi > 0, Q_1 = P_1^{-1} > 0, P_2 > 0$ and $R_i, i = 1, 2, 3$, the following inequality is made:*

$$\begin{bmatrix} \sigma_{11} & \sigma_{12} & \sigma_{13} & \bar{G} & 0 \\ * & \sigma_{22} & \sigma_{23} & 0 & P_2 \\ * & * & \mu^2 \Psi^{-1} - 2\mu Q_1 & 0 & 0 \\ * & * & * & -\mu_1^2 I & 0 \\ * & * & * & * & -\bar{M}^{-1} \end{bmatrix} < 0 \tag{23}$$

where

$$\begin{cases} \sigma_{11} = sym\left\{\bar{A}Q_1 + \sum_{i=1}^{2^m} \chi_i \bar{B}(D_i R_1 + D_i^- R_2)\right\} + \delta^2 \Psi + Q_1 \\ \sigma_{12} = -\bar{B}V - \left(\sum_{i=1}^{2^m} \chi_i R_3 \bar{B} D_i^-(R_2 - R_1)\right)^T \\ \sigma_{13} = -\sum_{i=1}^{2^m} \chi_i \bar{B}(D_i R_1 + D_i^- R_2) \\ \sigma_{22} = sym\{P_2 W + R_3 \bar{B}V\} + U_\sigma^T U_\sigma + \mu_2^{-2} I + P_2 \\ \sigma_{23} = \sum_{i=1}^{2^m} \chi_i R_3 \bar{B} D_i^-(R_2 - R_1) \end{cases}$$

is solvable, and the adaptive law of $\hat{M}(t)$ is designed by (14); then, both the controlled system (19) and the dynamical error system (20) will be stable and the augmented variable $\xi(t)$ will retain a small set $\Theta_{\xi(t)}$, where

$$\Theta_{\xi(t)} = \left\{ \xi(t) \mid \|\xi(t)\| \leq \sqrt{(\mu_1^2 y_d^2 + \kappa)/\lambda_{\min}(P_1)} \right\}.$$

Moreover, the gain matrices K, H_1, L and Ψ are, respectively, given by

$$K = R_1 Q_1^{-1},\ H_1 = R_2 Q_1^{-1},\ L = P_2^{-1} R_3,\ \hat{\Psi} = Q_1 \Psi Q_1$$

Proof. Select the Lyapunov functions as

$$V_1(\tilde{x}(t), t) = \tilde{x}^T(t) P_1 \tilde{x}(t) \tag{24}$$

and

$$V_2(e_\sigma(t), t) = e_\sigma^T(t) P_2 e_\sigma(t) + tr\left\{ \tilde{M}^T(t) \gamma^{-1} \tilde{M}(t) \right\} \tag{25}$$

Along the trajectory of (19), we have, from (24), that

$$\dot{V}_1 \leq \tilde{x}^T(t) sym\left\{ P_1 \bar{A} + \sum_{i=1}^{2m} \chi_i P_1 \bar{B}(D_i K + D_i^- H_1) \right\} \tilde{x}(t) + \tilde{x}^T(t) \left\{ \mu_1^{-2} P_1 \bar{G} \bar{G}^T P_1 + \delta^2 \Psi \right\} \tilde{x}(t)$$

$$- 2\tilde{x}^T(t) \sum_{i=1}^{2m} \chi_i P_1 \bar{B}(D_i K + D_i^- H_1) e_k(t) - 2\tilde{x}^T P_1 \bar{B} V e_\sigma(t) + \mu_1^2 z_d^2 - e_k^T(t) \Psi e_k(t) \tag{26}$$

The derivative of V_2 along (20) is deduced by

$$\dot{V}_2 \leq e_\sigma^T(t) \left(sym(P_2 W + P_2 L \bar{B} V) + P_2 \bar{M} P_2 + U_\sigma^T U_\sigma \right) e_\sigma(t)$$

$$- 2e_\sigma^T(t) \sum_{i=1}^{2m} \chi_i P_2 L \bar{B} D_i^-(H_1 - K) \tilde{x}(t) + 2e_\sigma^T(t) \sum_{i=1}^{2m} \chi_i P_2 L \bar{B} D_i^-(H_1 - K) \tilde{e}_k(t)$$

$$+ 2\|\hat{\sigma}(t)\| \|M^*\|_F^2 + \sqrt{\frac{2\theta_g n_1}{\lambda_{\min}(V^T V)}} \|P_2\| \left(\gamma \sqrt{n_1} P_2 + \sqrt{tr(\tilde{M})} \right) \tag{27}$$

Notice that

$$2\|\hat{\sigma}(t)\| \|M^*\|_F^2 \leq 2\|\sigma(t)\| \|M^*\|_F^2 + 2\|e_\sigma(t)\| \|M^*\|_F^2$$

$$\leq 2\sqrt{\frac{\theta_g}{\lambda_{\min}(V^T V)}} tr(\tilde{M}) + \mu_2^2 (tr(\tilde{M}))^2 + \mu_2^{-2} e_\sigma^T(t) e_\sigma(t) \tag{28}$$

Then, integrating (26) and (27) with (28) produces

$$\dot{V}_1 + \dot{V}_2 \leq \xi^T(t) \Omega \xi(t) + \mu_1^2 y_d^2 + \kappa \tag{29}$$

where the parameter κ is expressed as

$$\kappa = \sqrt{\frac{2\theta_g n_1}{\lambda_{\min}(V^T V)}} \|P_2\| \left(\gamma \sqrt{n_1} P_2 + \sqrt{tr(\tilde{M})} \right) + 2\sqrt{\frac{\theta_g}{\lambda_{\min}(V^T V)}} tr(\tilde{M}) + \mu_2^2 (tr(\tilde{M}))^2 \tag{30}$$

and

$$\Omega = \begin{bmatrix} \omega_{11} & \omega_{12} & \omega_{13} \\ * & \omega_{22} & \omega_{23} \\ * & * & -\Psi \end{bmatrix} \tag{31}$$

with

$$\begin{cases} \omega_{11} = sym\left\{P_1\bar{A} + P_1\sum_{i=1}^{2^m}\chi_i\bar{B}(D_iK + D_i^-H_1)\right\} + \mu_1^{-2}P_1\check{G}\check{G}^TP_1 + \delta^2\Psi \\ \omega_{12} = -P_1\bar{B}V - \left(P_2\sum_{i=1}^{2^m}\chi_iL\bar{B}D_i^-(H_1 - K)\right)^T \\ \omega_{13} = -P_1\sum_{i=1}^{2^m}\chi_i\bar{B}(D_iK + D_i^-H_1) \\ \omega_{22} = sym\{P_2(W + L\bar{B}V)\} + P_2\bar{M}P_2 + U_\sigma^TU_\sigma + \mu_2^{-2}I \\ \omega_{23} = P_2\sum_{i=1}^{2^m}\chi_iL\bar{B}D_i^-(H_1 - K). \end{cases}$$

Based on the Lemma 2, by multiplying the matrix $diag\{P_1, I, P_1, I, I, I\}$ to two sides of (23), we have

$$(23) \iff \Omega < diag\{P_1, P_2, 0\}$$

Then, (29) is expressible as

$$\dot{V}_1 + \dot{V}_2 \leq -\xi^T(t)\tilde{P}\xi(t) + \mu_1^2z_d^2 + \kappa \qquad (32)$$

where $\tilde{P} = diag\{P_1, P_2, \alpha I\}$, with α being a proper positive constant. If

$$\xi^T(t)\tilde{P}\xi(t) > \mu_1^2z_d^2 + \kappa$$

then it is easy to arrive at

$$\dot{V}_1 + \dot{V}_2 < 0.$$

Thus, for any $\bar{x}(t)$, $e_\sigma(t)$ and $e_k(t)$, we have

$$\xi^T(t)\tilde{P}\xi(t) \leq \max\left\{\xi^T(0)\tilde{P}\xi(0), \mu_1^2z_d^2 + \kappa\right\} = \pi \qquad (33)$$

which implies that the controlled system (21) is stable with the original state $\xi(0)$. Thus, the state $\xi(t)$ can be ensured to converge into $\Theta_{\xi(t)}$. The proof is complete. □

Theorem 3. *For given positive parameters μ_i, $i = 1, 2$ and δ, if there exists $P_1^{-1} = Q_1 > 0$, $P_2 > 0$, $\Psi > 0$ and R_i, $i = 1, 2, 3$ satisfying (23) and the conditions*

$$\begin{bmatrix} Q_1 & Q_1\bar{C}_i^T \\ * & (\pi^{-1}z_{di}^2)I \end{bmatrix} \geq 0, i = 1, 2, \cdots, p \qquad (34)$$

$$\begin{bmatrix} \pi^{-1} & R_2^l & V^l \\ * & Q_1 & 0 \\ * & * & P_2 \end{bmatrix} \geq 0, l = 1, 2, \cdots, m \qquad (35)$$

where $\bar{C}_i$ and z_{di}, respectively, represent the ith row of $\bar{C}$ and the ith component of z_d, R_2^l and V^l are, respectively, the ith row of R_2 and V and the adaptive regulation law of $\hat{M}(t)$ is designed by (14), the augmented system (21) will be stable and the tracking error of the output will astringe to zero; that is,

$$\lim_{t\to\infty} z(t) = z_d$$

Moreover, the state saturation constraint $\eta(t) \in L(H)$ will also be satisfied. In addition, the gain matrices K, H_1, L and Ψ are, respectively, given by

$$K = R_1Q_1^{-1}, H_1 = R_2Q_1^{-1}, L = P_2^{-1}R_3, \check{\Psi} = Q_1\Psi Q_1$$

Proof. Similar to the above Theorem, the stability of the augmented system (21) will be proved. From (34), it is not hard to deduce that

$$\tilde{C}_i^T \tilde{C}_i \leq \pi^{-1} z_{di}^2 P_1.$$

Thus, the inequality can be obtained by

$$z_i^2(t) = \tilde{x}^T(t)\tilde{C}_i^T \tilde{C}_i \tilde{x}(t) \leq \pi^{-1} z_{di}^2 \tilde{x}^T(t) P_1 \tilde{x}(t) \leq z_{di}^2 \tag{36}$$

On one hand, it can be known that the term $\int_0^t e(\tau)d\tau$ is a part of $\tilde{x}(t)$. Therefore, when $t \to +\infty$, it can be verified that the integral item must be bounded. Meanwhile, due to the constraint condition of each component of the output (36), the sign of $e(t)$ will stay the same for all $t \geq 0$. In general, it can be concluded that the tracking error satisfies $\lim_{t\to\infty} z(t) = z_d$.

On the other hand, according to the Theorem 1, the $\eta(t)$ will stay in the defined ellipsoid $\Omega(\bar{P}, \pi)$, where $\bar{P} = diag\{P_1, P_2\}$. In addition, by multiplying left and right sides of (35) with the matrix $diag\{I, Q_1^{-1}, I\}$, one has

$$\begin{bmatrix} \pi^{-1} & H_1^l & V^l \\ * & P_1 & 0 \\ * & * & P_2 \end{bmatrix} > 0 \tag{37}$$

Applying the Schur formula into (37) yields

$$\left(H^l \eta(t)\right)^T \left(H^l \eta(t)\right) \leq \pi^{-1} \eta^T(t) \bar{P} \eta(t) \leq 1 \tag{38}$$

Thus, it can be inferred that $\Omega(\bar{P}, \pi) \subset L(H)$ can be met for all $\eta(t)$. Therefore, $\eta(t) \in L(H)$ can be pledged for all $\eta(t) \in \Omega(\bar{P}, \pi)$. □

The next theorem is concerned with the problem of how to determine the minimum triggering time interval.

Theorem 4. *For the system (4), under the designed event-triggering format (8), the minimum triggering interval can be given by*

$$\tilde{T} = \min_k \{t_{k+1} - t_k\} = \frac{1}{a}\ln\left(1 + \frac{a}{b}\Delta(t)\right) > 0 \tag{39}$$

where

$$a = |\lambda_{\max}(\bar{A})|, \quad b = a\|\bar{B}\| \|\tilde{x}(t_k)\| + \|\bar{G}\| \|z_d\|, \quad \Delta(t) = \delta \frac{\lambda_{\max}(\Psi)}{\lambda_{\min}(\Psi)} \|\tilde{x}(t)\| \tag{40}$$

Proof. From (9), it is obtained that

$$\dot{e}_k(t) = \bar{A}\tilde{x}(t) + \bar{B}sat(u(t) + g(t)) + \bar{G}z_d$$

Furthermore, for all $t \in [t_k, t_{k+1})$, one has

$$\frac{d}{dt}\|e_k(t)\| \leq |\lambda_{\max}(\bar{A})| \|e_k(t)\| + |\lambda_{\max}(\bar{A})| \|\tilde{x}(t_k)\| + \|\bar{B}\| + \|\bar{G}\| \|z_d\| \tag{41}$$

By defining a and b as given in (40), the inequality (41) is described as

$$\frac{d}{dt}\|e_k(t)\| \leq a\|e_k(t)\| + b \tag{42}$$

It is easy to deduce that

$$\left\|\Psi^{\frac{1}{2}}e_k(t)\right\| \leq \frac{a}{b}\left(e^{a(t-t_k)}-1\right)$$

Based on the event-triggering condition, by solving $\Delta(t) = \frac{a}{b}\left(e^{a(t-t_k)}-1\right)$, we can achieve

$$\widetilde{T} = \frac{1}{a}\ln\left(1+\frac{a}{b}\Delta(t)\right)$$

which is the minimum triggering time interval. Based on the definition of $\bar{x}(t)$, $\|x(t)\| \neq 0$ is true. Thus, the minimum triggering time interval $\widetilde{T} > 0$ holds. In conclusion, the Zeno phenomenon will not happen in the designed event-triggered algorithm. □

Please note that (23) in Theorem 2 is not a standard LMI and is actually a BLMI. Generally, the BLMI can be solved by fixing the matrix R_3 or the matrices R_1 and R_2 beforehand. As such, the results in Theorem 2 really do not give a convex optimization algorithm. Therefore, the next theorem intends to further improve the results of Theorem 2.

Theorem 5. *Given parameters $\mu_i > 0$, $\alpha_i > 0$, $\delta > 0$ and $\delta_1 > 0$, if there are matrices $P_2 > 0$, $Q_1 = P_1^{-1} > 0$, $\check{\Psi} > 0$, $R > 0$ and R_i such that the conditions*

$$\begin{bmatrix} \psi_{11} & \psi_{12} & \psi_{13} & \bar{G} & 0 & Q_1 & 0 \\ * & \psi_{22} & 0 & 0 & P_2 & 0 & 0 \\ * & * & \psi_{33} & 0 & 0 & 0 & Q_1 \\ * & * & * & -\mu_1^2 I & 0 & 0 & 0 \\ * & * & * & * & -\bar{M}^{-1} & 0 & 0 \\ * & * & * & * & * & -\alpha_1^{-1}I & 0 \\ * & * & * & * & * & * & -\alpha_3^{-1}I \end{bmatrix} < 0 \quad (43)$$

with

$$\begin{cases} \psi_{11} = sym\left\{\bar{A}Q_1 + \sum_{i=1}^{2^m}\chi_i \bar{B}(D_i R_1 + D_i^- R_2)\right\} + \delta^2 \check{\Psi} \\ \psi_{12} = \bar{B}V \\ \psi_{13} = -\sum_{i=1}^{2^m}\chi_i \bar{B}(D_i R_1 + D_i^- R_2) \\ \psi_{22} = sym\{P_2 W + R_3 \bar{B} V\} + U_\sigma^T U_\sigma + \mu_2^{-2} I + \alpha_2 I \\ \psi_{33} = \mu^2 \check{\Psi}^{-1} - 2\mu Q_1 \end{cases}$$

and

$$\varepsilon < 2\sqrt{\frac{\alpha_1 \alpha_2 \alpha_3}{\alpha_1 + \alpha_2}} \quad (44)$$

are solvable, and the adaptive regulation law of $\hat{M}(t)$ is designed by (14), the augmented system (21) will be stable. The gain matrices K, H_1, L and Ψ are, respectively, given by

$$K = R_1 Q_1^{-1}, \; H_1 = R_2 Q_1^{-1}, \; L = P_2^{-1} R_3, \; \check{\Psi} = Q_1 \Psi Q_1$$

Proof. Similar to Theorem 2, by taking the derivative of the functions given in (24) and (25), inequalities (26) and (27) can still be satisfied. As for the coupling term

$$e_\sigma^T(t)\sum_{i=1}^{2^m}\chi_i L\bar{B}(D_i K + D_i^- H_1)\bar{x}(t_k)$$

in (34), we can conclude that, if the inequality (43) holds, then there must exist a parameter $\varepsilon > 0$, depending on L, K and H_1, such that

$$e_\sigma^T(t) \sum_{i=1}^{2^m} \chi_i L\bar{B}(D_i K + D_i^- H_1) \tilde{x}(t_k) \leq \varepsilon \|e_\sigma(t)\|(\|\tilde{x}(t)\| + \|e_k(t)\|) \quad (45)$$

By means of (43), (27) is translated as

$$\dot{V}_2 \leq e_\sigma^T(t)\left(sym(P_2 W + P_2 L\bar{B} V) + P_2 \bar{M} P_2 + U_\sigma^T U_\sigma\right) e_\sigma(t) + \varepsilon \|e_\sigma(t)\|(\|\tilde{x}(t)\| + \|e_k(t)\|)$$
$$+ \sqrt{\frac{2\theta_d n_1}{\lambda_{min}(V^T V)}} \|P_2\| \left(\gamma\sqrt{n_1} P_2 + \sqrt{tr(\bar{M})}\right) + 2\|\hat{\sigma}(t)\| \|\|M^*\|_F^2 \quad (46)$$

Furthermore, by using (26) and (46), we can obtain

$$\dot{V}_1 + \dot{V}_2 \leq \xi^T(t) \Omega_1 \xi(t) + \varepsilon \|e_\sigma(t)\| \|\tilde{x}(t)\| + \varepsilon \|e_\sigma(t)\| \|e_k(t)\| + \mu_1^2 y_d^2 + \kappa \quad (47)$$

where

$$\Omega_1 = \begin{bmatrix} \omega_{11} & P_1 \bar{B} V & \omega_{13} \\ * & \omega_{22} & 0 \\ * & * & -\Psi \end{bmatrix}.$$

By using the Schur lemma, we can attain

$$(43) \iff \Omega_1 < diag\{\alpha_1 I, \alpha_2 I, \alpha_3 I\}.$$

Then, (47) is inferred as

$$\dot{V}_1 + \dot{V}_2 = -\tilde{\xi}^T(t) Y \tilde{\xi}(t) + \mu_1^2 y_d^2 + \kappa \quad (48)$$

where

$$\tilde{\xi}(t) = [\|\tilde{x}(t)\|, \|e_\sigma(t)\|, \|e_k(t)\|]^T, Y = \begin{bmatrix} \alpha_1 & 0 & -\frac{\varepsilon}{2} \\ * & \alpha_2 & -\frac{\varepsilon}{2} \\ * & * & \alpha_3 \end{bmatrix} \quad (49)$$

It is noted that, if Y is a positive real matrix, the stability of system (21) can be pledged. Further, the characteristic polynomial of Y is described by

$$4\alpha_3 \lambda_i^2 - 4(\alpha_1 \alpha_3 + \alpha_2 \alpha_3)\lambda_i + \left(4\alpha_1 \alpha_2 \alpha_3 - \alpha_1 \varepsilon^2 - \alpha_2 \varepsilon^2\right) = 0 \quad (50)$$

where λ_i are the eigenvalues of Y. From (50), it is inferred that

$$\lambda_1 + \lambda_2 = \alpha_1 + \alpha_2 > 0$$
$$\lambda_1 \lambda_2 = \frac{4\alpha_1 \alpha_2 \alpha_3 - (\alpha_1 + \alpha_2)\varepsilon^2}{4\alpha_3}$$

If the condition (44) is met, then it is easy to conclude that $\lambda_1 \lambda_2 > 0$. To sum up, the matrix Y is a positive real matrix and thus the augmented system (21) is proved to be stable. □

6. Simulation

Consider the A4D aircraft model as the controlled system. In a flight environment of 16,000 ft altitude and 0.9 Mach, the dynamics of the A4D system can be modeled by (1), where $x(t) \in \mathbb{R}^4$ represents the state of the aircraft, $x_1(t)$ is the forward velocity $(ft \cdot s^{-1})$, $x_2(t)$ is the attack angle (rad), and $x_3(t)$ and $x_4(t)$ are the velocity of pitch $(rad \cdot s^{-1})$ and the angle of pitch (rad), respectively. $u(t)$ is the elevator deflection (deg) and the output $z(t)$

is selected as the forward velocity $x_1(t)$. Similar to [45,46], the dynamic model was modeled by using the principle of system identification. Based on the idea of sparse identification, the input and output data of A4D aircraft were identified by the generalized least squares method, and then the state parameter matrices A, B and C of the system were obtained as

$$A = \begin{bmatrix} -0.0605 & 32.38 & 0 & 32 \\ -0.0015 & -1.47 & 1 & 0 \\ -0.0111 & -34.72 & -2.793 & 0 \\ 0 & 0 & 1 & 0 \end{bmatrix}, B = \begin{bmatrix} 0 \\ -0.1064 \\ -33.8 \\ 0 \end{bmatrix}, C = \begin{bmatrix} 1 & 0 & 0 & 0 \end{bmatrix}$$

Next, we considered the anti-disturbance control for two types of irregular disturbances by selecting different excitation functions.

First, in order to describe attenuated harmonic (AH) disturbances, the DNN parameters of the disturbance model were selected as

$$W = \begin{bmatrix} 0 & 4 \\ -4 & 0 \end{bmatrix}, V = \begin{bmatrix} 0.7 & 0 \end{bmatrix}, M^* = \begin{bmatrix} -0.3 & -0.05 \\ 0.01 & 0.45 \end{bmatrix}, \Phi(t) = \begin{bmatrix} \arctan(t) \\ \arctan(t) \end{bmatrix}$$

Preselect the candidate value of R_3 as

$$R_3 = \begin{bmatrix} 0 & -30.3493 & -5.6140 & 0 & 0 \\ 0 & 50.1378 & 10.9620 & 0 & 0 \end{bmatrix}.$$

Meanwhile, by defining $\mu_1 = \mu_2 = 1$ and solving inequalities (23), (34), (35), we obtained

$$K = \begin{bmatrix} 0.0038 & -0.6781 & 0.0339 & 0.7744 & 0.0008 \end{bmatrix}$$

$$L = 10^{-6} * \begin{bmatrix} 0 & -0.1055 & -0.0195 & 0 & 0 \\ 0 & 0.1743 & 0.0381 & 0 & 0 \end{bmatrix}$$

$$H_1 = \begin{bmatrix} 0.0014 & -0.1912 & 0.0115 & 0.233 & 0.0005 \end{bmatrix}$$

Assume that the initial conditions of the augmented states and the desired output are selected as

$$x_0 = [2, -2, 3, -2]^T, \ \sigma_0 = [4, 4]^T, \ z_d = 18$$

Suppose that Ψ is an identity matrix, $\delta = 0.01$. Figure 1 reflects the triggered release time and the corresponding interval. The dynamics of the states are plotted in Figure 2, which can reflect the favorable stability. Both the attenuated harmonic disturbances and the disturbance estimated value together with the estimated error are displayed in Figure 3. Thus, the satisfactory capacities of disturbance modeling and estimation are fully embodied. Figures 4 and 5 depict the dynamical trajectories of the input and output, respectively, which verifies the favorable input constraint and dynamical tracking performance. The dynamics of the DNN weight are exhibited in Figure 6.

Second, sawtooth wave (STW) signals usually appear in some circuit or electromagnetism systems, and it is quite hard to monitor them using common epitaxial systems. For modeling STW disturbances, the specific parameters of DNNs are considered as

$$W = \begin{bmatrix} 0 & -6 \\ 2 & -0.01 \end{bmatrix}, V = \begin{bmatrix} -0.01 & -1 \end{bmatrix}, M^* = \begin{bmatrix} 0 & 0.02 \\ -0.2 & 0.45 \end{bmatrix}$$

$$\Phi(t) = \begin{cases} \begin{bmatrix} \frac{1}{1+e^{-0.5t}} \\ \frac{1}{1+e^{-0.5t}} \end{bmatrix} & t \geq 0 \\ \begin{bmatrix} -2.1 \\ -2.1 \end{bmatrix} & t < 0 \end{cases}$$

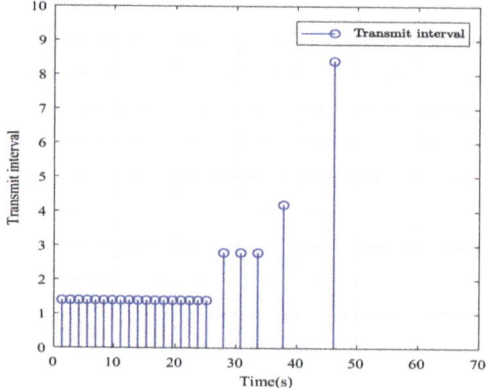

Figure 1. The event-triggered release times and intervals in the case of AH disturbances.

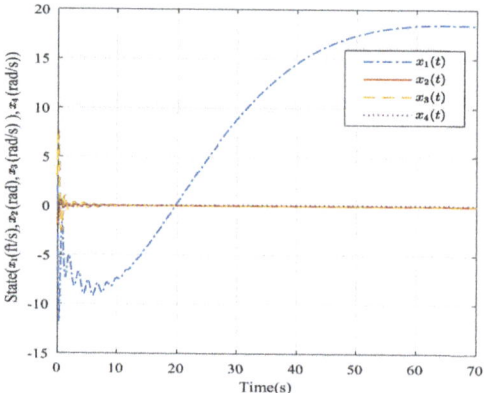

Figure 2. The trajectories of the system states in the case of AH disturbances.

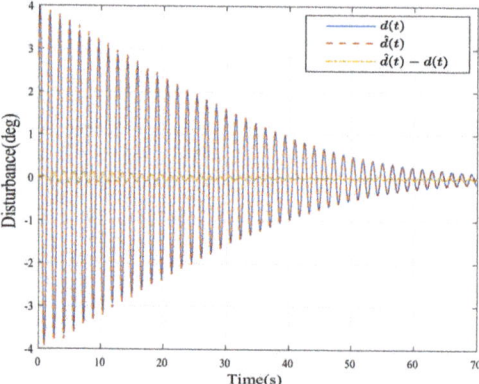

Figure 3. The disturbance estimates and estimation error in the case of AH disturbances.

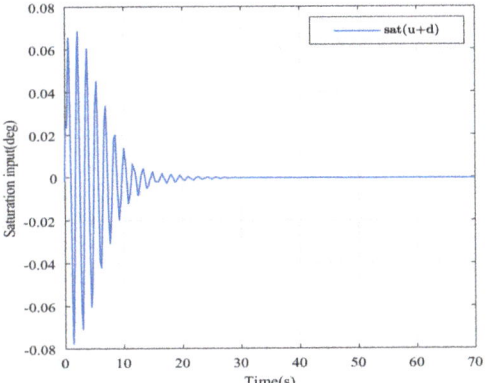

Figure 4. The dynamics of the saturated control input in the case of AH disturbances.

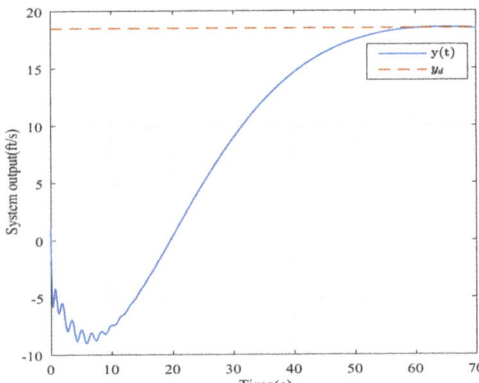

Figure 5. The trajectory of the system output in the case of AH disturbances.

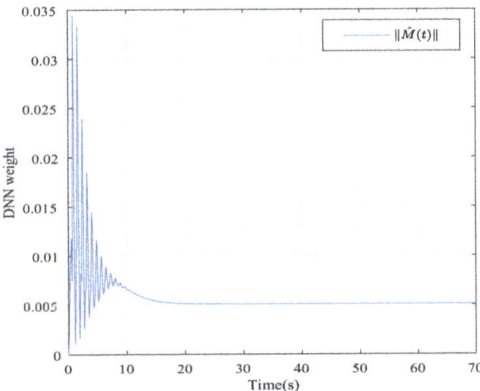

Figure 6. The trajectory of the dynamical weights in the case of AH disturbances.

By solving inequalities (23), (34) and (35), the gains K, L and H_1 can be found to be

$$K = \begin{bmatrix} 0.0231 & 0.2521 & 0.0184 & 0.9451 & 0.0431 \end{bmatrix}$$

$$L = 10^{-6} * \begin{bmatrix} 0 & -0.1441 & -0.0266 & 0 & 0 \\ 0 & 0.0792 & 0.0173 & 0 & 0 \end{bmatrix}$$

$$H_1 = \begin{bmatrix} -0.1475 & 0.0062 & 0.0230 & 0.1648 & 0.0430 \end{bmatrix}$$

Suppose that the initial values are, respectively, given by

$$x_0 = [2, -2, 3, -2]^T, \quad \sigma_0 = [3, 3]^T.$$

The desired output is defined as $z_d = 17$. The triggered release time and corresponding intervals are displayed in Figure 7. Figure 8 is the tracks of the states of the A4D system. Figure 9 exhibits the dynamics of STW and its estimates. Figures 10 and 11, respectively, present the saturated input and the system output. Figure 12 depicts the dynamics of the designed DNN weight. Figures 8–12 demonstrate that the designed event-triggered PI control input can obtain favorable control performances in the case of STW disturbances while saving a considerable amount of resources (see Figure 7).

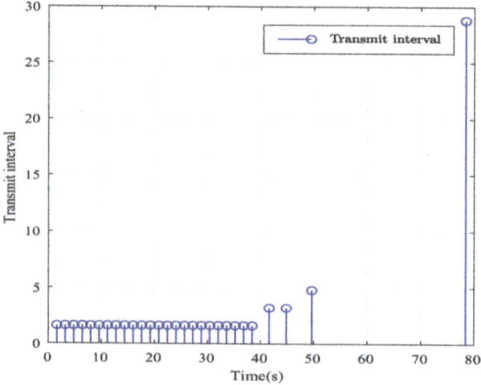

Figure 7. The event-triggered release times and intervals in the case of STW disturbances.

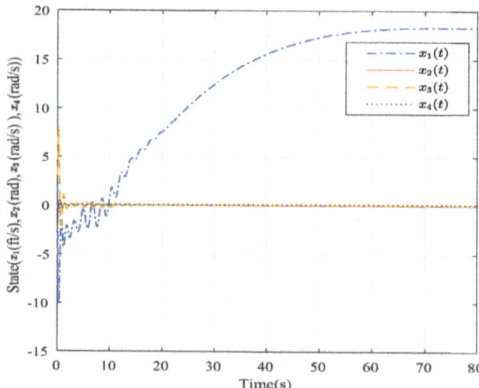

Figure 8. The trajectories of the system states in the case of STW disturbances.

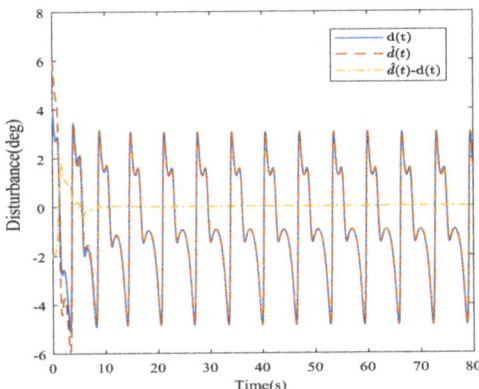

Figure 9. The disturbance estimates and estimation error in the case of STW disturbances.

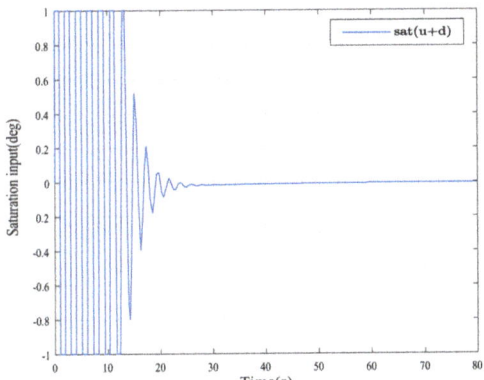

Figure 10. The dynamics of the saturated control input in the case of STW disturbances.

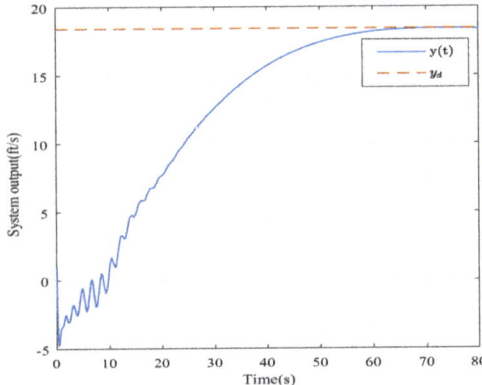

Figure 11. The trajectory of the system output in the case of STW disturbances.

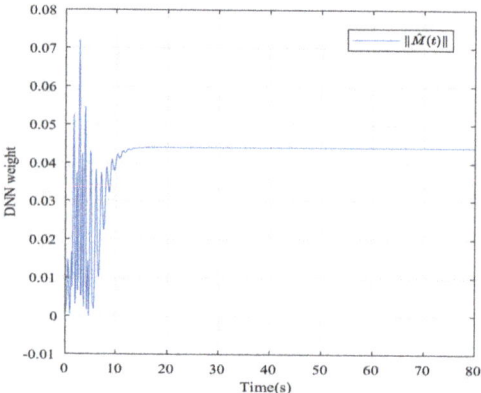

Figure 12. The trajectory of the dynamical weights in the case of STW disturbances.

By effectively estimating for AH and STW disturbances, respectively, a satisfactory anti-disturbance control frame can be embodied in the above simulation. Compared to those results that rely on constant or harmonic disturbances, the main advantages of the suggested method are reflected in wider anti-disturbance ranges, more objective control tasks and less data transfer. Of course, some existing disadvantages—for example, more conservative algorithms and higher real-time requirements—need to be fully considered in the future work.

7. Conclusions

In this paper, a valid anti-disturbance event-triggered control probelm is discussed for systems with multiple constraints under the frame of DNN disturbance modeling. Different from the usual time-triggered problem, the whole algorithm design was made with the event-triggered frame. After constructing the augmented event-triggering condition, a novel event-triggered DOBAC algorithm was designed by integrating the modified adaptive regulation law with the DNN disturbance models. Meanwhile, a composite event-triggered controller was successively designed with a polytopic description of the saturated actuator. By using the convex optimization theory, the relevant proofs were given to verify the stability of the closed-loop augmented system and to meet the multiple constraints regarding the augmented states, as well as the system output. Moreover, the dynamics of the tracking error can be displayed as converging to zero. Finally, the simulation results illustrate that the proposed scheme is effective in terms of desired control performances and significantly reduced resource utilization.

Author Contributions: Data curation: H.S.; resources: H.S. and Y.Y.; software: H.S. and Q.W.; writing original draft: H.S., Q.W. and Y.Y.; writing—review and editing: H.S. and Q.W. All authors have read and agreed to the published version of the manuscript.

Funding: This work was supported in part by NSFC under Grants 61973266, and the Project of Xuzhou Key Research and Development under Grant KC21080.

Data Availability Statement: Not applicable.

Conflicts of Interest: The authors declare no conflict of interest.

References

1. Chen, W.H.; Ballanceand, D.J.; Gawthrop, P.J. A Nonlinear Disturbance Observer for Robotic Manipulators. *IEEE Trans. Ind. Electron.* **2000**, *47*, 932–938. [CrossRef]
2. Zhang, H.F.; Wei, X.J.; Karimi, H.R.; Han, J. Anti-Disturbance Control Based on Disturbance Observer for Nonlinear Systems with Bounded Disturbances. *J. Frankl. Inst.* **2017**, *355*, 4916–4930. [CrossRef]
3. Nguyen, M.H.; Dao, H.V.; Ahn, K.K. Adaptive Robust Position Control of Electro-Hydraulic Servo Systems with Large Uncertainties and Disturbances. *Appl. Sci.* **2022**, *12*, 794. [CrossRef]
4. Abdul-Adheem, W.R.; Alkhayyat, A.; Al Mhdawi, A.K.; Bessis, N.; Ibraheem, I.K.; Abdulkareem, A.I.; Humaidi, A.J.; AL-Qassar, A.A. Anti-Disturbance Compensation-Based Nonlinear Control for a Class of MIMO Uncertain Nonlinear Systems. *Entropy* **2021**, *23*, 1487. [CrossRef]
5. Zhou, L.; Tse, K.T.; Hu, G.; Li, Y. Higher Order Dynamic Mode Decomposition of Wind Pressures on Square Buildings. *J. Wind. Eng. Ind. Aerodyn.* **2021**, *211*, 104545. [CrossRef]
6. Nguyen, M.H.; Dao, H.V.; Ahn, K.K. Extended Sliding Mode Observer-Based High-Accuracy Motion Control for Uncertain Electro-Hydraulic Systems. *Int. J. Robust Nonlinear Control* **2022**, *33*, 1351–1370. [CrossRef]
7. Zong, G.D.; Qi, W.H.; Karimi, H.R. L_1 Control of Positive Semi-Markov Jump Systems with State Delay. *IEEE Trans. Syst. Man Cybern. Syst.* **2021**, *51*, 7569–7578. [CrossRef]
8. Zhong, Z.X.; Wang, X.Y.; Lam, H.K. Finite-Time Fuzzy Sliding Mode Control for Nonlinear Descriptor Systems. *IEEE/CAA J. Autom. Sin.* **2021**, *8*, 1141–1152. [CrossRef]
9. Gao, Z. Active Disturbance Rejection Control for Nonlinear Fractional Order Systems. *Int. J. Robust Nonlinear Control* **2016**, *26*, 876–892. [CrossRef]
10. Chen, P.; Luo, Y.; Peng, Y.; Chen, Y. Optimal Fractional-Order Active Disturbance Rejection Controller Design for PMSM Speed Servo System. *Entropy* **2021**, *23*, 262. [CrossRef]
11. Aishwarya, A.; Ujjwala, T.; Vrunda, J. Disturbance Observer Based Speed Control of PMSM Using Fractional Order PI Controller. *IEEE/CAA J. Autom. Sin.* **2019**, *6*, 316–326.
12. Hua, Z.G.; Chen, M. Coordinated Disturbance Observer-Based Flight Control of Fixed-Wing UAV. *IEEE Trans. Circuits Syst. II Exp. Briefs* **2022**, *69*, 3545–3549.
13. Zhang, J.H.; Zheng, W.X.; Xu, H.; Xia, Y.Q. Observer-Based Event-Driven Control for Discrete-Time Systems with Disturbance Rejection. *IEEE Trans. Cybern.* **2021**, *51*, 2120–2130. [CrossRef] [PubMed]
14. Li, R.; Zhu, Q.; Yang, J.; Narayan, P.; Yue, X. Disturbance-Observer-Based U-Control (DOBUC) for Nonlinear Dynamic Systems. *Entropy* **2021**, *23*, 1625. [CrossRef] [PubMed]
15. Wang, X.Y.; Li, S.H.; Wang, G.D. Distributed Optimization for Disturbed Second-Order Multi-Agent Systems Based on Active Anti-Disturbance Control. *IEEE Trans. Neural Netw. Learn. Syst.* **2020**, *31*, 2104–2117. [CrossRef]
16. Yi, Y.; Zheng, W.X.; Sun, C.Y.; Guo, L. DOB Fuzzy Controller Design for Non-Gaussian Stochastic Distribution Systems Using Two-Step Fuzzy Identification. *IEEE Trans. Fuzzy Syst.* **2016**, *24*, 401–418. [CrossRef]
17. Zhao, Z.J.; Ahn, C.K.; Li, H.X. Boundary Anti-Disturbance Control of a Spatially Nonlinear Flexible String System. *IEEE Trans. Ind. Electron.* **2020**, *67*, 4846–4856. [CrossRef]
18. Hu, T.S.; Lin, Z. *Control Systems with Actuator Saturation: Analysis and Design*; Birkhäuser: Boston, MA, USA, 2001.
19. Tarbouriech, S.; Garcia, G.; Gomes, J.M.; Queinnec, I. *Stability and Stabilization of Linear Systems With Saturating Actuators*; Springer: London, UK, 2011.
20. Fridman, E.; Pila, A.; Shaked, U. Regional Stabilization and H_∞ Control of Time-Delay Systems with Saturating Actuators. *Int. J. Robust Nonlinear Control* **2003**, *13*, 885–907. [CrossRef]
21. Zhou, B.; Zheng, W.X.; Duan, G.R. An Improved Treatment of Saturation Nonlinearity with Its Application to Control of Systems Subject to Nested Saturation. *Automatica* **2011**, *47*, 306–315. [CrossRef]
22. Wei, Y.L.; Zheng, W.X.; Xu, S.Y. Anti-Disturbance Control for Nonlinear Systems Subject to Input Saturation via Disturbance Observer. *Syst. Control Lett.* **2015**, *85*, 61–69. [CrossRef]
23. Li, Y.L.; Lin, Z.L. A Complete Characterization of the Maximal Contractively Invariant Ellipsoids of Linear Systems Under Saturated Linear Feedback. *IEEE Trans. Autom. Control* **2015**, *85*, 179–185. [CrossRef]
24. Bai, W.W.; Zhou, Q.; Li, T.S.; Li, H.Y. Adaptive Reinforcement Learning Neural Network Control for Uncertain Nonlinear System with Input Saturation. *IEEE Trans. Cybern.* **2020**, *50*, 3433–3443. [CrossRef] [PubMed]
25. Wang, X.L.; Ding, D.R.; Dong H.L.; Zhang, X.M. Neural-Network-Based Control for Discrete-Time Nonlinear Systems with Input Saturation Under Stochastic Communication Protocol. *IEEE/CAA J. Autom. Sin.* **2021**, *8*, 766–778. [CrossRef]
26. Pan, H.H.; Sun, W.C.; Gao, H.J.; Jing, X.J. Disturbance Observer-Based Adaptive Tracking Control with Actuator Saturation and Its Application. *IEEE Trans. Autom. Sci. Eng.* **2016**, *13*, 868–875. [CrossRef]
27. Li, Z.J.; Zhao, J. Adaptive Consensus of Non-Strict Feedback Witched Multi-Agent Systems with Input Saturations. *IEEE/CAA J. Autom. Sin.* **2021**, *8*, 1752–1761. [CrossRef]
28. Tee, K.P.; Ren, B.B.; Ge, S.S. Control of Nonlinear Systems with Time Varying Output Constraints. *Automatica* **2011**, *47*, 2511–2516. [CrossRef]
29. Ngo, K.B.; Mahony, R.; Jiang, Z.P. Integrator Backstepping Using Barrier Functions for Systems with Multiple State Constraints. In Proceedings of the 44th IEEE Conference on Decision and Control, Seville, Spain, 15 December 2005; pp. 8306–8312.

30. Meng, W.C.; Yang, Q.M.; Si, S.N.; Sun, Y.X. Adaptive Neural Control of a Class of Output-Constrained Non-Affine Systems. *IEEE Trans. Cybern.* **2016**, *46*, 85–89. [CrossRef]
31. Liu, Y.J.; Ma, L.; Liu, L.; Tong, S.C.; Chen, C.L.P. Adaptive Neural Network Learning Controller Design for a Alass of Nonlinear Systems with Time-Varying State Constraints. *IEEE Trans. Neural Netw. Learn. Syst.* **2020**, *31*, 66–75. [CrossRef]
32. Astrom, K.; Bernhardsson, B. Comparison of Periodic and Event Based Sampling for First-Order Stochastic Systems. In Proceedings of the 14th IFAC World Congress, Beijing, China, 5–9 July 1999; pp. 301–306.
33. Sahoo, A.; Xu, H.; Jagannathan, S. Neural Network-Based Event Triggered State Feedback Control of Nonlinear Continuous-Time Systems. *IEEE Trans. Neural Netw. Learn. Syst.* **2016**, *27*, 497–509. [CrossRef]
34. Dolk, V.; Borgers, D.; Heemels, W.P.M.H. Output-Based and Decentralized Dynamic Event-Triggered Control with Guaranteed L_p-Gain Performance and Zeno-Freeness. *IEEE Trans. Autom. Control* **2017**, *62*, 34–49. [CrossRef]
35. Chen P.; Li, F.Q. A Survey on Recent Advances in Event-Triggered Communication and Control. *Inf. Sci.* **2018**, *457*, 113–125.
36. Wang, W.; Li, Y.M.; Tong, S.C. Neural-Network-Based Adaptive Event-Triggered Consensus Control of Nonstrict-Feedback Nonlinear Systems. *IEEE Trans. Neural Netw. Learn. Syst.* **2021**, *32*, 1750–1764. [CrossRef] [PubMed]
37. Li, Y.X.; Yang, G.H. Model-Based Adaptive Event-Triggered Control of Strict-Feedback Nonlinear Systems. *IEEE Trans. Neural Netw. Learn. Syst.* **2018**, *29*, 1033–1045. [CrossRef] [PubMed]
38. Wu, Z.G.; Xu, Y.; Pan, Y.J.; Su, S.H.; Tang, Y. Event-Triggered Control for Consensus Problem in Multi-Agent Systems with Quantized Relative State Measurements and External Disturbance. *IEEE Trans. Circuits Syst. I Reg. Pap.* **2018**, *65*, 2232–2242. [CrossRef]
39. Zhang, Y.H.; Sun, J.; Liang, H.J.; Li, H.Y. Event-Triggered Adaptive Tracking Control for Multiagent Systems with Unknown Disturbances. *IEEE Trans. Cyber.* **2020**, *50*, 890–901. [CrossRef]
40. Ren, C.E.; Fu, Q.X.; Zhang, J.G.; Zhao, J.S. · Adaptive Event-triggered Control for Nonlinear Multi-agent Systems with Unknown Control Directions and Actuator Failures. *Nonlinear Dyn.* **2021**, *105*, 1657–1672. [CrossRef]
41. Yang, L.W.; Liu, T.; Hill, D.J. Decentralized Event-Triggered Frequency Control with Guaranteed L_∞-Gain for Multi-Area Power Systems. *IEEE Control Syst. Lett.* **2021**, *5*, 373–378. [CrossRef]
42. Deng, Y.J.; Zhang, X.K.; Im, N.K.; Zhang, G.Q.; Zhang, Q. Model-Based Event-Triggered Tracking Control of Underactuated Surface Vessels with Minimum Learning Parameters. *IEEE Trans. Neural Netw. Learn. Syst.* **2020**, *31*, 4001–4014. [CrossRef]
43. Yu, W.; Rosen, J. Neural PID Control of Robot Manipulators with Application to an Upper Limb Exoskeleton. *IEEE Trans. Cybern.* **2013**, *43*, 673–684.
44. Han, H.G.; Zhang, L.; Hou, Y.; Qiao, J.F. Nonlinear Model Predictive Control Based on a Self-Organizing Recurrent Neural Network. *IEEE Trans. Neural Netw. Learn. Syst.* **2016**, *27*, 402–415. [CrossRef]
45. Guo, L.; Chen, W.H. Disturbance Attenuation and Rejection for Systems with Nonlinearity via DOBC Approach. *Int. J. Robust Nonlinear Control* **2005**, *15*, 109–125. [CrossRef]
46. McRuer, D.; Ashkenas, I.; Graham, D. *Aircraft Dynamics and Automatic Control*; Princeton University Press: Princeton, NJ, USA, 1976.

Disclaimer/Publisher's Note: The statements, opinions and data contained in all publications are solely those of the individual author(s) and contributor(s) and not of MDPI and/or the editor(s). MDPI and/or the editor(s) disclaim responsibility for any injury to people or property resulting from any ideas, methods, instructions or products referred to in the content.

Article

Robust Variable-Step Perturb-and-Observe Sliding Mode Controller for Grid-Connected Wind-Energy-Conversion Systems

Ilham Toumi [1], Billel Meghni [2], Oussama Hachana [3], Ahmad Taher Azar [4,5,*], Amira Boulmaiz [6], Amjad J. Humaidi [7], Ibraheem Kasim Ibraheem [8], Nashwa Ahmad Kamal [9], Quanmin Zhu [10], Giuseppe Fusco [11] and Naglaa K. Bahgaat [12]

1. Department of Electronics and Telecommunications, Faculty of New Technologies of Computing and Communication, University of Ouargla, Ouargla 30000, Algeria; toumi.ilham@univ-ouargla.dz
2. Algeria LSEM Laboratory, Department of Electrical Engineering, University Badji Mokhtar, Annaba 23000, Algeria; bilel.maghni@univ-annaba.dz
3. Department of Drilling and Rig Mechanics, Faculty of Hydrocarbons, Renewable Energies and Earth and Universe Sciences, University of Ouargla, Ouargla 30000, Algeria; oussama.hachana@gmail.com
4. College of Computer and Information Sciences, Prince Sultan University, Riyadh 11586, Saudi Arabia
5. Faculty of Computers and Artificial Intelligence, Benha University, Benha 13518, Egypt
6. Department of Electronics, University of Badji Mokhtar, Annaba 23000, Algeria; amira.boulmaiz@univ-annaba.org
7. Department of Control and Systems Engineering, University of Technology, Baghdad 10001, Iraq; amjad.j.humaidi@uotechnology.edu.iq
8. Department of Computer Techniques Engineering, Dijlah University College, Baghdad 10001, Iraq; ibraheemki@coeng.uobaghdad.edu.iq
9. Faculty of Engineering, Cairo University, Giza 12613, Egypt; nashwa.ahmad.kamal@gmail.com
10. Department of Engineering Design and Mathematics, Frenchy Campus Coldharbour Lane, University of the West of England, Bristol BS16 1QY, UK; quan.zhu@uwe.ac.uk
11. Department of Electrical and Information Engineering, Università degli Studi di Cassino e del Lazio Meridionale, 03043 Cassino, Italy; fusco@unicas.it
12. Department of Communications and Electronics Engineering, Faculty of Engineering, Canadian International College (CIC), Shiekh Zayed City, Egypt; naglaa_kamel@cic-cairo.com
* Correspondence: aazar@psu.edu.sa or ahmad.azar@fci.bu.edu.eg or ahmad_t_azar@ieee.org

Abstract: In order to extract efficient power generation, a wind turbine (WT) system requires an accurate maximum power point tracking (MPPT) technique. Therefore, a novel robust variable-step perturb-and-observe (RVS-P&O) algorithm was developed for the machine-side converter (MSC). The control strategy was applied on a WT based permanent-magnet synchronous generator (PMSG) to overcome the downsides of the currently published P&O MPPT methods. Particularly, two main points were involved. Firstly, a systematic step-size selection on the basis of power and speed measurement normalization was proposed; secondly, to obtain acceptable robustness for high and long wind-speed variations, a new correction to calculate the power variation was carried out. The grid-side converter (GSC) was controlled using a second-order sliding mode controller (SOSMC) with an adaptive-gain super-twisting algorithm (STA) to realize the high-quality seamless setting of power injected into the grid, a satisfactory power factor correction, a high harmonic performance of the AC source, and removal of the chatter effect compared to the traditional first-order sliding mode controller (FOSMC). Simulation results showed the superiority of the suggested RVS-P&O over the competing based P&O techniques. The RVS-P&O offered the WT an efficiency of 99.35%, which was an increase of 3.82% over the variable-step P&O algorithm. Indeed, the settling time was remarkably enhanced; it was 0.00794 s, which was better than for LS-P&O (0.0841 s), SS-P&O (0.1617 s), and VS-P&O (0.2224 s). Therefore, in terms of energy efficiency, as well as transient and steady-state response performances under various operating conditions, the RVS-P&O algorithm could be an accurate candidate for MPP online operation tracking.

Keywords: robust variable-step perturb and observe; normalization; second-order sliding mode controller; systematic step size; super-twisting algorithm

1. Introduction

Global energy consumption is mostly covered by fossil fuels that have a detrimental effect on the natural environment [1]. The increasing demand for energy with the consideration of global warming and environmental pollution has pushed interesting development of renewable energies. The wind system as an energy source has demonstrated important progression with a considerable production rate and maintenance cost [2]. It is the fastest-growing source, with a growing average of 20% per year in the energy sector [3]. A wind turbine (WT) can be categorized as variable- or fixed-speed. In the first configuration, the variable-speed wind turbine (VSWT), to permanently reach the maximum power point (MPP), its speed is constantly varied depending on the wind-speed fluctuations [4]. Hence, several generator types can be used, and the PMSG remains an attractive solution "without gearbox" in onshore and offshore applications, as it provides many advantages, such as: high energy production, good power/weight ratio, better reliability, and a high capacity to maximize energy production [5,6]. In addition, a variable speed, PMSG, horizontal axis, and direct drive without gearboxes are features that provide a positive impact on a WT system's mechanical framework design [5]. They permit the development of even larger VSWTs at greater heights.

In a machine-side converter (MSC), the VSWT should operate at the optimum speed during changes in wind speed to produce the maximum electrical power. This is realized by a fast and adequate MPPT algorithm. In order to enhance the dynamic performances, the MPPT techniques have recently gained considerable interest [7,8]. In the recent literature, there are three categories of MPPT algorithms; namely, the indirect power controller (IPC), direct power controller (DPC), and artificial intelligence (AI) [9,10].

The first category (IPC) involves the following techniques: optimal torque (OT) [11], power signal feedback (PSF) [12], tip speed ratio (TSR) [13], and sliding mode control (SMC) [14]. TSR-based MPPT is the simplest technique with a faster response time in which the wind speed data are recorded by means of anemometers. However, the availability of speed sensors increases the complexity of the wind power system, as well as the implementation and maintenance cost. In the OT and PSF techniques, prior knowledge of the turbine generator's mathematical model is necessary to predetermine the PMSG speed, TSR, and torque constant. However, it is difficult to precisely follow the MPP under a lower wind speed due to the relativity between the tracking speed and generator inertia. The SMC technique has been widely proposed in the literature [15] and is simple to implement, but it generates the well-known phenomenon of chattering, in which high-frequency oscillations around the MPP occur in a steady state caused by the sign function nature [16,17].

Furthermore, artificial intelligence (AI)-based MPPT control techniques, such as FLC [10] and ANN [18], have been proposed to track the MPP well, but their industrial applications are limited. The standard FLC-based MPPT technique requires many precise guidelines in the controller design, such as the quantity of choices to be measured, as well as the determination of fuzzification, inferences, and defuzzification [10]. In addition, a larger data memory space implies much more execution time to obtain the optimum solution, which is a significant drawback for online applications. The ANN-based MPPT technique is an expert knowledge strategy that requires a huge amount of data under various operating conditions. It usually needs a formal method to define the optimal network layout and number of neurons to place in the hidden layer. Indeed, choosing the initial values of the network weights and setting the learning step are of important concern [19].

The third family (DPC) allows tracking of the MPP by controlling the power fluctuation given by the mechanical speed under the wind speed variation. This category comprises P&O [8], incremental conductance (INC) [20,21], and optimum relation-based

(ORB) [11]. The conventional INC technique provides good results under constant wind-speed conditions [7]. Meanwhile, their performances are not ensured under sudden and faster wind-speed variations.

The P&O technique has been used effectively to follow the optimum rotor speed with interesting ease of implementation, which renders it the most common and applied algorithm in the literature [22]. It was developed in such a way to perturb the rotor speed at several steps and then observe the change in the extracted power until the power–speed curve slope becomes zero. The perturbation and observation actions are realized without using anemometers. The suitable choice of step size is the major concern of the P&O algorithm, as it directly affects the WECS performances [3,23]. The step size during the disturbance of the rotor speed can be fixed or variable. If a small step size is adopted by using the classical P&O algorithm, the tracking speed response becomes very slow, which causes more power losses [5]. Meanwhile, it shows small steady-state oscillations around the MPP. In contrast, a large step size engenders a faster tracking-speed response but with large steady-state oscillations that harm larger inertia WTs, and hence reduce the performance of the WECS [24,25].

To overcome the downsides of the fixed-step (FS)-P&O algorithms and efficiently achieve the optimum dynamic performance of a WECS, many modified P&O versions have been proposed [2,5]. They can be classified into two main groups: modified and adaptive P&O algorithms. By applying the modified P&O algorithms, the variable step (VS) sizes are attained by dividing the P/ω curve into several operating areas, with each one having a predefined step size based on a synthesized curve or ratio.

Adaptive (A)-P&O was presented in [26,27]. The step size was modified according to an objective function that relied upon various control variables and the wind speed. This method provided interesting results under uniform atmospheric conditions. However, the performances were reduced under a large random wind-speed variation when the P/ω curve included multiple peak points. In [3], the proposed algorithm combined the generation of adaptive step sizes with dividing the P/ω curve into several sections. The authors of [28,29] used a modified (M)-P&O algorithm based on the comparison of several P/ω curves and the sector's intersection points. It employed a forward large step and a small step around the MPP. Meanwhile, the larger step induced oscillations at steady state with no structured relation to select the required step length and WT properties. The authors of [30] proposed a robust MPPT control scheme for a grid-connected PMSG-WT using a P&O-based nonlinear adaptive control. This approach used many assumptions that decreased the system efficiency caused by unwanted fluctuations around the MPP. A VS-P&O algorithm was developed in [2,22] in which the step size was determined by observing the distance between the operating point and the MPP in the P/ω curve. The authors subdivided the P/ω curve into modular operating sectors using predefined ratios. However, the performances of this approach remained poor under rapid climatic variations, as it needed to calculate a specific ratio at each wind-speed value. In [18,31], the authors suggested a hybrid P&O algorithm to eliminate the disadvantages of the conventional FS-P&O. Based on the error observation between the instantaneous and reference rotor speeds, the hybridized algorithm, while usually employing FLC, ANN, PSO, and ANFIS, etc., ensured the subdivision of the P/ω curve into several sectors. The simulation results showed the efficiency of the hybrid techniques in spite of the algorithm complexity.

Motivated by the above discussion, this paper presents a recently developed robust variable-step perturb-and-observe (RVS-P&O)-based MPPT algorithm to eliminate the drawbacks of the classical P&O technique, such as "slower time response, influence of the WT inertia and the step size selection concerns". The control method proposed can realize stability in the system to maximize the power extraction in the WT system under rapid wind speed changes. Regarding the P/ω curve, it is more appropriate to adjust the speed reference step as a function of the MPP error. Hence, it was proposed to adapt this step by a proportional factor to reach the MPP. To measure this action, the normalized power level was subdivided into a finite number of sectors. For each sector, the corresponding step size was determined as the optimum speed percentage. The objective of this work was to design the adaptive control in order to achieve the best performances of the MPPT operation. The main contributions of this paper can be summarized as outlined below.

- Normalization of the observation measurement and the speed variation allows the controller optimality to be maintained during the use of WTs with different dimensions. In addition to the normalization of the power measurement and the set-point speed increment:
- For more robustness, correction of the observation measurement was carried out by compensating for the wind speed effect;
- The RVS-P&O optimization strategy was based on subdividing the P/w curve into several modular operating sectors according to the distance in the ratio measurement between the actual and desired MPP;
- The RVS-P&O method improved the performance and the efficiency of VS-P&O algorithm variants while eliminating the drawbacks of the traditional FS-P&O ones;
- In terms of accelerated dynamic response capacity, the RVS-P&O algorithm tracked the MPP well during rapid climate variations, with a fast response time of 0.00794 s;
- The RVS-P&O approach enhanced the efficiency of the WECS by 3.82% compared to the conventional algorithms (FS-P&O and VS-P&O);
- The RVS-P&O algorithm showed a high level of stability with a small variation around the MPP, where the mean energy loss was estimated as 13.1826 W regardless of the operating conditions;
- The novel proposed approach was simple and easy to implement in practice;
- A DPC-SOSMC–STA controller was utilized in the grid-side converter (GSC) to obtain a smooth setting of the active and reactive power-quantity interchange between the generator and grid based on grid demand during realistic variable wind speeds.

To verify the performances of the proposed RVS-P&O algorithm, it was fairly compared to small step (SS)-P&O, large step (LS)-P&O, and VS-P&O techniques. The proposed algorithm was tested in different environmental conditions. This was on the basis of multiple data sets of wind speeds: gradual changes and experimental random variations. At the grid-side converter, a control strategy based on the DPC-SOSMC–super-twisting algorithm (STA) controller was utilized to realize the smooth setting of active and reactive power-quantity interchange between the generator and grid according to the real power request and variable wind speeds.

The rest of this paper consists of five sections that are organized as follows: Section 2 provides the mathematical modeling of the PMSG-based VSWT principal parts. The converter controller architecture is discussed in Sections 3 and 4. Simulation tests using MATLAB/Simulink and comparison results are provided in Section 5 to validate the effectiveness of the proposed RVS-P&O algorithm under several operating scenarios of real wind-speed variation. Finally, Section 6 concludes by providing the obtained results and perspectives.

2. Mathematical Modeling of the WECS

In order to establish the system control, the examined WECS is introduced in this section. Figure 1 depicts a representative topology of the considered WECS, which included a three-bladed turbine with a horizontal axis, with the rotor of the VSWT connected directly without a gearbox to the PMSG shaft [26]. The electronic power devices (two back-to-back AC/DC/AC IGBT bridges) supplied power from the used generator to the grid via a common DC bus [5,32].

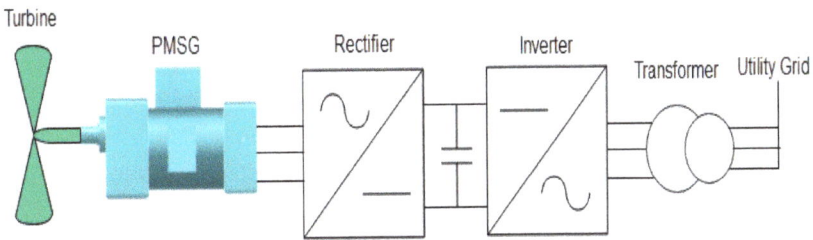

Figure 1. Configuration of the studied wind-generation system.

2.1. Wind Turbine Model

A wind sail converts a quantity of air mass energy into movement; during the circulation of the wind in an active surface (S), the power of the air mass (P_ω) is given by Equation (1) [33]:

$$P_\omega = \frac{1}{2}\rho \cdot S \cdot v^3 \tag{1}$$

It will be transmitted to the generator shaft as aerodynamic power or turbine power, as expressed by Equation (2) [34]:

$$P_k = \frac{1}{2}\rho \cdot S C_p(\lambda, \beta) \cdot V_k^3 \tag{2}$$

where λ is the relation between the turbine angular speed and the wind speed. This denominates the tip speed ratio (TSR), and is given by Equation (3) [35]:

$$\lambda = \frac{R \times \Omega_k}{V_k} \tag{3}$$

The aerodynamic efficiency varies according to λ. In other words, the maximum $C_{p\ max}$ is reached when λ is optimal (λ_{opti}). Figure 2 presents the resultant C_p according to the λ variation when β is fixed [36].

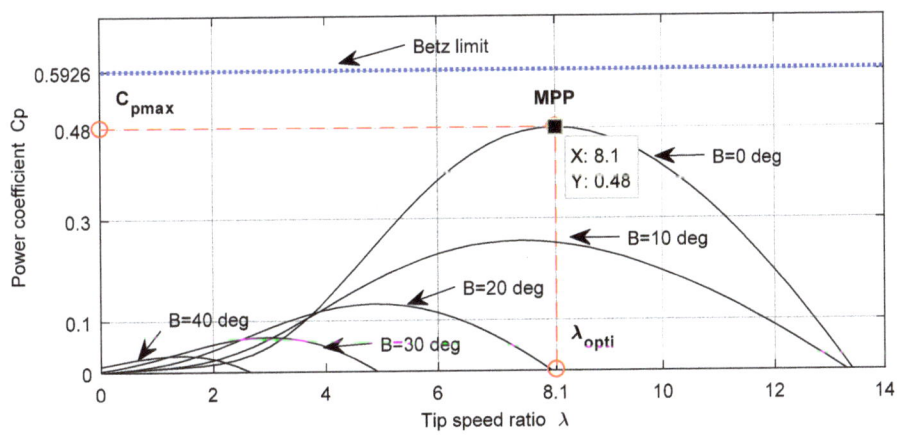

Figure 2. Model power coefficient (C_p) with tip speed ratio (λ) curve.

2.2. PMSG Model

To enhance the controlling procedure of the electric generator's dynamic performance, the model was based on the stator voltage within the Parck model, which is defined by using Equation (4) [37]:

$$\begin{cases} V_d = R_s I_d + L_d \frac{dI_d}{dt} - \omega L_q I_q \\ V_q = R_s I_q + L_q \frac{dI_q}{dt} + \omega (L_d I_d + \psi_f) \end{cases} \quad (4)$$

where R_s, L_d, L_q, and ψ_f are given in Tables A1 and A2 in Appendix A.

2.3. Grid Model

The grid model in the d-q plane is given by Equation (5) [4]:

$$\begin{cases} V_{dg} = V_{di} - R_g I_{dg} - L_{dg} \frac{dI_{dg}}{dt} + L_{qg} w_g I_{qg} \\ V_{qg} = V_{qi} - R_g I_{qg} - L_{qg} \frac{dI_{qg}}{dt} - L_{dg} w_g I_{dg} \end{cases} \quad (5)$$

where R_g, L_{dg}, and L_{qg} are given in Table A3 in Appendix A.

3. Converter Controller Architecture

3.1. General Description

The energetic and environmental constraints of the WT-PMSG presented above required the application of a sophisticated supervision system and an adequate energy-management system. The control scheme is described in Figure 3, in which the control strategy was divided into two main parts: MSC and GSC.

- *Machine-Side Converter:* An advanced controller based on an RVS-P&O-based MPPT algorithm and the SOSMC were applied to control the PMSG speed and torque, thus extracting the MPP for each sampled wind speed value.
- *Grid-Side Converter:* While the wind speed fluctuated, the amplitude of the energy produced and the electrical frequency were constantly changing, which was not a perspective appropriate for grid integration. To resolve this problem, the GSC was usually employed to ensure the wind system's connection to the electrical grid with better active and reactive powers. After that, the active and reactive power of the reference voltage generation was directly controlled by means of the DPC-SOSMC-STA-SVM strategy, unlike the traditional vector method.

3.2. Machine-Side Converter Controller

A WT is usually characterized by the P/w curve showing the relationship between the rotor speed and the generated mechanical energy amount. Given the limits of (v_{cut-in}) and $(v_{cut-out})$ as shown in Figure 4, this work focused mainly on region (2) [31].

In this region (2), the maximum speed of the rotor could be reached by adjusting the electromagnetic torque to extract the highest mechanical power; this was done by keeping the power coefficient (C_p) at the maximum $(C_{p\,max})$. To achieve this goal, the field-oriented control (FOC) strategy was used to control the PMSG. It comprised two control loops, an external one for the speed and an internal one for the current, as depicted in Figure 5.

The RVS-P&O-MPPT-SOSMC algorithm was utilized in the first control loop to reach a reference optimal speed for each wind speed in order to generate an electromagnetic torque reference. The current control loop was exploited to control the stator currents of the d-q axis separately based on Equation (5). The PI controller was applied to adjust the three-phase currents by generating the commutation pulses by means of the space vector pulse-width modulation (SVPWM) technique [5].

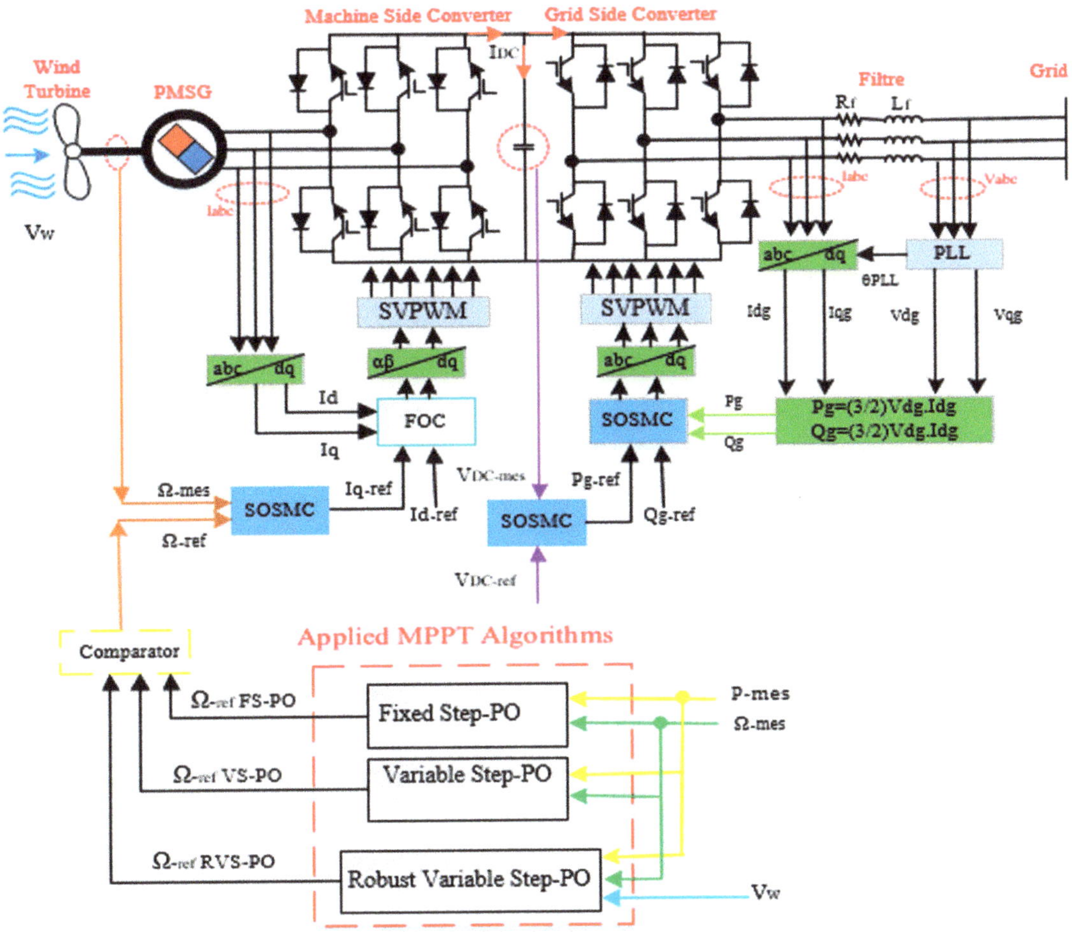

Figure 3. The complete control system description.

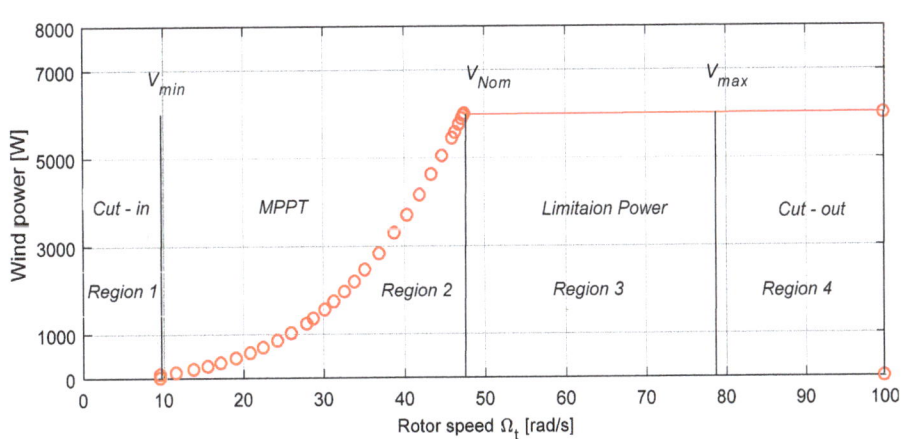

Figure 4. Power/speed curve showing the various operation regions of the VSWT.

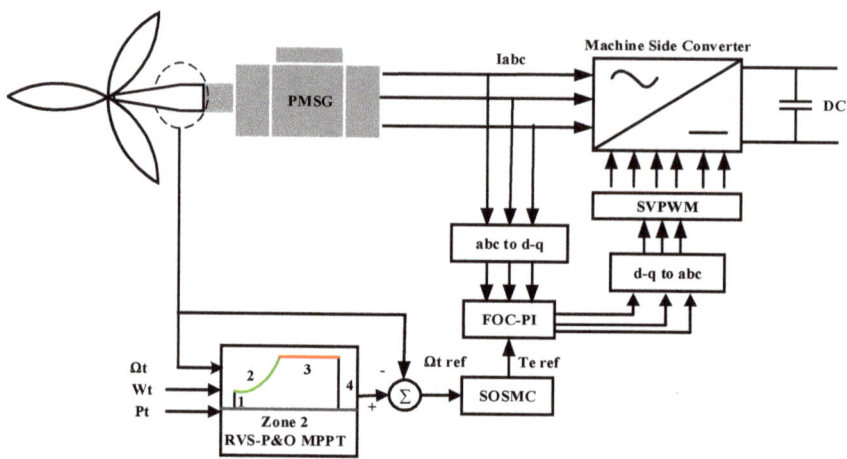

Figure 5. Block diagram of MSC contoller.

3.3. Grid-Side Converter Controller

The utmost challenge in wind power generation is the inherently sporadic nature of the wind, which can deviate quickly [37]. Its intermittent availability is the main impediment to power quality and flow control. Wind-speed variations lead to a fluctuating injected power; therefore, the stability and power quality of the grid operation is affected. Consequently, the fluctuations in wind power should be reduced to prevent a degradation of the grid's performance [38].

For this reason, we proposed a GSC to provide and arrange the energy required by the user regardless of operational conditions [37]. For controlling the active and reactive power supplied into the electrical grid, a DPC-SVM-based SOSMC-STA was recommended at this stage. The schematic diagram of the GSC control approach is shown in Figure 6. In contrast to the traditional vector technique [38], the DPC-SVM-based SOSMC-STA approach provided the grid voltage directly to the GSC.

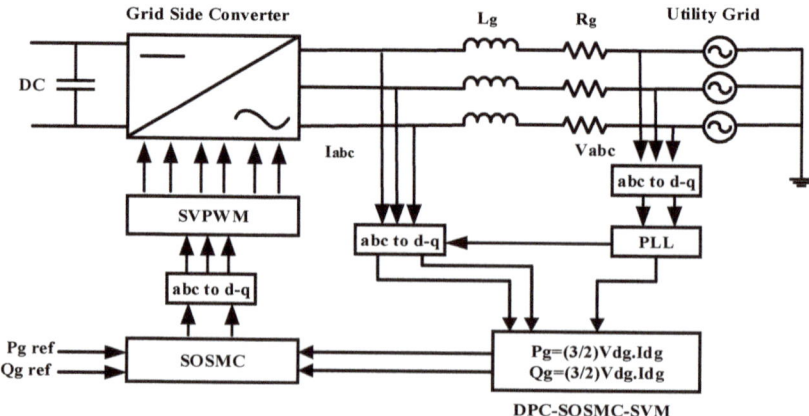

Figure 6. Block diagram of GSC-based DPC-SVM with SOSMC-STA controller.

A. Higher Order SMC Based DPC-SVM Design

The first-order SMC engenders the chattering phenomena, which is a major inconvenience in practical operating implementation. To avoid such an issue, higher-order

SMC application is a feasible solution that significantly reduces the multiple undesirable oscillations by maintaining the performances of the first-order controller [39,40].

The active and reactive grid powers are given by Equation (6) as follows:

$$\begin{cases} P_g = \frac{3}{2} V_{dg} I_{dg} \\ Q_g = \frac{3}{2} V_{dg} I_{qg} \end{cases} \quad (6)$$

In order to establish a null operating-power factor, the optimal reactive power was set to be $Q_{g\,ref} = 0$, while the optimal active power $P_{g\,ref}$ depended on the grid requirement. The SOSMC block diagram is shown in Figure 6. The sliding surfaces of the active and reactive powers (S_P and S_Q) were determined using Equation (7):

$$\begin{cases} s_P = P_{gref} - P_g \\ s_Q = Q_{gref} - Q_g \end{cases} \quad (7)$$

The first derivatives of the sliding surfaces are given by Equation (8):

$$\begin{cases} \dot{s}_P = \dot{P}_{gref} - \frac{1.5\,V_{dg}}{L_g}(-V_{dg} - R_g I_{dg} + L_g w_g I_{qg}) - \frac{V_{id}}{L_g} \\ \dot{s}_Q = \dot{Q}_{gref} - \frac{1.5\,V_{qg}}{L_g}(-V_{qg} - R_g I_{qg} - L_g w_g I_{dg}) - \frac{V_{iq}}{L_g} \end{cases} \quad (8)$$

Equation (9) gives the second derivative of both surfaces:

$$\begin{cases} \ddot{s}_P = \dot{G}_P - \frac{\dot{V}_{id}}{L_g} \\ \ddot{s}_Q = \dot{G}_Q - \frac{\dot{V}_{iq}}{L_g} \end{cases} \quad (9)$$

where G_P and G_Q are defined by Equation (10):

$$\begin{cases} G_P = \dot{P}_{gref} - \frac{1.5\,V_{dg}}{L_g}(-V_{dg} - R_g I_{dg} + L_g w_g I_{qg}) \\ G_Q = \dot{Q}_{gref} - \frac{1.5\,V_{qg}}{L_g}(-V_{qg} - R_g I_{qg} - L_g w_g I_{dg}) \end{cases} \quad (10)$$

The SOSMC defines two main parts, either for V_P^{ref} or V_Q^{ref}, as given by Equation (11):

$$\begin{cases} V_P^{ref} = V_P^N + V_P^{eq} \\ V_Q^{ref} = V_Q^N + V_Q^{eq} \end{cases} \quad (11)$$

where V^N is determined by Equation (12):

$$\begin{cases} \dot{w}_1 = -K \cdot sign(s_P) \\ w_2 = -M \cdot \sqrt{|s_P|} sign(s_P) \\ V_P^N = w_1 + w_2 \\ V_Q^N = w_1 + w_2 \end{cases} \quad (12)$$

The STA introduced by Levant [41] can be determined using Equation (13):

$$\begin{cases} V_P^{ref} = V_P^{eq} - M\sqrt{|s_P|} sign(s_P) - K \int sign(s_P) \\ V_Q^{ref} = V_Q^{eq} - M\sqrt{|s_Q|} sign(s_Q) - K \int sign(s_Q) \end{cases} \quad (13)$$

where K and M are unknown parameters to maintain the sliding manifolds' convergence to zero in finite time [42]. Both parameters could be limited as determined by Equation (14):

$$\begin{cases} K > \frac{C_0}{K_m} \quad 0 < \rho < 0.5 \\ M^2 \geq \frac{4C_0 K_M(K-C_0)}{K_m^2 K_m(K-C_0)} \; if\, \rho = 0.5 \end{cases} \quad (14)$$

where C_0, K_m, and K_M are positive constants.

4. MPPT-Based Control Algorithms

To enhance the overall efficiency of the WTs by capturing the highest energy output of the VSWT, an accurate MPPT algorithm should usually be implemented. Less-transient response oscillations, rapid dynamics, and a low design cost are the important requirements for an efficient MPPT technique. The VSWT is regulated to extract the highest generated power below the nominal wind speed. Therefore, to place the WT blades in front of the wind, the pitch angle should be zero. The MPP was determined by achieving the ideal values of λ_{opt} and $C_{p\,opt}$, which were 8.1 and 0.48, respectively.

4.1. Classical P&O Algorithm

The P&O algorithm is determined by the introduction of a small speed perturbation of $(+\Delta\Omega - ref / +\Delta\Omega - ref)$, as illustrated in Figure 7. The effect of this disturbance is subsequently noticed in the PMSG output power.

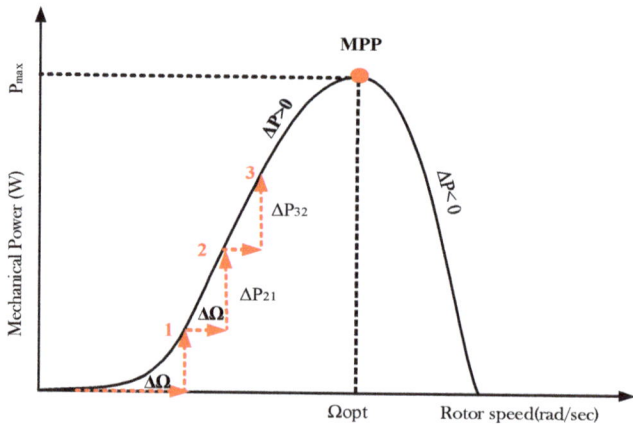

Figure 7. Working principle of the P&O-based MPPT technique.

A P&O algorithm is an iterative approach that needs just two sensors for sensing the power and the speed of the WT. Its operating principle, as depicted in Table 1, is based on perturbing the speed in small increments and comparing the power with that of the preceding perturbation cycle. If the perturbation leads to an increase (decrease) in wind power, the succeeding perturbation is made in the same (opposite) direction. In this manner, the MPP tracker incessantly seeks to find the maximum power location.

Table 1. Effect of wind-speed variation on the conventional P&O algorithm's convergence rate.

	Operating Point is on the Left Side of P_{max}		Operating Point is on the Right Side of P_{max}	
	$\Delta\Omega < 0$	$\Delta\Omega > 0$	$\Delta\Omega < 0$	$\Delta\Omega > 0$
Increase in wind speed	Moves away to the left side of P_{max}	Converges toward the best P_{max}	Moves away to the left side of P_{max}	Moves away to the right side of P_{max}
Decrease in wind speed		Very slow convergence to reach P_{max}		

The behavior of the conventional P&O technique under varying climatic conditions was evaluated. In a basic analysis, this technique showed remarkable drawbacks, such as:

- The P&O algorithm step size was usually fixed and lacked any clarification regarding how it was determined;

- Through the observation of the P/w curve, it was more convenient to adjust the speed reference step according to the MPP error;
- The P&O algorithm was developed on the basis of a constant or slowly varying wind speed, which is not practical. In reality, the convergence rate is strongly affected by the rapid variation in the wind speed;
- The output power displayed several oscillations with a large magnitude permanently, even during fixed wind speeds.

To overcome these concerns, we proposed a robust variable-step P&O.

4.2. Proposed Robust Variable-Step P&O Algorithm

The RVS-P&O was based on the standardization of the generator speed and the mechanical power variables. Algorithm characteristic parameters are summarized in Table 2. A correction of the power-variation calculation was introduced by canceling the effect of wind disturbances.

Table 2. Algorithm characteristic parameters.

Sector	β_{L-1}	α_L
l = 1	0.6	0.03
l = 2	0.4	0.02
l = 3	0.01	0.01
l = L	0	0.0001

4.2.1. Power Normalization

To provide a systematic method for sizing the reference step size, a WT-PMSG system operating under the wind speed v_k at instant k was considered. The maximum mechanical power P_k^{max} is given by Equation (15):

$$P_k^{max} = \frac{1}{2} \rho \cdot s \cdot v_k^3 \cdot C_{p\,max} \qquad (15)$$

To maintain the optimal controller dynamics with turbines of different sizes, a standardization of the power measurement and the set-point speed increment is suggested [43].

The normalized power P_k^N is instantaneously defined as the ratio of the actual absorbed power to the maximum available one using Equation (16):

$$P_k^N = \frac{P_k}{P_k^{max}} \times 100 \qquad (16)$$

4.2.2. Speed Step Selection

If the speed reference step is taken to be constant, for considerable variation in wind speed, the controller will take more time to reach the MPP, as a nonadaptive step will provide the same action as that taken in a small variation in the wind speed case. Therefore, to avoid the slow reaction, an adaptation of this step size by a proportional amount to the correction signal to reach the MPP was proposed [42]. To subdivide the range of the normalized power in finite number of sectors ($l = 1 \ldots L$), it is required to define $(L-1)$ level as the delimiter. For that, let us consider a maximum power level in each sector, denoted by P_{max}^l as a ratio (β_l) of the maximum mechanical actual power P_k^{max}, which is defined by means of Equation (17):

$$P_{max}^l = \beta_l \cdot P_k^{max} \qquad (17)$$

where the ratio β_l is in the range of [0,1], while $l = 1, \ldots, L-1$.

For each sector, the corresponding step size is defined by the weighting factor (α_l) from the actual optimal speed (Ω_k^{opt}) at instant k using Equation (18).

$$\Delta\Omega_k^{ref} = \alpha_l \times \Omega_k^{opt} \tag{18}$$

In addition, the normalized actual speed is defined by Equation (19):

$$\Omega_k^N = \frac{\Omega_k}{\Omega_k^{opt}} \times 100 \tag{19}$$

while the optimal speed (Ω_k^{opt}) is given by Equation (20):

$$\Omega_k^{opt} = \frac{\lambda_{opt} \times V_k}{R} \tag{20}$$

where $l = 1, \ldots, L$ denotes the sector index and α_l is in the range of [0,1]. The weighting factor reflects the amount of the speed adjustment relative to the optimal speed. Since a fine adjustment is needed near the MPP, this factor should be decreased when moving from a sector to the upper one. Figure 8 shows an example of the normalized P/w curve with three modular operating sectors.

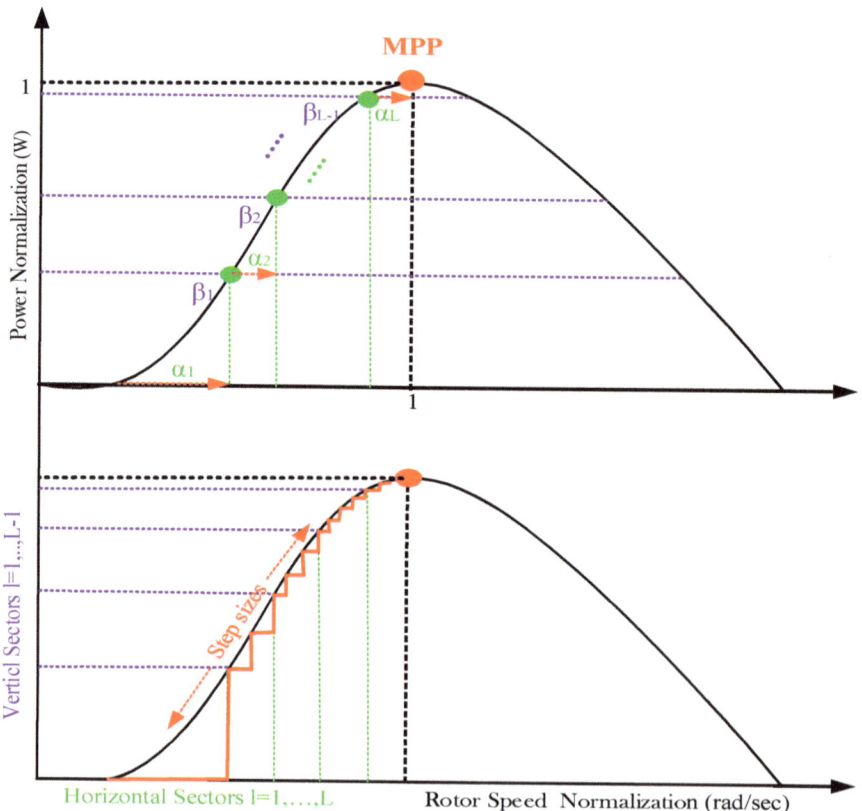

Figure 8. Operation principal of the RVS-P&O-based MPPT controller.

4.2.3. Compensation for Wind-Speed Variation

The P&O algorithm is based essentially on the product sign of the power variation and the speed step increment. If positive, the speed reference step will be increased, and in the negative case, it will be decreased. The power variation also depends on the wind-speed variation, as it introduces a perturbation of the power variation ΔP_k, and the algorithm will behave with less efficiency. This is one of the reasons we proposed the RVS-P&O strategy. It is necessary to eliminate this perturbation while taking into account only the part of ΔP_k induced by the speed adjustment in the previous step. This makes the control algorithm more robust against perturbations of the variation in wind speed. It is known that the power at instant k depends on the turbine speed and the wind speed as given by Equation (21):

$$P_k = f(\Omega_k, v_k) \tag{21}$$

Hence, the power variation at instant k is given by Equation (22):

$$\Delta P_k = P_k(\Omega_k, v_k) - P_k(\Omega_{k-1}, v_{k-1}) \tag{22}$$

The development in the first order of Equation (22) is given by Equation (23):

$$\Delta P_k \simeq f(\Omega_k, v_{k-1}) + \left.\frac{\partial f}{\partial v}\right|_{(\Omega_k, v_{k-1})} \Delta v_k - f(\Omega_{k-1}, v_{k-1}) \tag{23}$$

The second term in Equation (23) represents the perturbation of the wind-speed variation. When the wind speed is constant, this term goes to zero. So, the corrected power variation ΔP_k^ω without wind disturbance is determined by Equation (24):

$$\Delta P_k^\omega = \Delta P_k - \left.\frac{\partial f}{\partial v}\right|_{(\Omega_k, v_{k-1})} \Delta v_k \tag{24}$$

Practically, ΔP_k is easily deduced through successively finding the error between the calculated powers (P_k and P_{k-1}) at different instances. The RVS-P&O algorithm flowchart is illustrated in Figure 9.

It should be mentioned that the arbitrary parameter β_i defines the corresponding sector size, since a coarse action should be taken when the operating point is located far from the MPP, and conversely. A fine step-size adjustment must be applied around the MPP, where the condition determined by Equation (24) is suggested:

$$\beta_l = \begin{cases} 0.5 & l = 1 \\ 1.5\,\beta_{l-1} & 1 < l < L \end{cases} \tag{25}$$

This will ensure an initial fast response in the presence of perturbation at steady state.

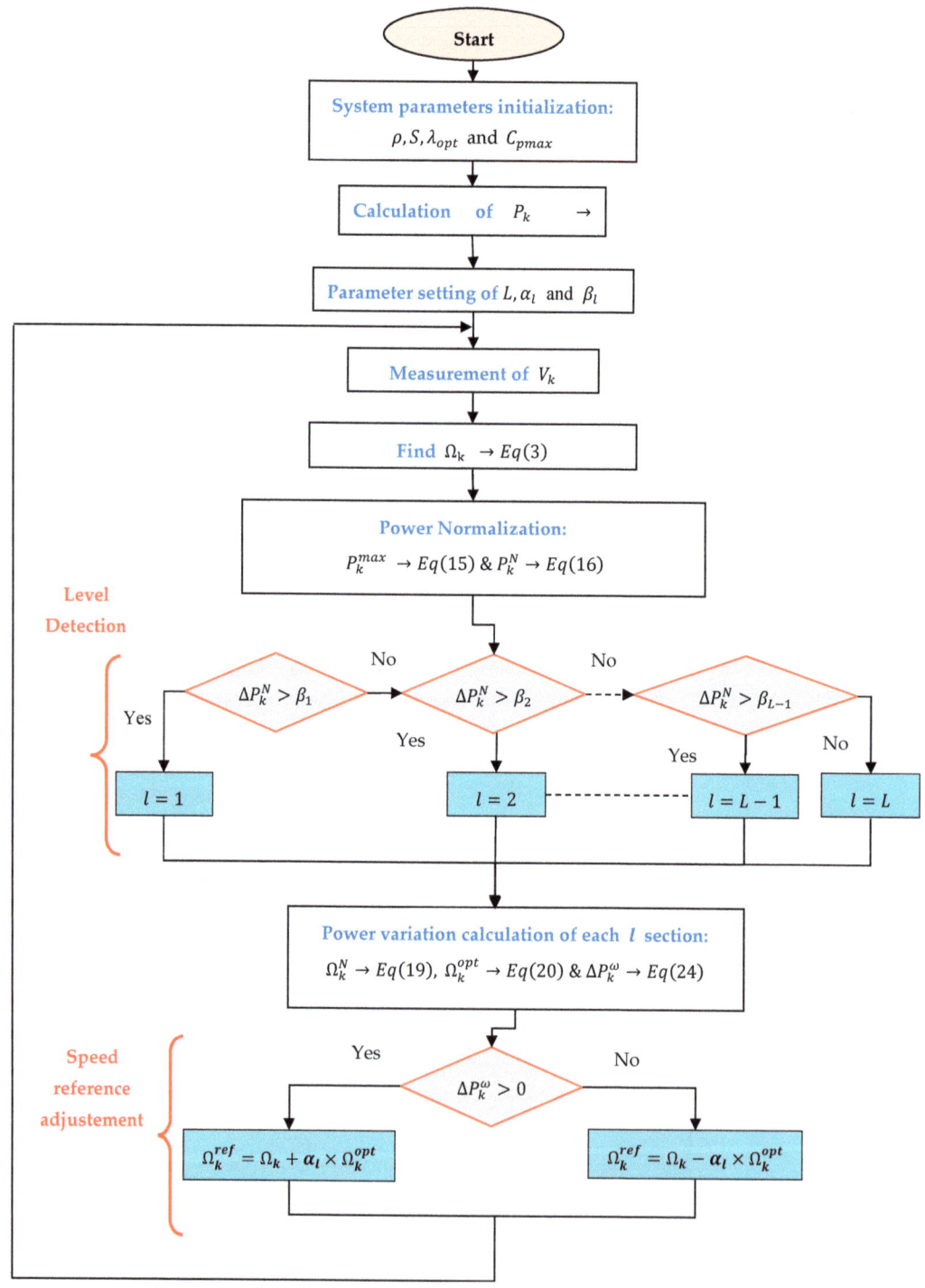

Figure 9. Detailed flowchart of the RVS-P&O-based MPPT technique.

5. Simulation Results and Discussion

To verify the effectiveness of the proposed algorithm, as well as its robustness compared to other existing MPPT algorithms, several simulations using MATLAB/Simulink were performed. This was done by using two different case studies of wind-speed profiles. Tables A1–A3 in Appendix A provide the control parameters of the WT, PMSG, and grid, respectively. The organic ranking cycles (ORCs) and the overall efficiency of the WECS were calculated. The results are summarized in Table 3.

Table 3. Overall performance assessment of the competing algorithms under variable fluctuations in wind speed.

MPPT Method	Average Value			Average Error Value			Efficiency
	P_t (w)	C_p	λ	P_t (w)	C_p	Ω_t (rad/s)	η(%)
SS-P&O	1.9935×10^3	0.4710	8.0282	30.9802	0.0090	6.5315×10^{-4}	98.47
LS-P&O	1.9679×10^3	0.4690	8.0029	49.1288	0.0110	0.0696	97.57
VS-P&O	1.8868×10^3	0.4616	7.8862	88.2430	0.0184	2.3098×10^{-4}	95.53
RVS-P&O	2.0097×10^3	0.4770	8.0483	13.1826	0.0030	8.9757×10^{-4}	99.35

5.1. Gradual Variations in Wind Speed

Figure 10 depicts the machine-side results of four algorithms—SS-P&O, LS-P&O, VS-P&O, and RVS-P&O—under gradual variations in the wind speed. This was to well assess the transient and steady-state performances of the RVS-P&O as shown in Figure 10a. As can be observed, the predicted wind speed based on the step change profile was utilized to analyze the suggested P&O algorithms, in which the wind speed was varied by 6.6 m/s, 7.5 m/s, 9.5 m/s, 11.4 m/s, and 10 m/s every 5 s of samples. The obtained results were compared with the standard method (FS-P&O) and VS-P&O. The most important criteria to verify the effectiveness of the proposed technique were the optimal values of C_p and λ. The behavior of the values is shown in Figure 10b,c. As shown in Figure 10b, the suggested algorithm (RVS-P&O) followed the ideal C_p value faster than the SS-P&O, LS-P&O, and VS-P&O techniques, where the 5% settling time of 7.94 ms was compared to 161.7 ms, 84.1 ms, and 222.4 ms for the SS-P&O, LS-P&O, and VS-P&O techniques, respectively. In the transient response, during an abrupt variation in wind speed (9.5 m/s to 11.4 m/s) at 15 s, at this moment the SS-P&O and LS-P&O algorithms showed large oscillations around the MPP, with settling times of 1.5 s and 0.55 s, respectively. Meanwhile, RVS-P&O had an interesting settling time of 0.2 s compared to the VS-P&O algorithm, which had a time 0.37 s, as depicted in the zoomed part of Figure 10b.

The tip speed ratio was kept at the most optimal value (8.1) with all competing algorithms, as described in Figure 10c. Nevertheless, the RVS-PO effectively preserved the operation with an optimal TSR, and followed it with a lower settling time and without any overshooting as compared to the other algorithms during the fast wind change. At 15 s, the overshoot values of SS-P&O, LS-P&O, and VS-P&O were 8.939, 8.697, and 8.104, respectively. However, the RVS-P&O technique provided a better rapidity performance of 8.101, as depicted in the zoomed section of Figure 10c. Meanwhile, the rotor speed settling time was about 9.7 ms when using the RVS-P&O algorithm, as compared to the SS-P&O, LS-P&O and VS-P&O algorithms, which had times of 689.2, 330.4, and 310.2 ms, respectively, as it can be seen in the zoomed section of Figure 10d. Furthermore, it was clear that the RVS-P&O and VS-P&O algorithms had no remarkable overshoot on the tracking of the rotor speed compared to SS-P&O and LS-P&O.

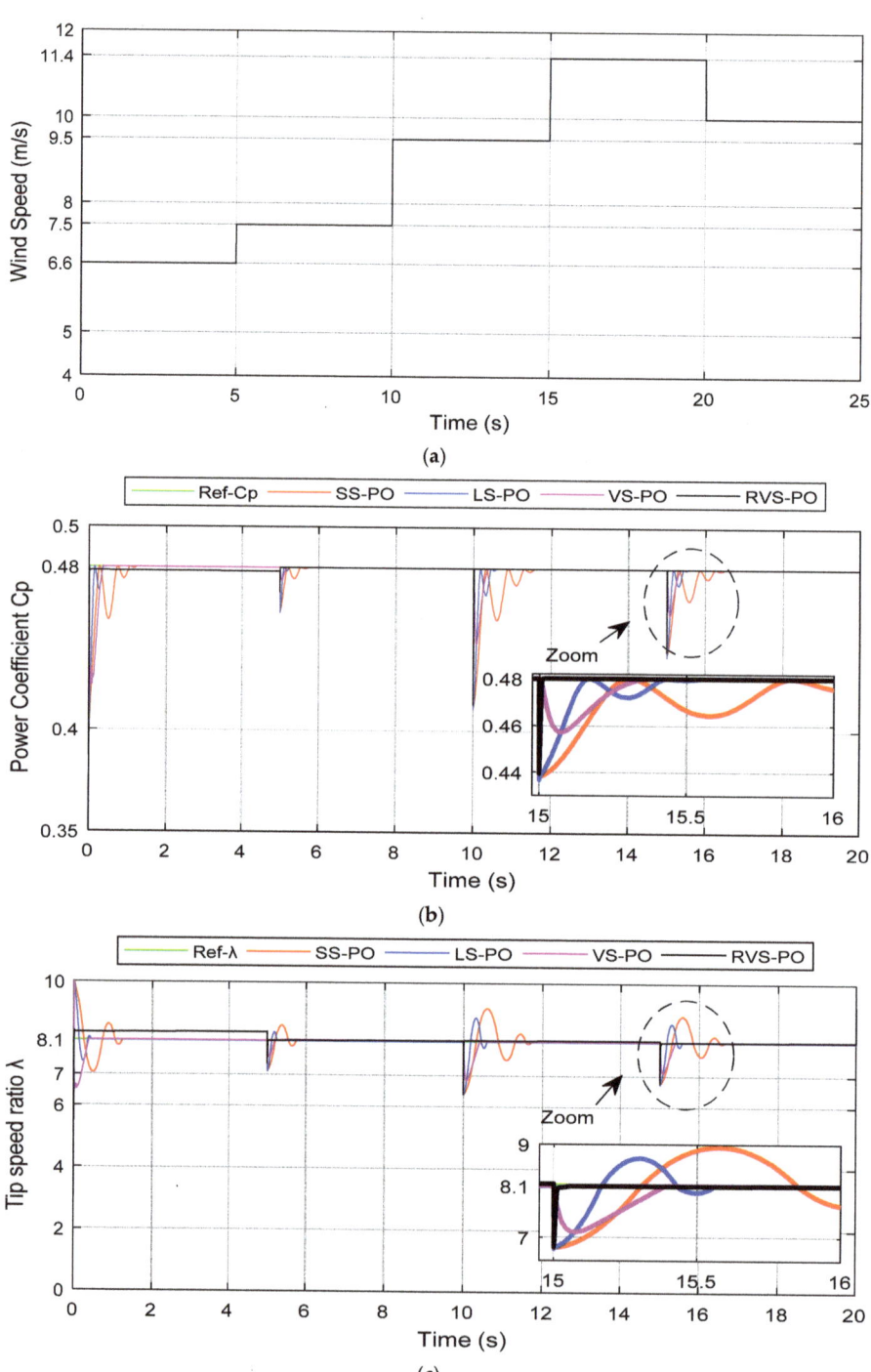

Figure 10. Cont.

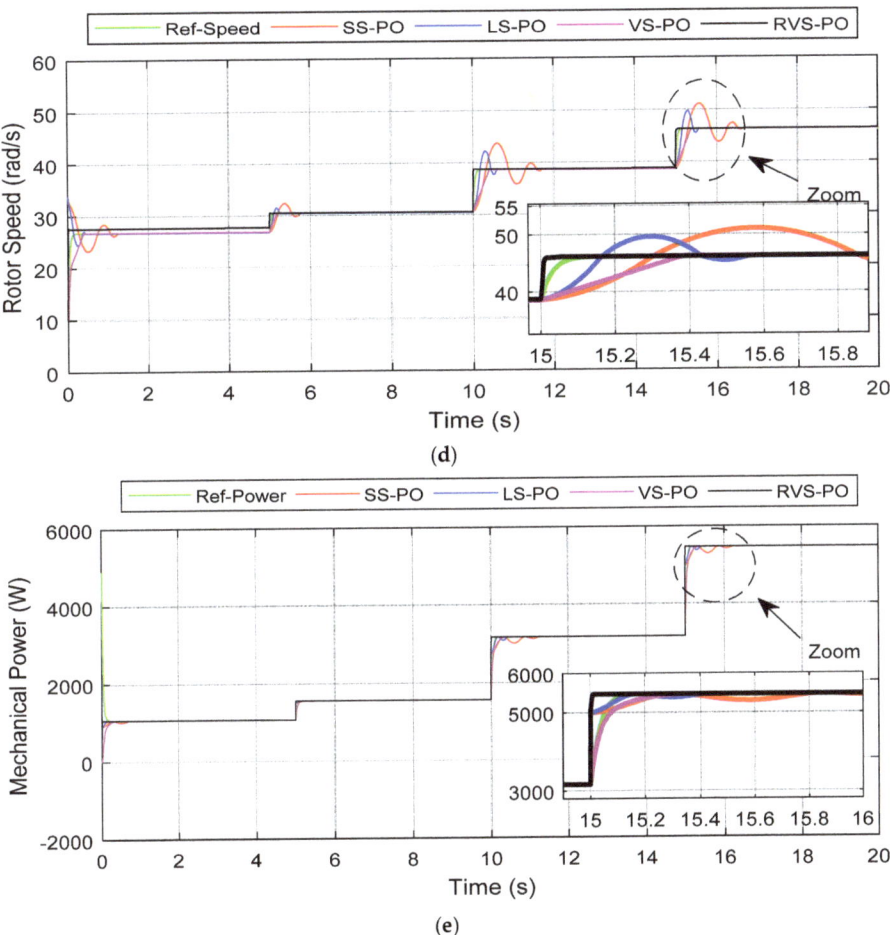

Figure 10. Machine-side results under gradual variations in wind speed. (**a**) Wind-speed profile; (**b**) Power coefficient; (**c**) Tip speed ratio; (**d**) Rotor speed; (**e**) Mechanical power.

Figure 10e depicts the mechanical power by means of the competing algorithms in order to verify the optimal power-extraction performances' quality. The power oscillations of both algorithms, VS-P&O and LS-P&O, at steady state were lower around the extracted MPP. Meanwhile, the proposed RVS-P&O did not show any power oscillations for rapid variations in the wind speed. Simultaneously, the RVS-P&O algorithm took less time than the SS-P&O, LS-P&O, and VS-P&O algorithms to reach the new MPP under rapid fluctuations in the wind speed. For instance, during an abrupt variation of 9.5 m/s to 11.4 m/s at 15 s, the RVS-P&O algorithm required only 0.1 s, which was better than the time needed by the other algorithms (SS-P&O = 0.7 s, LS-P&O = 0.15 s, and VS-P&O = 0.3 s), as depicted in the zoomed part of Figure 10e. Therefore, RVS-P&O showed the best the power-extraction performances, as illustrated in Table 3.

5.2. Variable Fluctuations in Wind Speed

Figure 11 demonstrates the machine-side results of the algorithms in competition under variable fluctuations in wind speed. In order to check the performance of the suggested RVS-P&O algorithm under variable environmental conditions, the system was simulated using a wind speed with an average value of 9 m/s, as shown in Figure 11a. The proposed RVS-P&O algorithm reached the optimal power coefficient (C_p = 0.48) more rapidly than the SS-P&O, LS-P&O, and VS-P&O algorithms, as depicted in the zoomed part of Figure 11b. It can be observed that the SS-P&O, LS-P&O, and VS-P&O algorithms were not able to efficiently track the MPP during these rapid operating conditions. Furthermore, they took more time to track the MPP due to the perturbation misdirection problem. In contrast, the proposed RVS-P&O sustained the optimal C_p efficiently, with a mean value of 0.4770 during the 10 s wind speed variation. It can be mentioned that the mean C_p values shown by the SS-P&O, LS-P&O, and VS-P&O algorithms were 0.4710, 0.4690, and 0.4616, respectively. The four algorithms preserved the optimal value of the TSR, as depicted in Figure 11c. However, RVS-P&O did not show any overshoot compared to the others, which presented relatively considerable ones. The rotor-speed tracking results are shown in Figure 11d. It is remarkable that the RVS-P&O was able to quickly regulate the generator speed under the rapid variation conditions, with very small ripples compared to SS-P&O, LS-P&O, and VS-P&O. Regarding the convergence aspect, the proposed algorithm quickly tracked the reference, with a lower speed error of 8.9757×10^{-5} rad/s compared to the competing algorithms, as displayed in Figure 11e and Table 3. An efficient speed tracking significantly increased the power-extraction quality, as the extracted power during 10 s in the same conditions was estimated at 2.0097×10^3 W for RVS-P&O, while it was 1.9935×10^3, 1.9679×10^3, and 1.8868×10^3 for SS-P&O, LS-P&O, and VS-P&O, respectively.

Figure 11f,g shows the RVS-P&O algorithm efficiency quality when tracking the MPP with small oscillations during random fluctuations in the wind speed. In addition, the waveforms of the mechanical power when using LS-P&O, SS-P&O, and VS-P&O showed some oscillations that affected the energy quality. This can be explained by their inability to track the MPP. The operating step sizes of the proposed RVS-P&O algorithm are depicted in Figure 11h.

5.3. Optimal Rotational Speed

The organic ranking cycles also denominated the optimal rotational speed; evolutions by means of the four algorithms under variable fluctuations in wind speed are depicted in Figure 12. The wind energy system operated around the ORC while maintaining the MPP for each variation in the wind speed. Figure 12 illustrates the ORC profiles of the MPPT methods. The results were obtained by applying a mean wind profile of 11.55 m/s. It appears clearly that RVS-P&O was more efficient than the competing algorithms in the ORC smooth tracking, as shown in Figure 12e. The produced energy quality was better in terms of oscillation frequency and power loss, with an overall estimated efficiency of 99.35% by means of the proposed technique.

The dynamical behavior of the RVS-P&O-MPPT applied to the MSC was analyzed, as depicted in Figure 13, in terms of settling time, rise time, and undershoot. Whatever the instantaneous variations in the wind speed, the power extracted was the maximum value. The settling time (s) given by RVS-P&O was 0.00794; meanwhile, it was 0.2224 and 0.0841 for VS-P&O and LS-P&O, respectively. However, it was 0.1617 when using SS-P&O. Furthermore, the rise time was 0.0068 by means of RVS-P&O and VS-P&O, but it was 0.0496 and 0.0905 by using LS-P&O and SS-P&O, respectively. The undershoot (%) was 0.0481 when using RVS-P&O, which was better than for VS-P&O (0.0845), LS-P&O (0.0902), and SS-P&O (0.0902).

5.4. Grid-Side Converter DPC-SOSMC-STA Controller

When guaranteeing to supply the energy demanded, the quality of that energy is determined by the control tactics used in the regulation of parameters associated with the electrical grid. To achieve such an objective, a novel direct power control DPC–SVM that employed a nonlinear control SOSMC was developed. To evaluate the effectiveness of the proposed technique, a comparison with the FOSMC classical controller and the SOSMC was carried out; the results were validated by the harmonic analysis of each controller. The interchange of electric power between the PMSG and the grid is only assured if the DC bus is set to a constant value, regardless of the momentary variation in available power from the wind. The DC-link voltage of 800 V should be maintained around its nominal value by the machine-side converter, as depicted in Figure 14a. The electrical power injected into the grid was controlled by the DPC-SVM and two regulator types, as shown in Figure 14b,c. The required value could be accurately tracked by the FOSMC and SOSMC control units. However, there was a difference in the quality of the active and reactive powers. The simulation results revealed the superiority of the suggested regulator (SOSMC) based on the "super-adaptive convolution" algorithm that ensured high efficiency and a smooth desired slip path without the phenomenon of chatter or oscillations.

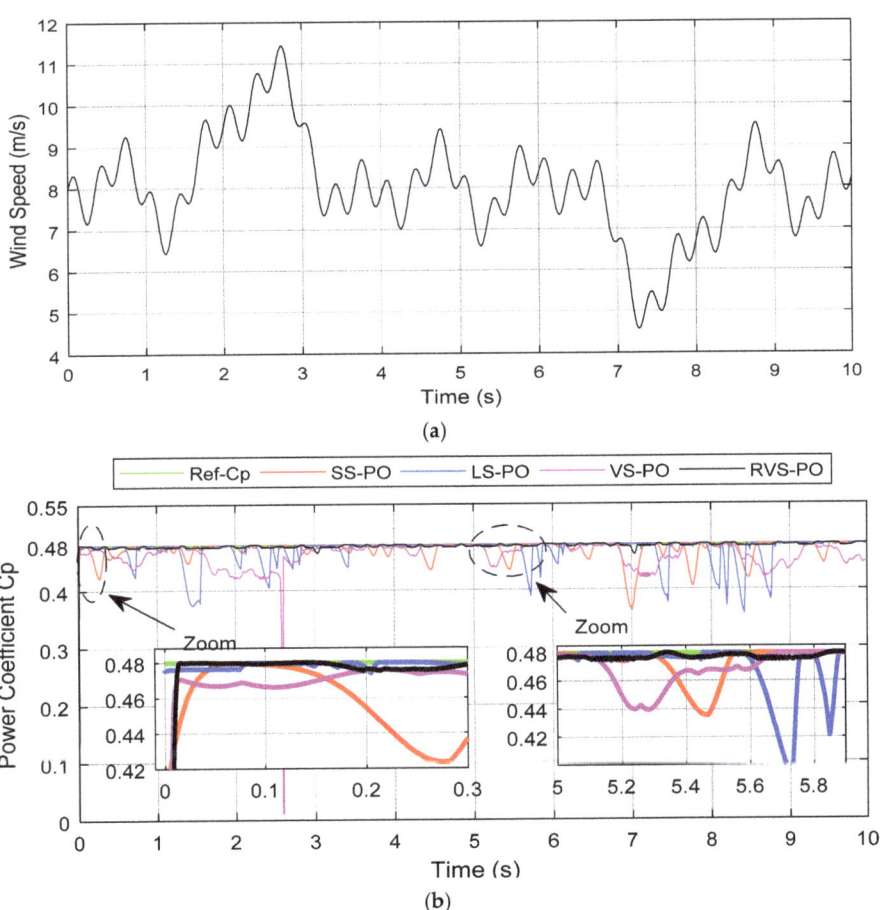

Figure 11. *Cont.*

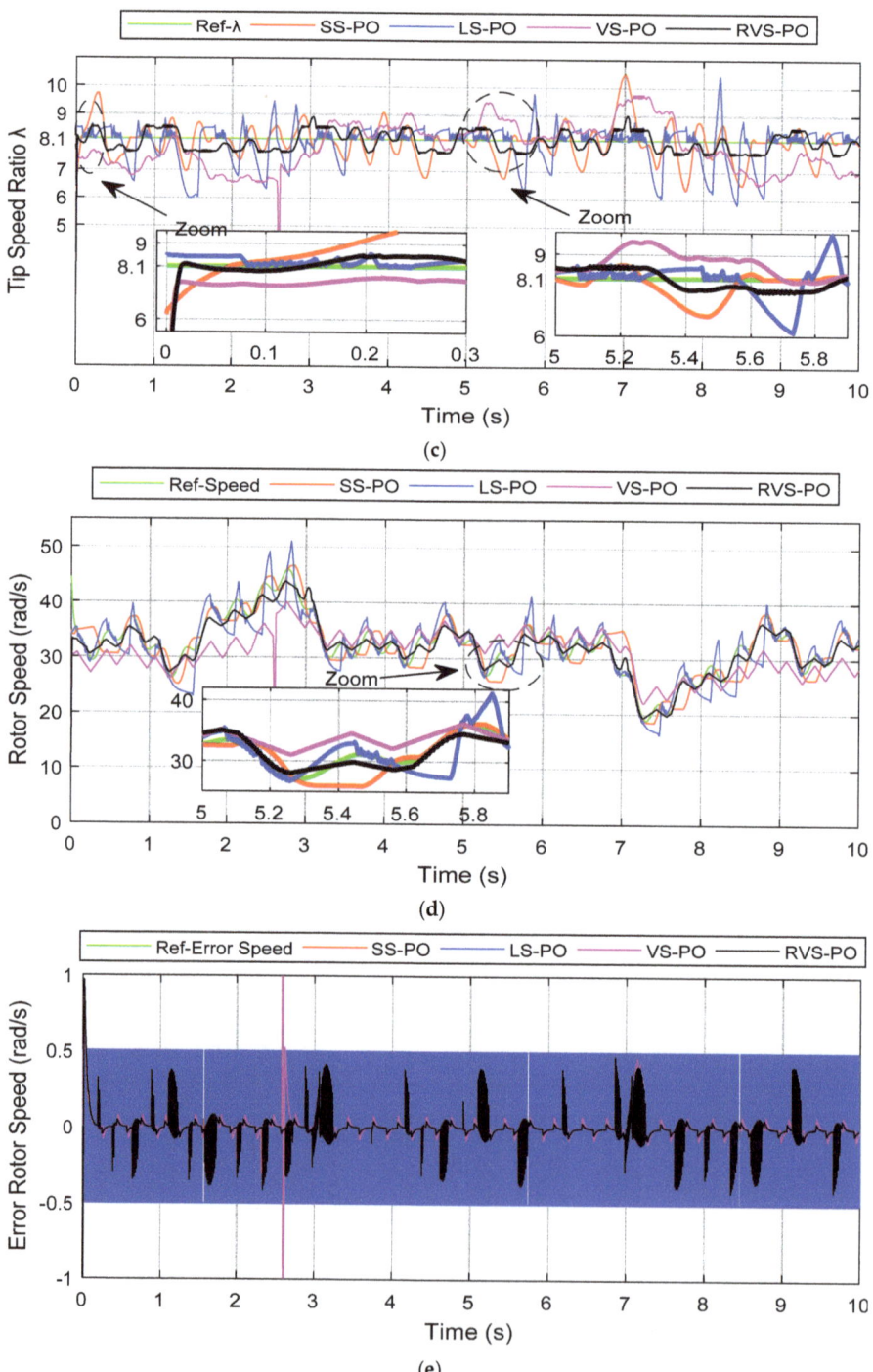

Figure 11. Cont.

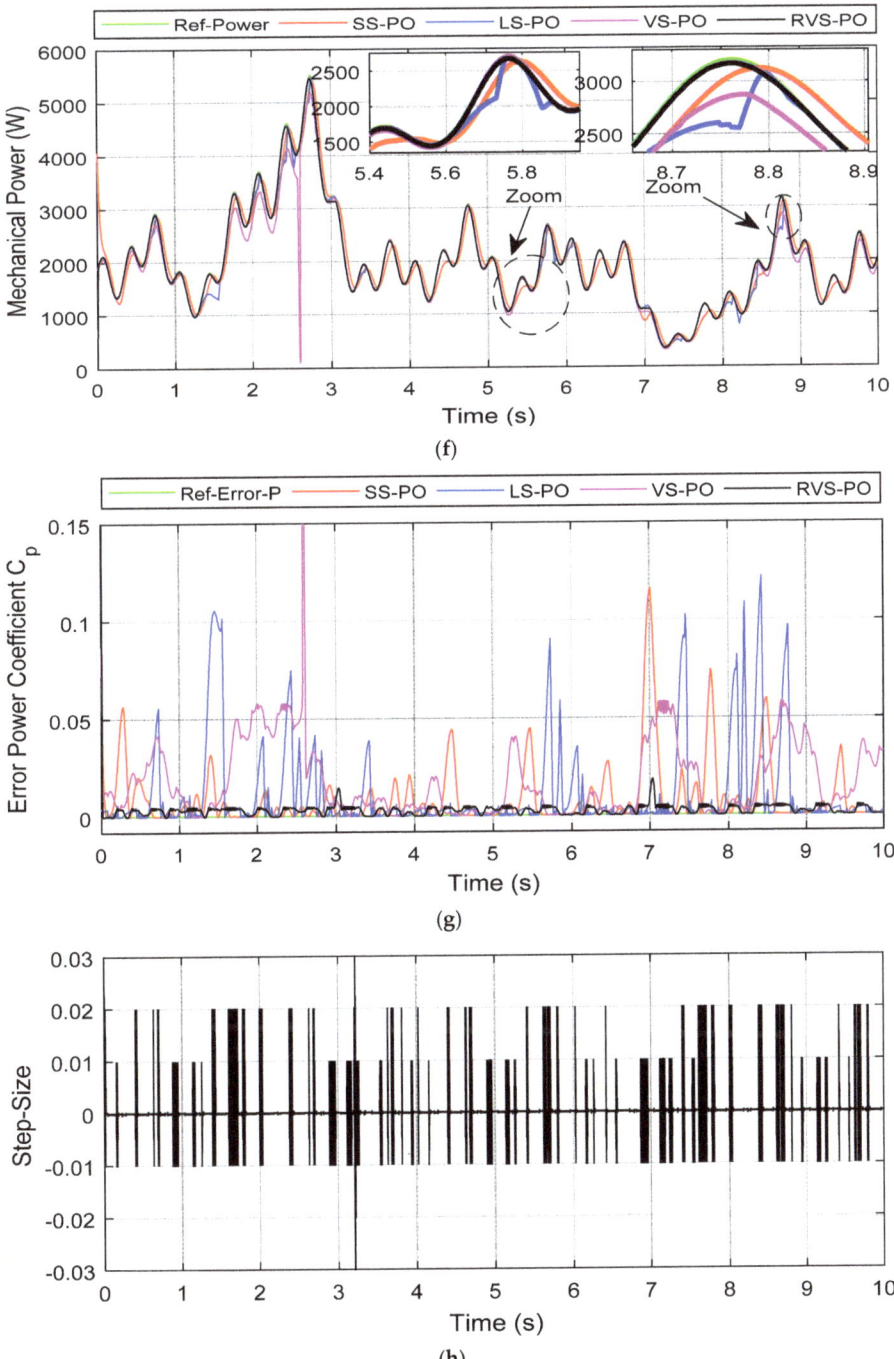

Figure 11. Machine-side results under variable fluctuations in wind speed. (**a**) Wind-speed profile; (**b**) Power coefficient; (**c**) Tip speed ratio; (**d**) Rotor speed; (**e**) Error rotor speed; (**f**) Mechanical power; (**g**) Extracted power error; (**h**) Step size.

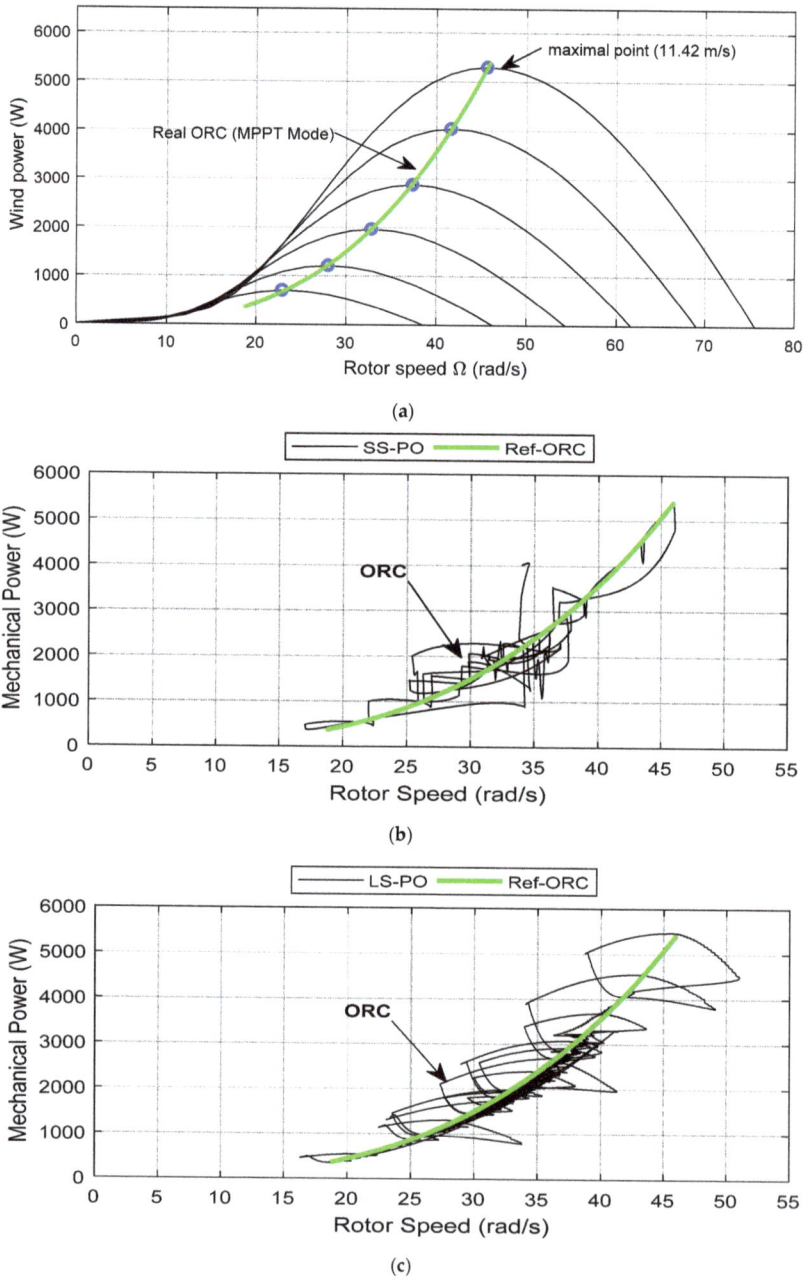

Figure 12. Cont.

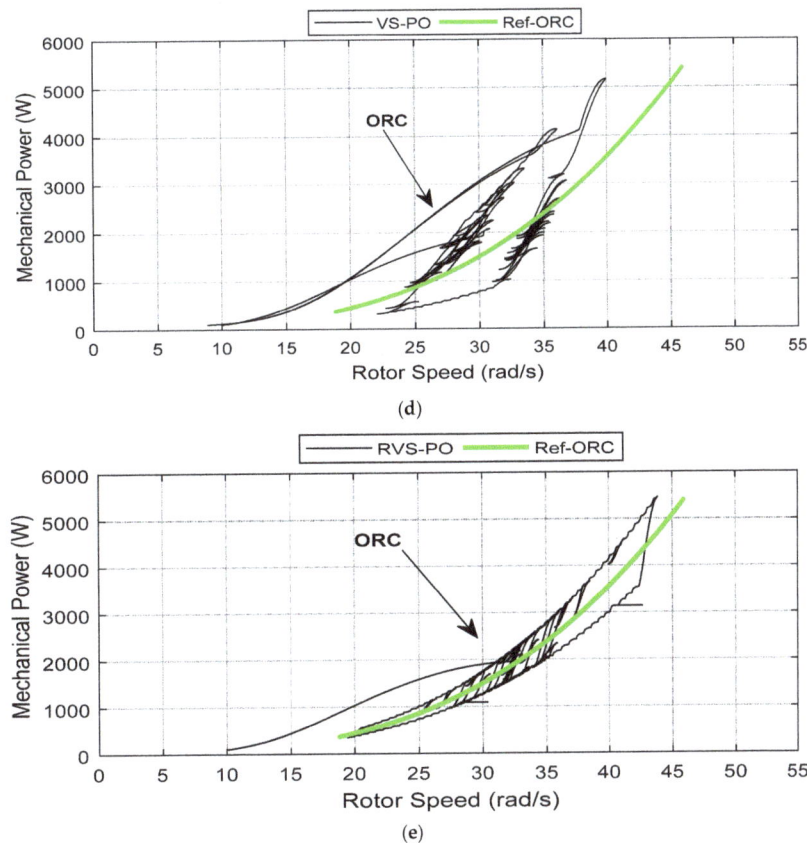

Figure 12. Optimal rotational-speed profile. (**a**) Real tracking of the optimal rotational speed ORC in region "II"; (**b**) ORC for SS-P&O; (**c**) ORC for LS-P&O; (**d**) ORC for VS-P&O; (**e**) ORC for RVS-P&O.

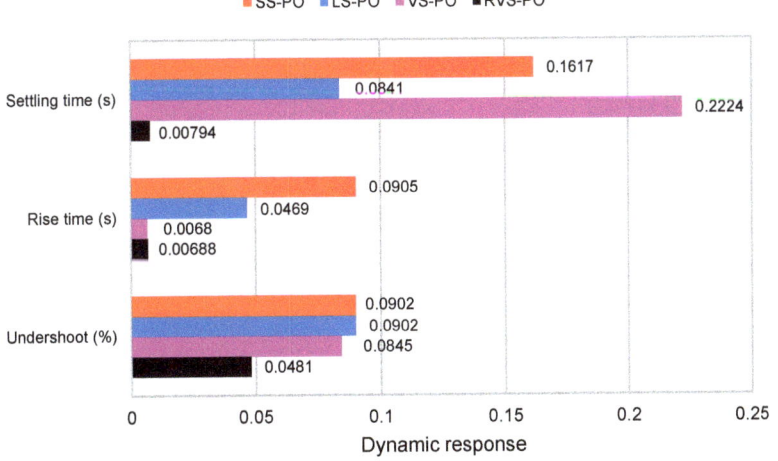

Figure 13. Dynamic response of the competing algorithms (SS-P&O, LS-P&O, VS-P&O, and RVS-P&O).

To illustrate the performance of the proposed control strategy (DPC–SVM) and the effectiveness of the SOSMC used in this work, an evaluation and a comparison with the conventional technique (FOSMC) was conducted. Figure 14d,e represent the grid injected current into phase A for both controllers. Furthermore, the THD of the current (phase A) was higher, at 1.38%. In Figure 14f, a distorted version of a highly unwanted current (phase A) can be seen during simulations in which the use of the FOSMC led to a poor quality of the grid's electrical power. Through the smooth shape of the current, the superiority of the SOMSC was evident, as illustrated in Figure 14e. In addition, the decrease in the best current distortion reached 0.98%, as depicted in Figure 14g. The THD reduction, filtering, and the elimination of odd harmonics all showed considerable improvements [44,45]. Using the illustrated results, we deduced that the SOSMC approach attenuated 30% to 70% of the odd harmonics presented when using FOSMC.

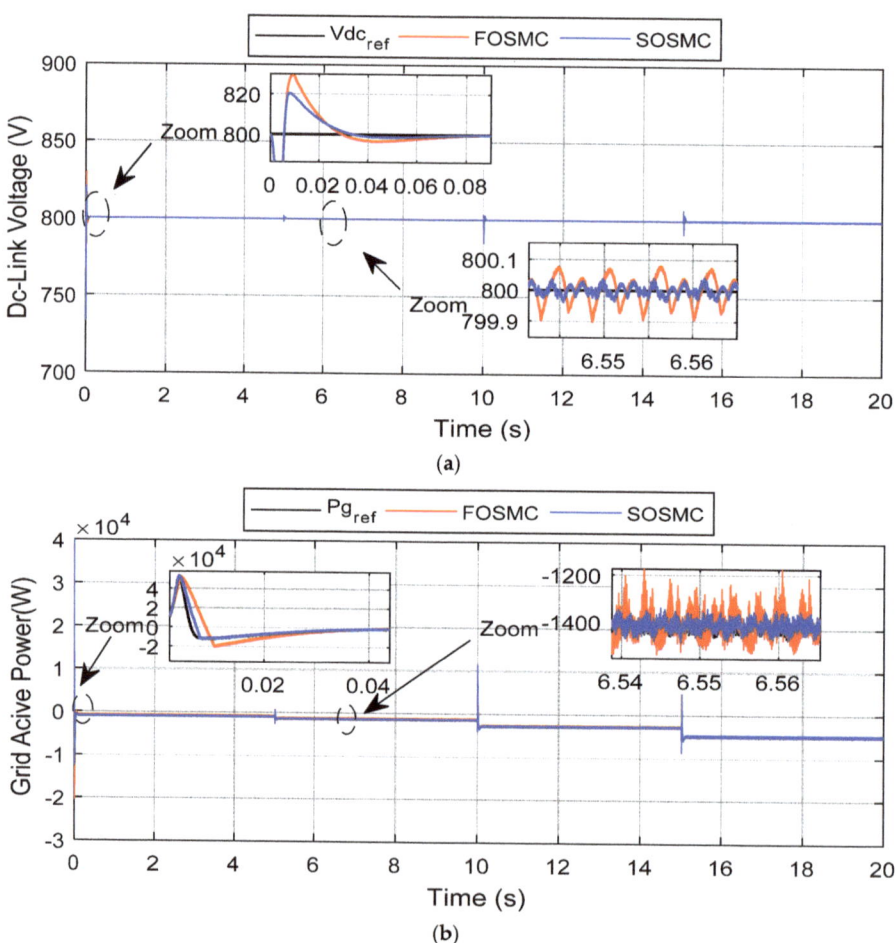

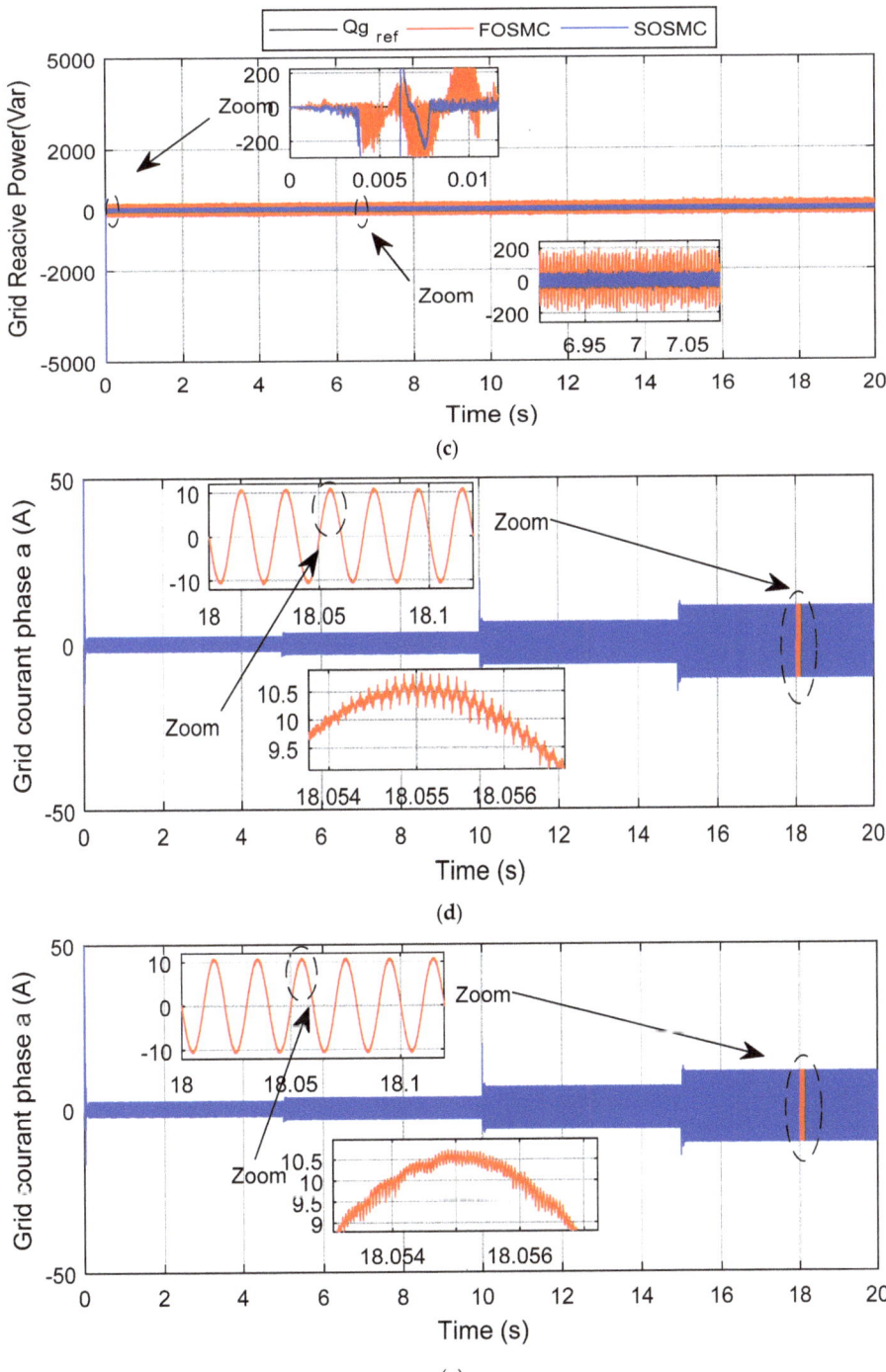

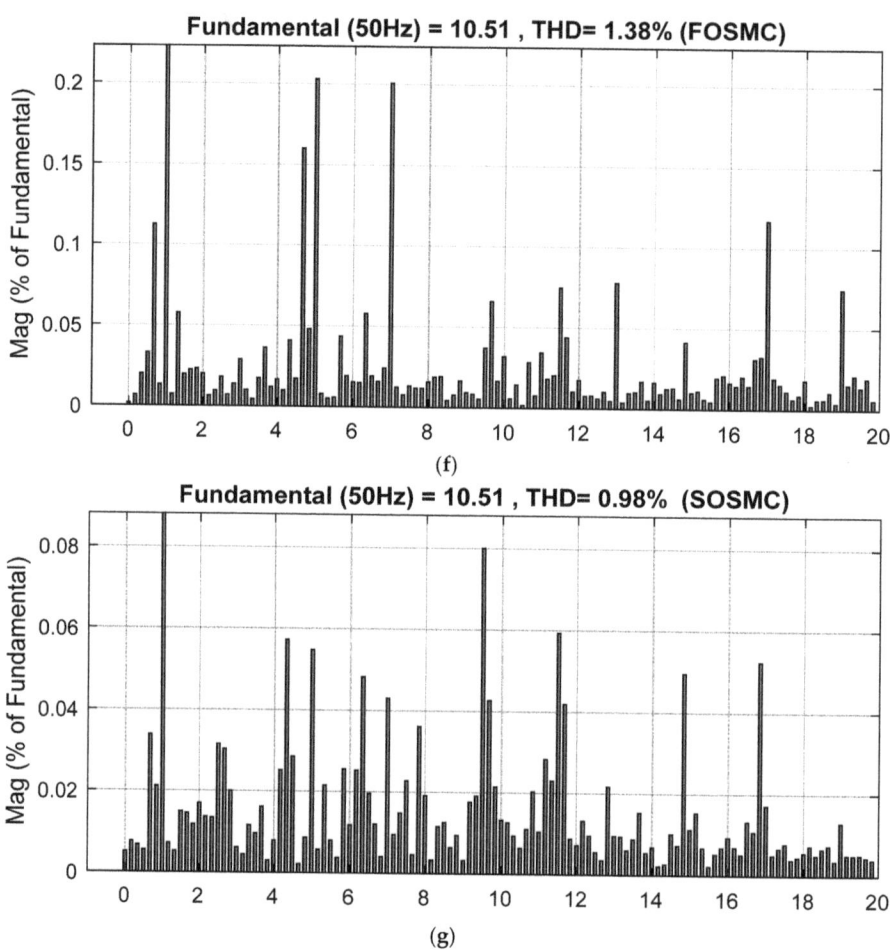

Figure 14. Grid-side results for FOSMC and SOSMC algorithms. (**a**) DC-link voltage; (**b**) Grid active power; (**c**) Grid reactive power; (**d**) Grid current phase "A" for FOSMC; (**e**) Grid current phase "A" for SOSMC; (**f**) THD for FOSMC algorithm; (**g**) THD for SOSMC algorithm.

6. Conclusions

To obtain an optimal and beneficial behavior in a wind turbine installation, an efficient MPPT technique to extract the wind power should be carried out. In this work, to eliminate the drawbacks of the existing conventional MPPT algorithms, particularly FS-P&O and VS-P&O, as they are highly used in current industrial applications, a new Robust variable-step P&O-based MPPT algorithm was proposed and validated under variable operating conditions of wind speed. The proposed RVS-P&O approach was based on the subdivision of the P/w curve into several horizontal modular operating sectors by comparing a newly synthesized ratio with another one related to the required power accuracy. To ensure an initial fast response in the presence of perturbations, the adjustment of arbitrary parameters (β_l) defined the corresponding sector size with a smooth alignment at steady state. In addition, to verify the performances of the proposed RVS-P&O algorithm, it was fairly compared to SS-P&O, LS-P&O, and VS-P&O techniques. The tracking-loss concern and the misdirection of the other techniques were avoided, and the step-size value was accurately estimated in each modular operating sector to reach the appropriate MPP.

In transient conditions, the proposed RVS-P&O algorithm reacted quickly to rapid fluctuations in wind speeds, with an interesting setting time of 7.94 ms and without any overshoot. In terms of steady-state stability, the RVS-P&O was more accurate than the competing algorithms. In both regimes, the proposed RVS-P&O algorithm combined lower oscillations with a power loss of 0.65% at 10 s variation, and a competitive tracking quality under limited speed fluctuations of 8.9757×10^{-5} rad/s. Furthermore, it provided an adeptly better quality of the extracted power during rapid changes in the wind speed, since the overall efficiency was 99.35% which was increased by 0.88%, 1.78%, and 3.82% compared to SS-P&O, LS-P&O, and VS-P&O, respectively. In fact, not only was the loss-of-tracking problem avoided, but the dynamic tracking performances also were improved in either transient or steady-state regimes under several operating conditions. The specified high-order SMC was built to manage the active and reactive powers exchanged between the generator and the grid in the GSC. The grid power values given by the SOSMC method, on the other hand, displayed smooth waveforms with acceptable tracking indices and low THD, as well as unwanted current distortion. In the case of the FOSMC control, the chattering phenomenon was ruled out.

Author Contributions: Conceptualization, I.T., B.M., O.H., A.T.A., A.B. and N.A.K.; Data curation, A.B., A.J.H., I.K.I., Q.Z., G.F. and N.K.B.; Formal analysis, I.T., B.M, O.H., A.T.A., A.B., A.J.H., I.K.I., N.A.K., Q.Z., G.F. and N.K.B.; Investigation, B.M., A.T.A., A.B., A.J.H., I.K.I., N.A.K., Q.Z., G.F. and N.K.B.; Methodology, I.T., B.M, O.H., A.T.A., A.B., A.J.H., I.K.I., N.A.K., Q.Z., G.F. and N.K.B.; Resources, I.T., B.M., O.H., A.T.A., A.B., A.J.H., I.K.I., Q.Z., G.F. and N.A.K.; Software, I.T., O.H., A.B, G.F. and N.K.B.; Supervision, A.T.A.; Validation, I.T., B.M., O.H., A.T.A., A.J.H., I.K.I., N.A.K., Q.Z., G.F. and N.K.B.; Visualization, A.T.A., A.J.H., I.K.I., N.A.K., Q.Z. and N.K.B.; Writing—original draft I.T., B.M., O.H. and A.B.; Writing—review & editing I.T., B.M., O.H., A.T.A. and N.A.K.; Data curation, A.B., A.J.H., I.K.I., Q.Z., G.F. and N.K.B. All authors have read and agreed to the published version of the manuscript.

Funding: This research received no external funding.

Institutional Review Board Statement: Not applicable.

Informed Consent Statement: Not applicable.

Data Availability Statement: Not applicable.

Acknowledgments: The authors would like to acknowledge the support of Prince Sultan University for supporting this publication. Special acknowledgement is given to the Automated Systems & Soft Computing Lab (ASSCL), Prince Sultan University, Riyadh, Saudi Arabia. In addition, the authors wish to acknowledge the editor and anonymous reviewers for their insightful comments, which have improved the quality of this publication.

Conflicts of Interest: The authors declare no conflict of interest.

Nomenclature

Variables

C_P	Coefficient power
F	Simplex
f_g	Grid frequency
I_d	d-axis current
I_{dg}	Grid d-axis current
I_q	q-axis current
I_{qg}	Grid q-axis current
K	First unknown gain
l	Sector index
L_d	d-axis inductance
L_q	q-axis inductance
M	Second unknown gain
m	Complex size

N		Normalization index
P		Number of independent complexes
P_g		Grid active power
P_t		Power of the air mass
P_k		Turbine power
Q_g		Grid reactive power
R_s		Stator resistance
S		Surface
S_P		Sliding surface of the active power
S_Q		Sliding surface of the reactive power
T_e		Electromagnetic torque
α_L		Weighting factor
V		Wind speed
V_d		d-axis voltage
V_{dg}		Grid d-axis voltage
V_{di}		Inverter d-axis voltage
V_q		q-axis voltage
V_{qg}		Grid q-axis voltage
V_{qi}		Inverter q-axis voltage
W		Selection factor

Subscripts and superscripts

d	stator axis
e	Electromagnetic
f	Flux
g	Grid
i	Element (solution)
j	Point index
k	Complex index
max	Maximum
mes	Measure
opti	Optimum
p	Power
q	Stator axis
ref	Reference
s	Stator

Greek letters

α	Number of iteration for each simplex
	Blade pitch angle
λ	Tip speed ratio
ρ	Air density
τ	Number of offspring
ω	Electric pulsation
ψ_f	Magnetic flux

Abbreviations

AC	Alternating current
AI	Artificial intelligence
ANFIS	Adaptive neuro fuzzy inference system
ANN	Artificial neural network
DC	Firect current
DFIG	Doublyfed induction generator
DPC	Direct power control
FLC	Fuzzy logic control
FOSMC	First-order sliding mode controller
FOC	Field-oriented control
FS	Fixed step
GSC	Grid-side converter
INC	Incremental conductance
IPC	Indirect power controller

LS		Large step	
MPPT		Maximum power point tracking	
MSC		Machine-side converter	
ORB		Optimum relation-based	
ORC		Optimal rotational cycle	
OTC		Optimal torque control	
P&O		Perturb and observe	
PMSG		Permanent magnet synchronous generator	
PSO		Particle swarm optimizer	
PSF		Power signal feedback	
RVS		Robust variable ste	
SCIG		Squirrel-cage induction generator	
SS		Small step	
SOSMC		Second-order sliding mode controller	
STA		Super-twisting algorithm	
SVM		Support vector machine	
SVPWM		Space vector pulse-width modulation	
THD		Total harmonic distortion	
VS		Variable step	
VSWT		Variable-speed wind turbine	
WECS		Wind-energy control system	
WSE		Wind speed estimated	
WT		Wind turbine	

Appendix A

Table A1. PMSG setting parameters.

Rated power	$P_e = 10$ kw	Permanent magnet flux	$\psi_m = 0.071$ wb
Stator resistance	$R_s = 0.00829\ \Omega$	Number of pole pairs	$n_p = 6$
Stator direct inductance	$L_d = 0.174$ mH	Inertia	$J_t = 0.089$ kg·m^2
Stator quadrature inductance	$L_q = 0.174$ mH	Friction	$f = 0.005$ N·m

Table A2. WT setting parameters.

Radius of the turbine	$R_t = 2$ m	Optimal tip speed ratio	$\lambda_{opti} = 8.1$
Air density	$\rho = 1.225$ kg·m^3	power Coefficient	$C_p = 0.48$
Pitch angle	$\beta = 0°$		

Table A3. DC bus and grid setting parameters.

Grid resistance	$R_g = 0.02\ \Omega$	Grid quadrature inductance	$L_{qg} = 0.005$ H
Grid direct inductance	$L_{dg} = 0.005$ H	DC-Link-Voltage	$V_{dc} = 800$ v

References

1. Dahbi, A.; Reama, A.; Hamouda, M.; Nait-Said, N.; Nait-Said, M.S. Control and study of a real wind turbine. *Comput. Electr. Eng.* **2019**, *80*, 106492. [CrossRef]
2. Mousa, H.H.; Youssef, A.R.; Mohamed, E.E. Variable step size P&O MPPT algorithm for optimal power extraction of multi-phase PMSG based wind generation system. *Int. J. Electr. Power Energy Syst.* **2019**, *108*, 218–231. [CrossRef]
3. Mousa, H.H.; Youssef, A.R.; Mohamed, E.E. Modified P&O MPPT algorithm for optimal power extraction of five-phase PMSG based wind generation system. *SN Appl. Sci.* **2019**, *1*, 838. [CrossRef]
4. Meghni, B.; Saadoun, A.; Dib, D.; Amirat, Y. Effective MPPT technique and robust power control of the PMSG wind turbine. *IEEJ Trans. Electr. Electron. Eng.* **2015**, *10*, 619–627. [CrossRef]
5. Meghni, B.; Ouada, M.; Saad, S. A novel improved variable-step-size P&O MPPT method and effective supervisory controller to extend optimal energy management in hybrid wind turbine. *Electr. Eng.* **2020**, *102*, 763–778. [CrossRef]
6. Youssef, A.R.; Mousa, H.H.; Mohamed, E.E. Development of self-adaptive P&O MPPT algorithm for wind generation systems with concentrated search area. *Renew. Energy* **2020**, *154*, 875–893. [CrossRef]

7. Mohapatra, A.; Nayak, B.; Das, P.; Mohanty, K.B. A review on MPPT techniques of PV system under partial shading condition. *Renew. Sustain. Energy Rev.* **2017**, *80*, 854–867. [CrossRef]
8. Kumar, D.; Chatterjee, K. A review of conventional and advanced MPPT algorithms for wind energy systems. *Renew. Sustain. Energy Rev.* **2016**, *55*, 957–970. [CrossRef]
9. Karami, N.; Moubayed, N.; Outbib, R. General review and classification of different MPPT Techniques. *Renew. Sustain. Energy Rev.* **2017**, *68*, 1–18. [CrossRef]
10. Eltamaly, A.; Farh, H.M. Maximum power extraction from wind energy system based on fuzzy logic control. *Electr. Power Syst. Res.* **2013**, *97*, 144–150. [CrossRef]
11. Tripathi, S.M.; Tiwari, A.N.; Singh, D. Grid-integrated permanent magnet synchronous generator-based wind energy conversion systems: A technology review. *Renew. Sustain. Energy Rev.* **2015**, *51*, 1288–1305. [CrossRef]
12. Castelló, J.; Espí, J.M.; García-Gil, R. Development details and performance assessment of a wind turbine emulator. *Renew. Energy* **2016**, *86*, 848–857. [CrossRef]
13. Ganjefar, S.; Ghassemi, A.A.; Ahmadi, M.M. Improving efficiency of two-type maximum power point tracking methods of tip-speed ratio and optimum torque in wind turbine system using a quantum neural network. *Energy* **2014**, *67*, 444–453. [CrossRef]
14. Mule, S.M.; Sankeshwari, S.S. Sliding mode control based maximum power point tracking of PV system. *IOSR J. Electr. Electron. Eng. Ver. II* **2015**, *10*, 2278-1676.
15. Zhu, Q.; Fusco, G.; Na, J.; Zhang, W.; Azar, A.T. Special Issue Complex Dynamic System Modelling, Identification and Control. *Entropy* **2022**, *24*, 380. [CrossRef]
16. Azar, A.T.; Serrano, F.E. Stabilization of Port Hamiltonian Chaotic Systems with Hidden Attractors by Adaptive Terminal Sliding Mode Control. *Entropy* **2020**, *22*, 122. [CrossRef] [PubMed]
17. Beltran, B.; Ahmed-Ali, T.; Benbouzid, M.E.H. Sliding mode power control of variable-speed wind energy conversion systems. *IEEE Trans. Energy Convers.* **2008**, *23*, 551–558. [CrossRef]
18. Meharrar, A.; Tioursi, M.; Hatti, M.; Stambouli, A.B. A variable speed wind generator maximum power tracking based on adaptive neuro-fuzzy inference system. *Expert Syst. Appl.* **2011**, *38*, 7659–7664. [CrossRef]
19. Liu, L.; Ma, D.; Azar, A.T.; Zhu, Q. Neural Computing Enhanced Parameter Estimation for Multi-Input and Multi-Output Total Non-Linear Dynamic Models. *Entropy* **2020**, *22*, 510. [CrossRef]
20. Yu, K.N.; Liao, C.K. Applying novel fractional order incremental conductance algorithm to design and study the maximum power tracking of small wind power systems. *J. Appl. Res. Technol.* **2015**, *13*, 238–244. [CrossRef]
21. Necaibia, S.; Kelaiaia, M.S.; Labar, H.; Necaibia, A.; Castronuovo, E.D. Enhanced auto-scaling incremental conductance MPPT method, implemented on low-cost microcontroller and SEPIC converter. *Sol. Energy* **2019**, *180*, 152–168. [CrossRef]
22. Belkaid, A.; Colak, I.; Kayisli, K. A novel approach of perturb and observe technique adapted to rapid change of environmental conditions and load. *Electr. Power Compon. Syst.* **2020**, *48*, 375–387. [CrossRef]
23. Mousa, H.H.; Youssef, A.R.; Mohamed, E.E. Adaptive P&O MPPT algorithm based wind generation system using realistic wind fluctuations. *Int. J. Electr. Power Energy Syst.* **2019**, *112*, 294–308. [CrossRef]
24. Mousa, H.H.; Youssef, A.R.; Hamdan, I.; Ahamed, M.; Mohamed, E.E. Performance assessment of robust P&O algorithm using optimal hypothetical position of generator speed. *IEEE Access* **2021**, *9*, 30469–30485. [CrossRef]
25. Morimoto, S.; Nakayama, H.; Sanada, M.; Takeda, Y. Sensorless output maximization control for variable-speed wind generation system using IPMSG. *IEEE Trans. Ind. Appl.* **2005**, *41*, 60–67. [CrossRef]
26. Putri, R.I.; Pujiantara, M.; Priyadi, A.; Ise, T.; Purnomo, M.H. Maximum power extraction improvement using sensorless controller based on adaptive perturb and observe algorithm for PMSG wind turbine application. *IET Electr. Power Appl.* **2018**, *12*, 455–462. [CrossRef]
27. Ahmed, J.; Salam, Z. An enhanced adaptive P&O MPPT for fast and efficient tracking under varying environmental conditions. *IEEE Trans. Sustain. Energy* **2018**, *9*, 1487–1496. [CrossRef]
28. Linus, R.M.; Damodharan, P. Maximum power point tracking method using a modified perturb and observe algorithm for grid connected wind energy conversion systems. *IET Renew. Power Gener.* **2015**, *9*, 682–689. [CrossRef]
29. Youssef, A.R.; Ali, A.I.; Saeed, M.S.; Mohamed, E.E. Advanced multi-sector P&O maximum power point tracking technique for wind energy conversion system. *Int. J. Electr. Power Energy Syst.* **2019**, *107*, 89–97. [CrossRef]
30. Sitharthan, R.; Karthikeyan, M.; Sundar, D.S.; Rajasekaran, S. Adaptive hybrid intelligent MPPT controller to approximate effectual wind speed and optimal rotor speed of variable speed wind turbine. *ISA Trans.* **2020**, *96*, 479–489. [CrossRef]
31. Mousa, H.H.; Youssef, A.R.; Mohamed, E.E. Hybrid and adaptive sectors P&O MPPT algorithm-based wind generation system. *Renew. Energy* **2020**, *145*, 1412–1429. [CrossRef]
32. Go, S.I.; Ahn, S.J.; Choi, J.H.; Jung, W.W.; Yun, S.Y.; Song, I.K. Simulation and analysis of existing MPPT control methods in a PV generation system. *J. Int. Counc. Electr. Eng.* **2011**, *1*, 446–451. [CrossRef]
33. Pathak, D.; Gaur, P. A fractional order fuzzy-proportional-integral-derivative based pitch angle controller for a direct-drive wind energy system. *Comput. Electr. Eng.* **2019**, *78*, 420–436. [CrossRef]
34. Zhou, F.; Liu, J. Pitch controller design of wind turbine based on nonlinear PI/PD control. *Shock. Vib.* **2018**, *2018*, 7859510. [CrossRef]

35. Belmokhtar, K.; Ibrahim, H.; Lamine Doumbia, M.A. *Maximum Power Point Tracking Control Algorithms for a PMSG-based WECS for Isolated Applications: Critical Review*; Aissaoui, A.G., Tahour, A., Eds.; Wind Turbines—Design, Control and Applications; IntechOpen: London, UK, 2016; p. 199.
36. García-Sánchez, T.; Mishra, A.K.; Hurtado-Pérez, E.; Puché-Panadero, R.; Fernández-Guillamón, A. A controller for optimum electrical power extraction from a small grid-interconnected wind turbine. *Energies* **2020**, *13*, 5809. [CrossRef]
37. Meghni, B.; Dib, D.; Azar, A.T.; Saadoun, A. Effective supervisory controller to extend optimal energy management in hybrid wind turbine under energy and reliability constraints. *Int. J. Dyn. Control.* **2018**, *6*, 369–383. [CrossRef]
38. Meghni, B.; Dib, D.; Azar, A.T. A second-order sliding mode and fuzzy logic control to optimal energy management in wind turbine with battery storage. *Neural Comput. Appl.* **2017**, *28*, 1417–1434. [CrossRef]
39. Benamor, A.; Benchouia, M.T.; Srairi, K.; Benbouzid, M.E.H. A novel rooted tree optimization apply in the high order sliding mode control using super-twisting algorithm based on DTC scheme for DFIG. *Int. J. Electr. Power Energy Syst.* **2019**, *108*, 293–302. [CrossRef]
40. Kelkoul, B.; Boumediene, A. Stability analysis and study between classical sliding mode control (SMC) and super twisting algorithm (STA) for doubly fed induction generator (DFIG) under wind turbine. *Energy* **2021**, *214*, 118871. [CrossRef]
41. Benbouzid, M.; Beltran, B.; Amirat, Y.; Yao, G.; Han, J.; Mangel, H. Second-order sliding mode control for DFIG-based wind turbines fault ride-through capability enhancement. *ISA Trans.* **2014**, *53*, 827–833. [CrossRef]
42. Rafiq, M.; Rehman, S.U.; Rehman, F.U.; Butt, Q.R.; Awan, I. A second order sliding mode control design of a switched reluctance motor using super twisting algorithm. *Simul. Model. Pract. Theory* **2012**, *25*, 106–117. [CrossRef]
43. Azar, A.T.; Serrano, F.E.; Zhu, Q.; Bettayeb, M.; Fusco, G.; Na, J.; Zhang, W.; Kamal, N.A. Robust Stabilization and Synchronization of a Novel Chaotic System with Input Saturation Constraints. *Entropy* **2021**, *23*, 1110. [CrossRef] [PubMed]
44. Abdou, E.H.; Youssef, A.R.; Kamel, S.; Aly, M.M. Sensor less proposed multi sector perturb and observe maximum power tracking for 1.5 MW based on DFIG. *J. Control Instrum. Eng.* **2020**, *6*, 1–13. [CrossRef]
45. Alizadeh, M.; Kojori, S.S. Small-signal stability analysis, and predictive control of Z-source matrix converter feeding a PMSG-WECS. *Int. J. Electr. Power Energy Syst.* **2018**, *95*, 601–616. [CrossRef]

Article

An Incremental Broad-Learning-System-Based Approach for Tremor Attenuation for Robot Tele-Operation

Guanyu Lai [1], Weizhen Liu [1], Weijun Yang [2,*], Huihui Zhong [1], Yutao He [1] and Yun Zhang [1]

[1] School of Automation, Guangdong University of Technology, Guangzhou 510006, China; lgy124@foxmail.com (G.L.); lwz0906@foxmail.com (W.L.); 3221001032@gdut.edu.cn (H.Z.); heyutao81@foxmail.com (Y.H.); yun@gdut.edu.cn (Y.Z.)

[2] School of Mechanical and Electrical Engineering, Guangzhou City Polytechnic, Guangzhou 510405, China

* Correspondence: ywj@gcp.edu.cn; Tel.: +86-020-32648840

Abstract: The existence of the physiological tremor of the human hand significantly affects the application of tele-operation systems in performing high-precision tasks, such as tele-surgery, and currently, the process of effectively eliminating the physiological tremor has been an important yet challenging research topic in the tele-operation robot field. Some scholars propose using deep learning algorithms to solve this problem, but a large number of hyperparameters lead to a slow training speed. Later, the support-vector-machine-based methods have been applied to solve the problem, thereby effectively canceling tremors. However, these methods may lose the prediction accuracy, because learning energy cannot be accurately assigned. Therefore, in this paper, we propose a broad-learning-system-based tremor filter, which integrates a series of incremental learning algorithms to achieve fast remodeling and reach the desired performance. Note that the broad-learning-system-based filter has a fast learning rate while ensuring the accuracy due to its simple and novel network structure. Unlike other algorithms, it uses incremental learning algorithms to constantly update network parameters during training, and it stops learning when the error converges to zero. By focusing on the control performance of the slave robot, a sliding mode control approach has been used to improve the performance of closed-loop systems. In simulation experiments, the results demonstrated the feasibility of our proposed method.

Keywords: hand physiological tremors; incremental broad learning system; tele-operation robot system; sliding mode controller

1. Introduction

With the rapid advancements of tele-operation techniques, robots have gradually improved in performance and have been applied for various areas such as medical and space exploration, see [1–3] for examples, where they are used to complete difficult and complicated scenes with greater precision and efficiency. The stability of tele-operating systems is susceptible to various factors, such as human hand tremors and transmission time delays. Hand tremors in tele-operation lead to suboptimal task tracking. The physiological tremors in human hands are natural, and not pathological [4,5]. These tremors exist in every part of the human body with an amplitude range between 50 and 100 μm in each principal axis, and their dominant frequency is usually distributed in the range of 8–12 Hz [6,7]. Note that physiological tremors are intolerant in the tele-operation scene requiring highly precise manual positioning [8–10], since they can make a remote robot generate motion deviations. Hence, it is imperative to compensate for these tremor signals to enhance the effectiveness of tele-robotic operation systems. To eliminate this influence of tremors, various related methods have been proposed; see [11–17] for examples.

Since physiological tremors exhibit a high-frequency characteristic, while human hand motion is low frequency, some scholars have proposed utilizing the linear low-pass filter [11], which can filter out high-frequency signals and retain low-frequency signals. However, digital filters usually require caching and data processing, which can cause time delays and affect the response speed of systems. The literature show the results of the implementation of canceling tremors on tele-operation systems using the low-pass filter, thereby demonstrating that it is fundamental to set the filter frequency threshold, wherein the optimal frequency threshold still loses some information [12]. To address the limitations of digital filters, in [13], T. A. Wei and P. K. Khosla proposed that the Kalman filter (KF) can combine sensor measurements and dynamic system modeling, as well as estimate the state of the system using the Kalman filter principle, to eliminate the tremor. Furthermore, in [14], Y. Wang proposed an innovative band-limited multiple Fourier linear combiner (BMFLC)-based KF approach (BMFLC-KF) to offer the decomposition of band-limited signals in the time frequency, thereby facilitating effective filtering and compensation. In [15], the authors proposed an autoregressive-based KF model (AR-KF) aimed at the real-time estimation of oscillatory patterns by leveraging past output data. In [16], the authors proposed algorithms based on multi-step (MS) prediction to address the phase delays in the sensors and filters, and they accurately eliminated real-time tremors. Since the fusion of various algorithms leads to computational cost increases, a reduced-order Kalman-enhanced-based BMFLC model (RKE-BMFLC) was proposed in [17], which can reduce the computational complexity of the system and improve real-time performance. Despite their promising performance in predicting tremors, the algorithms mentioned above still have certain limitations. First, the AR-KF method utilizes a linear prediction model to represent the tremor signal as a linear Gaussian distribution. To achieve accurate results, the KF approach must consider the specific characteristics of the tremor being analyzed. Second, when applying the BMFLC-KF, one may need to select parameters carefully, and as shown in [14]; the results clearly show that a minor frequency gap can lead to the infeasibility of such an approach for accurate estimation.

To remove the two limitations above, many machine learning-based approaches have been proposed, e.g., the small-scale sample learning method has been employed widely in physiological tremor elimination applications for tele-operation systems. In [18], Luo, J. proposed the support vector machine (SVM) algorithm as the model for tremor cancellation, wherein it demonstrated good generalization ability and excellent computational performance. In [19], Z. Liu made the SVM algorithm more adaptable to remote operating system tasks and proposed an adaptive fuzzy SVM-based algorithm filter, which filters time series signals and is capable of more accurately modeling tremor signals. In addition to the small-scale sample approach, strong deep learning models [20–22] can further learn the characteristics of tremor signals and achieve high-precision tremor elimination. Despite the merits of the machine-learning-based approaches such as those mentioned above, they still have some restrictions: (1) for a small-scale sample learning method, a loss in prediction accuracy may occur, because learning energy cannot be accurately assigned, and it may be insensitive to small amplitude signals, which may result in poor performance; (2) for deep learning models, a large number of hyperparameters is an unavoidable problem, regardless of their ability to process data efficiently.

Motivated by the observation above, in this study, an incremental broad learning system filter (I-BLSF) has been proposed to predict and cancel hand physiological tremors. In summary, the main novelties and contributions of this work are listed as follows:

- Unlike high-complexity deep learning networks, a simple and efficient network, broad learning system (BLS), is applied in tele-operation systems as a tremor filter, which overcomes the shortcomings of traditional deep neural networks by using the pseudo-inverse calculation. Due to the ill-posed problem, we combine the BLS with the ridge regression approach.
- Traditional batch-learning algorithms require a lot of time and computing resources, and they are limited in dealing with mass data. To solve the problem, incremental learning algorithms are introduced to rebuild the network model online, which can improve the model performance.

- A novel sliding mode controller is raised. The previous work [23] combined with the PD controller to achieve tremor canceling, and there was still room for improvement in tracking accuracy and robustness. Thus, in this paper, we apply a superior controller to control the slave robot.

The rest of this paper is structured as follows. First, Section 2 points out the research problem that needs to be addressed. Then, Section 3 describes the control strategies designed for teleoperation. The proposed broad-learning-system-based filter (BLSF) approach is introduced in Section 4. Section 5 shows experiment parameters and Section 6 validates by simulation that our proposed method has the capability for canceling tremors. Finally, Section 7 makes a summary of the paper.

2. Problem Description

The background problem of this study will be described in this section. First, we introduce the tele-operation system, including the master and slave devices. Then, in the master part, the joints of the master device are analyzed. Finally, the workspace relationship between the master and the slave is given.

2.1. Tele-Operated Robot System

In Figure 1, it shows components of a tele-operation robot system, which is composed of the following parts: (1) the master part (involving a haptic device and a sampling device); (2) the bluetooth communication channel; and (3) the slave part (containing a slave robot manipulator).

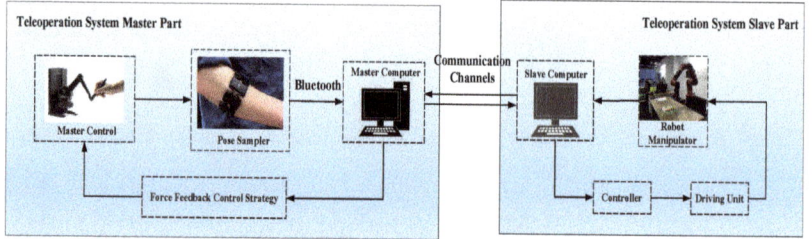

Figure 1. Tele-Operation robot system elements.

- **Haptic device and sampling device:** The haptic device contains a six degrees of freedom (DOFs), where the first three are used to describe the position of the haptic device, and the last three are used to describe the orientation of the haptic device. The sampling device (Myo armband) has eight electromyography (EMG) electrodes and one nine-axis inertial measurement unit (IMU), which can obtain the change in human arm muscle bioelectricity versus time.
- **Communication channels:** Bluetooth technology eliminates the need for wires between master devices and slave devices through wireless connections. Master–slave computers can communicate with each other at a certain distance through a wireless receiver on the chip.
- **Slave robot manipulator:** A multi-DOFs robot manipulator is used as the slave control object, which is equipped with force sensors and electric servers on each joint, where electric servers include the control circuit, direct current (DC) motor, and reduction gear set.

2.2. Master Joints Analysis

Although physiological tremors are normal signals in our daily life, they are a non-negligible issue for meeting about 10 μm range position accuracy [24]. These tremor signals affect each joint of the master device by yielding disturbance signals. In this paper, the modified D-H notation [25] has been adopted to express a haptic device with tremors, and we have the following equation:

$$\begin{cases} \theta_{ori_i} + \theta_{h_i} + \Delta\theta_i = \theta_{new_i}, & i = 1,2,3,4,5,6, \\ d_{ori_i} + d_{h_i} + \Delta d_i = d_{new_i}, & i = 1,2,3,4,5,6, \end{cases} \quad (1)$$

where θ_{ori} and d_{ori} are the original joint information, θ_h and d_h are the desired values from the human hand, and i represents the i-th joint. $\Delta\theta$ and Δd are the disturbed values influenced by tremors, and θ_{new} and d_{new} are actual joint information. And then, the homogeneous transformation matrix can be described in the following form:

$$_{i-1}^{i}\hat{T} = \begin{bmatrix} c\theta_{new_i} & -s\theta_{new_i} & 0 & a_{i-1} \\ s\theta_{new_i}c\alpha_{i-1} & c\theta_{new_i}c\alpha_{i-1} & -s\alpha_{i-1} & -s\alpha_{i-1}d_{new_i} \\ s\theta_{new_i}s\alpha_{i-1} & s\alpha_{i-1}c\theta_{new_i} & c\alpha_{i-1} & c\alpha_{i-1}d_{new_i} \\ 0 & 0 & 0 & 1 \end{bmatrix}, \quad (2)$$

$$s = sin(\cdot), c = cos(\cdot), i = 2, \ldots, 6,$$

where α is the kinematic link twist. To obtain the joint transformation matrix for the end effector, the six matrixes $_1^0\hat{T}, _2^1\hat{T}, _3^2\hat{T}, _4^3\hat{T}, _5^4\hat{T}$, and $_6^5\hat{T}$ can be multiplied in a sequence as follows:

$$_6^0\hat{T} = _1^0\hat{T}\,_2^1\hat{T}\,_3^2\hat{T}\,_4^3\hat{T}\,_5^4\hat{T}\,_6^5\hat{T} = \begin{bmatrix} \hat{n}_{11} & \hat{n}_{12} & \hat{n}_{13} & \hat{p}_x \\ \hat{n}_{21} & \hat{n}_{22} & \hat{n}_{23} & \hat{p}_y \\ \hat{n}_{31} & \hat{n}_{32} & \hat{n}_{33} & \hat{p}_z \\ 0 & 0 & 0 & 1 \end{bmatrix}, \quad (3)$$

where $\hat{n}_{ij}, i = 1,2,3, j = 1,2,3$, and $\hat{p}_x, \hat{p}_y, \hat{p}_z$ are the rotational factors and the position vector, respectively.

2.3. Workspace Description

In a tele-opeation robot system, the description of the coordinate system between the master and the slave is different due to their different physical characteristics, and their workspace relationship is shown as follows:

$$S_s = Z_\delta \times (\theta S_m + b), \quad (4)$$

where S_s defines the coordinates of the slave robot manipulator, and S_m defines the coordinates of the master device. Z_δ is the rotational matrix about the z-axis, and Equation (4) has the following form:

$$\begin{bmatrix} x_s \\ y_s \\ z_s \end{bmatrix} = \begin{bmatrix} \cos\delta & -\sin\delta & 0 \\ \sin\delta & \cos\delta & 0 \\ 0 & 0 & 1 \end{bmatrix} \times \left(\begin{bmatrix} \theta_x & 0 & 0 \\ 0 & \theta_y & 0 \\ 0 & 0 & \theta_z \end{bmatrix} \begin{bmatrix} x_m \\ y_m \\ z_m \end{bmatrix} + \begin{bmatrix} b_x \\ b_y \\ b_z \end{bmatrix} \right), \quad (5)$$

where δ is a rotation angle. θ_x, θ_y, and θ_z are the scale factors in the three-axis direction, and b_x, b_y, and b_z are the translation factors the three-axis direction. The parameters of Equation (5) are provided as below:

$$\begin{cases} \delta = \dfrac{\pi}{4} \\ [\theta_x \ \theta_y \ \theta_z]^T = [0.041 \ 0.040 \ 0.041]^T \\ [b_x \ b_y \ b_z]^T = [0.701 \ 0.210 \ 0.129]^T. \end{cases} \quad (6)$$

The presence of tremors in the homogeneous transformation matrix of a haptic device leads to changes in the master coordinates X_m, which in turn causes differences in the slave coordinates X_s. As the error in the master–slave position increases, the accuracy of the system decreases. To address this issue, a tremor attenuation filter can be designed to reduce the effects of tremors on the performance of tele-operation systems.

3. Control Strategies

In the paper, we integrated the force feedback module, the controller module, and the tremor filter module into the teleoperation system. An initial torque τ_o from a human hand is sent to the haptic device, and the haptic sends an operation trajectory S_{m1} to the filer unit. S_{m2} is a filtered trajectory, S_s is an actual trajectory of the slave manipulator, and S_e is an error trajectory, i.e., $S_e = S_{m2} - S_s$. The error signal is sent to the force feedback module to obtain the master and slave control variables $\dot{q}_{md}$ and $\dot{q}_{sd}$, respectively. In the controller module, the controller exports the master and slave torque, which are τ_m and τ_s, respectively. Figure 2 shows this process in tele-operation robot systems. Here, we provide more details on the control strategies as follows.

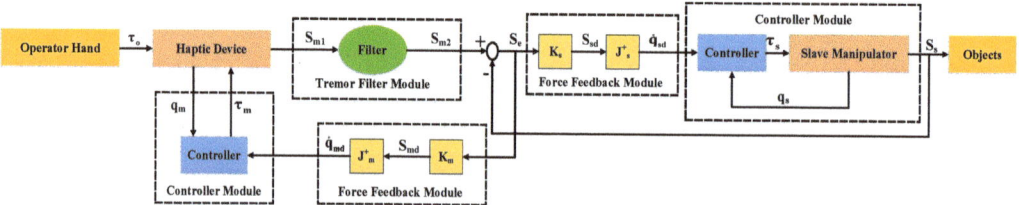

Figure 2. Control mode in teleoperation systems.

3.1. Force Feedback Control

When the end effector of the slave robot arm follows the motion of the master device, the master device can receive feedback information from the force sensors of the slave robot joints. Force feedback can achieve the integration of visual perception and tactile sensation, thus ensuring that the operator can perceive the remote environment and manipulate the robot more naturally [26]. The strength of the feedback force is expressed as [27]:

$$F_f = K_f \sqrt{(x_s - x_m)^2 + (y_s - y_m)^2 + (z_s - z_m)^2}, \tag{7}$$

where x_m, y_m, and z_m and x_s, y_s, and z_s are the coordinate values of the master device and the slave device, respectively. K_f is a feedback force parameter. With the optimization of a received feedback force from the slave part, the desired trajectories of the master and slave devices, S_{md} and S_{sd}, respectively, can be obtained. The pose information can be turned into the joint velocity information by the Jacobian matrixes J_m^+ and J_s^+ as follows:

$$\begin{cases} \dot{q}_{md} = J_m^+(S_m)\dot{S}_{md}, \\ \dot{q}_{sd} = J_s^+(S_s)\dot{S}_{sd}. \end{cases} \tag{8}$$

3.2. Sliding Mode Controller

The sliding mode controller, known for its ability to overcome system uncertainties and achieve robust control characteristics [28], employs different structures on both sides of the sliding surface. This nonlinear controller is particularly effective in dealing with the complexities of uncertain dynamic systems. For a typical second-order nonlinear uncertain dynamic system with a single input, its general state space expression can be described as follows [29]:

$$\dot{x}_1(t) = x_2(t) \tag{9}$$

$$\dot{x}_2(t) = \sum_{i=1}^{n}(a_i + \Delta_i(t))f_i(x_1, x_2, t) + b(x_1, x_2, t)u(t) + d(t) \tag{10}$$

$$x_1(t_0) = x_{1,0}, x_2(t_0) = x_{2,0}, \qquad (11)$$

where $x_1(t)$ and $x_2(t)$ are the state variables, a_i ($i = 1, \ldots, n$) are the constant parameters of the system, and Δ_i and $d(t)$ are the uncertain perturbations and the known disturbance, respectively. $u(t)$ represents the input signals, and $f_i(x_1, x_2, t)$ and $b(x_1, x_2, t)$ are derived by the system characteristics; $x_{1,0}$ and $x_{2,0}$ are initial conditions given at the initial time t_0. The main objective of this controller is to satisfy that $X(t) = [x_1(t), x_2(t)]^T$ can track the desired trajectory $X_d(t) = [x_{d1}(t), x_{d2}(t)]$. Hence, the control law should be designed to make the tracking error asymptotically arrive at zero. Since the above considered system is single input, there exists only one sliding surface $s(x_1, x_2) = 0$ for second order systems, and it is defined as follows:

$$s(x_1, x_2) = err_2(t) + c \times err_1(t), \qquad (12)$$

where c is a strictly positive real number, and the tracking errors $err_1(t)$ and $err_2(t)$ are written as follows:

$$\begin{aligned} err_1(t) &= x_1(t) - x_{d1}(t) \\ err_2(t) &= x_2(t) - x_{d2}(t). \end{aligned} \qquad (13)$$

Assume that $\dot{err}_1 = err_2$, and denote $E(t) = [err_1(t), err_2(t)]$. To obtain a unique solution of a homogeneous differential equation $err(t) = 0$, $s(x_1, x_2)$ is set as zero. Thus, the tracking error will asymptotically reach zero with a proper control law that can keep the trajectory on the sliding surface. The control law is designed as follows:

$$\begin{aligned} \dot{s} &= \dot{err}_2 + c \times err_2 = -b sgn(s), b > 0 \\ \dot{err}_2 &= -c \times err_2 - b sgn(s), b > 0, \end{aligned} \qquad (14)$$

where b is a positive number.

Remark 1. *Traditional controllers often rely on control algorithms such as the PID controller, which are simple and easy to implement but have limited accuracy and anti-interference capabilities. In contrast, the sliding mode control is effective in reducing the effects of uncertainties and external disturbances that are common in practical systems, which is achieved by designing a sliding surface that drives the system towards a stable equilibrium point, regardless of the uncertainties and disturbances. Furthermore, the sliding mode control provides a fast response and high tracking accuracy.*

3.3. Tremor Attenuation Filter

To provide a more detailed explanation of the flow of signals in the tremor filter, we provide the mathematical model of the designed tremor filter in Figure 3. The model illustrates the various flows involved in the filtering process and how they interact with each other.

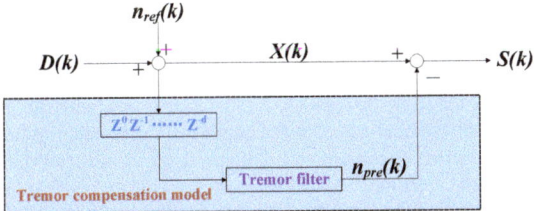

Figure 3. Mathematical expression of the tremor complementation model.

Since the data sampled by the sampling unit are hand trajectories and tremor disturbance signals, we denote the input of the tremor filter mathematical model as actual signals $X(k)$ and the tremor disturbance signals as $n_{ref}(k)$. The actual signals with tremors can be written as follows:

$$X(k) = D(k) + n_{ref}(k), \tag{15}$$

where k is the sampling point, and $D(k)$ is the desired signal without tremors. Through the prediction of the tremor filter, the output of the tremor filter mathematical model is

$$S(k) = X(k) - n_{pre}(k) = D(k) + n_{ref}(k) - n_{pre}(k), \tag{16}$$

where $n_{pre}(k)$ is the prediction signal of the tremor filter.

We denote the error as $\Delta n = n_{ref}(k) - n_{pre}(k)$ and aim for it to equal to zero. Theoretical predictions suggest that the model error ideally should be zero. In practical scenarios, there might exist a small residual deviation, thus resulting in a prediction error that is slightly larger than zero.

4. Design of Broad-Learning-System-Based Tremor Filter

While deep learning algorithms are efficient at processing large amounts of data, they often involve a large number of hyperparameters, which can be problematic. The broad learning system is a novel and efficient network architecture that avoids the complex and redundant structures found in traditional deep learning networks [30,31]. As a result, it provides a more efficient, interpretable, and scalable solution for processing data.

4.1. Broad Learning System

The proposed network architecture was developed by C. L. Philip Chen and is referred to as the broad learning system, which is depicted in Figure 4. This novel network architecture differs from deep learning neural networks, as it does not require backpropagation to update weights. The speed of the broad learning system is attributed to the fact that weights can be obtained via pseudo-inverse formulas. Moreover, the network weights are continuously updated as the system is trained with data, as the system employs an incremental learning algorithm to adjust nodes without reinputting previous data.

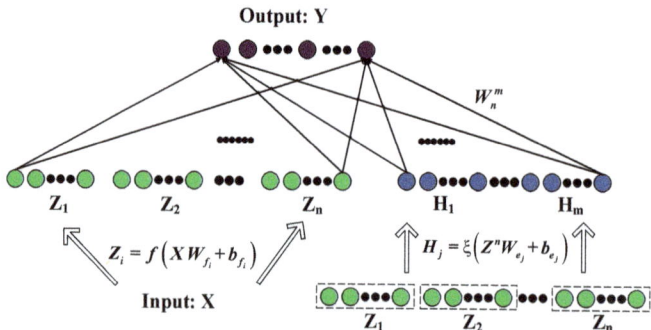

Figure 4. Broad learning system network model architecture.

In Figure 4, we denote $X \in \mathbb{R}^{M \times N}$ as the input into the BLS, and we denote $Y \in \mathbb{R}^{M \times C}$ as the output, where M, N, and $\mathbb{C}$ represent the number of samples, the number of features, and the number of output nodes, respectively. The input data X is randomly mapped to n sets of feature window nodes, thereby generating the feature layer of the network, which can be expressed in the following form:

$$Z_i = \phi\left(XW_{f_i} + \beta_{f_i}\right), \quad i = 1, \ldots, n, \tag{17}$$

where the variables W_{f_i} and β_{f_i} correspond to randomly generated weights and biases, respectively. $\phi(\cdot)$ refers to a random mapping function. Each mapping group contains k feature nodes. All the feature nodes can be represented as $Z^n \equiv [Z_1, Z_2, \ldots, Z_n]$, and we denote j-th group enhancement nodes as the following:

$$H_j = \xi\left([Z_1, Z_2, \ldots, Z_n] W_{hj} + \beta_{hj}\right) = \xi\left(Z^n W_{e_j} + \beta_{e_j}\right), \quad j = 1, \ldots, m, \tag{18}$$

where W_{hj} and β_{hj} are random weight coefficients, and the function $\xi(\cdot)$ is a nonlinear activation function, $f = tanh(\cdot)$. We denote all nodes as $L_n^m \equiv [Z_1, Z_2, \ldots, Z_n | H_1, H_2, \ldots, H_m]$. The output of the broad learning model is represented as follows:

$$\begin{aligned} Y &= [Z_1, \ldots, Z_n | \xi(Z^n W_{h1} + \beta_{h1}), \ldots, \xi(Z^n W_{hm} + \beta_{hm})] W_n^m \\ &= [Z_1, \ldots, Z_n | H_1, \ldots, H_m] W_n^m = L_n^m W_n^m. \end{aligned} \tag{19}$$

Here, W_n^m represents the connection weight of the network with n feature windows and m groups of enhancement nodes. For all nodes L_n^m, the pseudoinverse is equal to the following:

$$(L_n^m)^+ = [(L_n^m)^T L_n^m]^{-1} (L_n^m)^T, \tag{20}$$

and the weights can be represented as follows:

$$W_n^m = (L_n^m)^+ Y = [(L_n^m)^T L_n^m]^{-1} (L_n^m)^T Y. \tag{21}$$

Due to the ill-posed nature of the problem, where no stable or unique solution to the inverse matrix exists, the ridge regression algorithm is employed to obtain the connection weights of the structure. As a result, Equation (20) can be rewritten as follows:

$$(L_n^m)^+ = \lim_{\lambda \to 0} \left([\lambda E + (L_n^m)^T L_n^m]^{-1} (L_n^m)^T \right), \tag{22}$$

and we have

$$W_n^m = (L_n^m)^+ Y = \lim_{\lambda \to 0} \left([\lambda E + (L_n^m)^T L_n^m]^{-1} (L_n^m)^T \right) Y. \tag{23}$$

Remark 2. *The broad learning model (BLM) is a computational framework that offers a fast and efficient solution for various supervised and unsupervised machine learning tasks. The BLM has been developed to overcome the limitations of traditional deep learning architectures, which typically require a large number of layers and a large amount of computational resources to achieve high predictive performance. The singular value decomposition technique is used to simplify the complexity of the model, and incremental learning modes can be integrated to form the broad learning system.*

4.2. Incremental Learning Methods

In the broad learning system, to improve the system performance, an incremental learning approach is integrated. This incremental learning method has three updating forms, which contain the increment of the feature nodes, the increment of the enhancement nodes, and the increment of the input data. Since the input data is enough for our experiment, in this paper, the first two methods are considered. The details are given as follows.

4.2.1. Increment of Additional Enhancement Nodes

Denote $L_n^m = [Z_n | H_m]$ and denote the group of additional enhancement nodes as $H^* = \xi(Z^n W_{h(m+1)} + \beta_{h(m+1)})$. Hence, the new input matrix is written as follows:

$$L^{m+1} \equiv [L_n^m | \xi(Z^n W_{h(m+1)} + \beta_{h(m+1)})], \tag{24}$$

where $W_{h(m+1)}$ and $\beta_{h(m+1)}$ are random weights and random biases from n groups of features mapping to p additional enhancement nodes, respectively. The pseudo-inverse of the new matrix can be written as follows:

$$(L^{m+1})^{+} = \begin{bmatrix} (L_n^m)^{+} - DB^T \\ B^T \end{bmatrix}, \tag{25}$$

where $D = (L_n^m)^{+} \xi(Z^n W_{h(m+1)} + \beta_{h(m+1)})$,

$$B^T = \begin{cases} (C)^{+} & if\ C \neq 0 \\ (1 + D^T D)^{-1} D^T (A_n^m)^{+} & if\ C = 0, \end{cases} \tag{26}$$

and $C = \xi(Z^n W_{h(m+1)} + \beta_{h(m+1)}) - A_n^m D$; the new weights are denoted as the following:

$$W^{m+1} = \begin{bmatrix} W^m - DB^T Y \\ B^T Y \end{bmatrix}. \tag{27}$$

Remark 3. *When the trained network fails to achieve the desired accuracy, additional enhancement nodes can be added to improve the accuracy. By adding extra enhancement nodes into the network, the nonlinear capability can be enhanced. As shown in the equations above, the algorithm only requires the calculation of the pseudo-inverse of the new nodes rather than the entire matrix, thereby enabling the network to be rapidly restructured.*

4.2.2. Increment of Additional Feature Mapping Nodes

We point out that the dynamic increment of the enhancement nodes method cannot improve the current network performance, as it may fall into a locally optimal solution. The increment of additional feature mapping nodes is an effective learning method for neural networks, which only needs to calculate the pseudo-inverse of the new nodes and does not need to retrain the whole network. This method provides the benefits of saving time for improving the feature extraction capability.

Assume that the initial nodes are constructed by n groups of feature mapping nodes and m groups of enhancement nodes, and denote the additional (n+1)-th group feature mapping nodes as $Z_{n+1} = \phi(XW_{e(n+1)} + \beta_{e(n+1)})$, where $W_{e(n+1)}$ and $\beta_{e(n+1)}$ are randomly generated. The corresponding enhancement nodes generated by the additional (n+1)-th group feature mapping nodes are defined as follows:

$$H_j^* = \left[\xi(Z_{n+1} W_{e1}^* + \beta_{e1}^*), \xi(Z_{n+1} W_{e2}^* + \beta_{e2}^*), \ldots, \xi(Z_{n+1} W_{ej}^* + \beta_{ej}^*) \right], \quad j = 1, \ldots, m, \tag{28}$$

where W_{ei}^* and β_{ei}^* are random parameters. Here, we denote $L_{n+1}^m = [L_n^m \mid Z_{n+1} \mid H_j^*]$, and its pseudo-inverse matrix is defined as follows:

$$(L_{n+1}^m)^{+} = \begin{bmatrix} (L_n^m)^{+} - db^T \\ b^T \end{bmatrix}, \tag{29}$$

where $d = (L_{n+1}^m)^{+} [Z_{n+1} \mid H_j^*]$,

$$b^T = \begin{cases} (c)^{+} & if\ c \neq 0 \\ (1 + d^T d)^{-1} d^T (L_n^m)^{+} & if\ c = 0, \end{cases} \tag{30}$$

and $c = [Z_{n+1} \mid H_j^*] - L_n^m d$; the new weights are denoted as follows:

$$W_{n+1}^m = \begin{bmatrix} W_n^m - db^T Y \\ b^T Y \end{bmatrix}. \tag{31}$$

4.3. Sparse Autoencoder

Obtaining a good feature representation of input data is a critical step in machine learning. Traditionally, complex mathematical derivations have been used to derive features, or a set of features has been generated through random initialization. However, random features suffer from unpredictability and uncertainty, which may lead to incomplete feature extraction. As the dimensionality or size of the input data increases, it becomes necessary to remove redundant features.

To address these issues, the sparse autoencoder (SAE) model has been proposed, which fine-tunes random features into a set of sparse and compact features [32–34]. The SAE model structure is illustrated in Figure 5, and then the details of the sparse feature learning algorithm are described below.

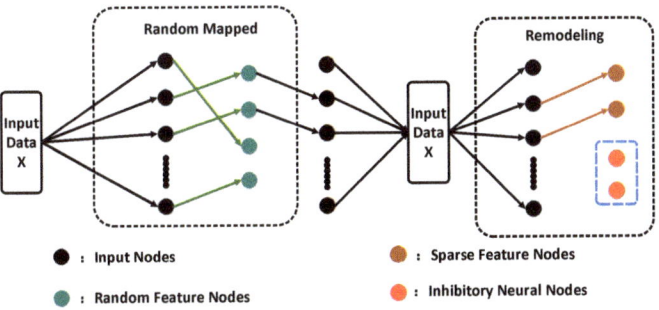

Figure 5. Sparse autoencoder structure diagram.

The extraction of sparse features is considered to be an optimization problem that requires addressing. Lasso regression (l_1 regularization) represents a convex optimization problem, as stated in [35]. To obtain the solution W^* for the sparse autoencoder, the following optimization problem can be used to obtain it:

$$f(W^*) = \arg\min_{W^*} \|ZW^* - X\|_2^2 + C\|W^*\|_1, \tag{32}$$

where C is the regularization parameter, and Z is the output of the linear random mapping equation, as shown in Equation (17). It is well-known that l_1 regularization is often used to solve linear inverse problems. A common approach is the alternating direction method of multipliers (ADMM), which is used to obtain the solution by minimizing one function at a time. To apply the ADMM algorithm, we first reformulate Equation (32) as follows:

$$f(W^*) = f(w,v) = \arg\min_{w,v} h(w) + g(v) \tag{33}$$
$$s.t\ w - v = 0,$$

where $h(w) = \|Zw - X\|_2^2$, $g(v) = c\|v\|_1$. In the augmented Lagrangian with a penalty form, we have the following:

$$\arg\min_{w,v} h(w) + g(v) + \lambda(w-v) + \frac{\rho}{2}\|w-v\|_2^2 \tag{34}$$
$$s.t\ w - v = 0,$$

and then the solution of the original problem can be obtained as follows:

$$w_{k+1} = \left(Z^T Z + \frac{\rho}{2}I\right)^{-1}\left[Z^T X + \frac{\rho}{2}(v_k - u_k)\right],$$

$$v_{k+1} = S_k(w_{k+1} + u_k), \quad k = \frac{c}{\rho}, \tag{35}$$

$$u_{k+1} = u_k + (w_{k+1} - v_{k+1}), \quad u_k = \frac{\lambda^T}{\rho},$$

where ρ is a positive penalty factor. $S(\cdot)$ is the soft thresholding operator, and it is defined as follows:

$$S_k(w_{k+1} + u_k) = \begin{cases} w_{k+1} + u_k - k & if(w_{k+1} + u_k > k) \\ 0 & if(|w_{k+1} + u_k| \leq k) \\ w_{k+1} + u_k + k & if(w_{k+1} + u_k < -k) \end{cases}. \tag{36}$$

4.4. Physical Model Structure of BLSF

A novel BLSF was proposed to address the effects of physiological tremors, and it consists of three main components: the sampling unit, the tremor filter unit, and the control unit. In the sampling unit, an internal measurement unit (IMU) captures real-time hand movements by measuring the three-axis position acceleration $\ddot{x}$, $\ddot{y}$, and $\ddot{z}$ and the three-axis joint angular velocity $\dot{\theta}_x$, $\dot{\theta}_y$, and $\dot{\theta}_z$. The tremor filtering unit utilizes the BLS network algorithm to forecast and compensate for tremor signals, thereby effectively neutralizing them in the actual signals. The control unit incorporates inverse kinematics calculations, single joint drivers, and motion feedback from deflection sensors to convert inverse kinematics into motion control variables for the robot manipulator.

By incorporating the BLS network algorithm, the proposed BLSF effectively forecasts tremor signals, as depicted in Figure 6. The three-axis compensation signals, namely, x_p, y_p, and z_p, and θ_{xp}, θ_{yp}, and θ_{zp}, exhibit equal magnitudes but opposite phases in comparison to the tremor signals. This unique characteristic allows them to effectively neutralize the tremor signals present in the actual signals x, y, and z.

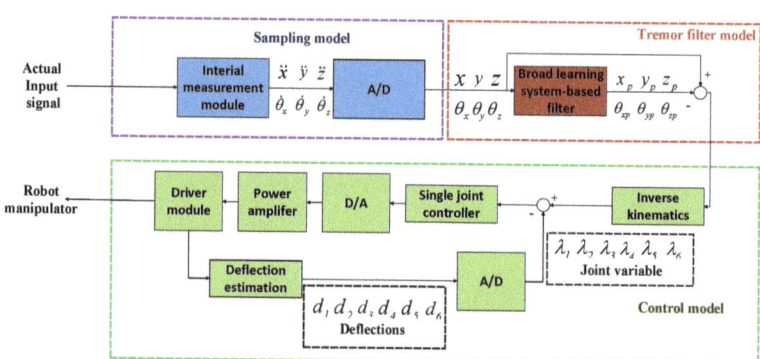

Figure 6. Block diagram of the broad-learning-system-based tremor filter.

5. Simulation Experiments

A mainstream SVM algorithm in the field of machine learning was used. Its model was built based on solving convex optimization problems in optimization problems. At the same time, kernel functions were used to replace the nonlinear mapping of high-dimensional space to realize the role of processing high-dimensional space data in the low-dimensional calculation. The desired classification flat was only related to the support vector samples, thereby enabling the SVM-based algorithm to have the ability of small sample learning. However, this algorithm suffers from poor performance in canceling tremors because of their characteristics.

Hence, in this subsection, we compared the broad-learning-system-based algorithm model with the support vector machine algorithm for the tele-operation systems, which were mainly based on the MATLAB Robotics and Libsvm toolbox.

5.1. Model Evaluation Metrics

Determing whether a model has the ability of classification or regression can be judged by some evaluation metrics, such as the Euclidean distance error. In this topic, the (1) sum square error (SSE); (2) root mean square error (RMSE); and (3) regression determination coefficient (R^2) were used as the evaluation strategies of the various network models.

$$SSE = \sum_{t=1}^{T} (n_{ref}(t) - n_{pre}(t))^2 \qquad (37)$$

$$RMSE = \sqrt{\frac{\sum_{t=1}^{T} (n_{ref}(t) - n_{pre}(t))^2}{N}} \qquad (38)$$

$$R^2 = 1 - \frac{\sum_{t=1}^{T} n_{ref}(t) - n_{pre}(t)}{\sum_{t=1}^{T} n_{ref}(t) - \tilde{n}_{ref}(t)}, \qquad (39)$$

where T represents the periods, and N is the number of samples. $\tilde{n}_{ref}(t)$ is the mean values of n_{ref}.

5.2. Data Pre-Processing

In order to satisfy the same distribution of the input data and to prevent the difference of varying data from being large, we pre-processed the input data. Specifically, we used z-score normalization processing, which allows the input data to be adjusted to present a standard normal distribution, i.e., the Gaussian distribution, which satisfies the zero mean and the one variance.

$$s = \frac{s_i - min(s)}{max(s) - min(s)}, \qquad (40)$$

where s represents the input vectors on the three-axis directions in the original input data. After standardization, all the elements in the vector are normalized to [0, 1], which can accelerate the training and learning speed of the network.

5.3. Parameter Settings

The simulation experiment had a sampling time of 200 s with an interval time of 1 s, thus resulting in a total of 200 sampling points. In MATLAB, we built a simulation robot, and its joint parameters are given in Table 1. Additionally, other simulation parameters, such as network nodes and activation functions, were set according to the following specifications.

Table 1. The MATLAB-based simulation robot arm joint parameters.

i	Theta	d	a	Alpha	Offset
1	q1	105	0	$\pi/2$	0
2	q2	0	−174	0	$-\pi/2$
3	q3	0	−174	0	0
4	q4	76	0	$\pi/2$	$-\pi/2$
5	q5	80	0	$-\pi/2$	0
6	q6	44	0	0	0

Noise signals were used to simulate the tremor signals of the human hand. We added these simulated signals into the trajectory, and they were set as two parts: (1) the low-frequency part and (2) the high-frequency part. The physiological tremor signal is defined as the joint angle signal of the haptic device that comes into contact with the human hand. Assuming the absence of any accompanying physiological tremor in the human body, the joint angle information of the joystick is denoted by q_d. However, after being affected by tremors, the joint angle is updated as the following:

$$q(t) = q_d(t) + n(t). \tag{41}$$

In Figure 7, the presence of a physiological tremor can cause the operating human hand to deflect during operations, particularly during slow movements. This amplifies the effect of the physiological tremor, and the subplot in Figure 7 demonstrates that the operator's actual trajectory deviated significantly from the desired target trajectory.

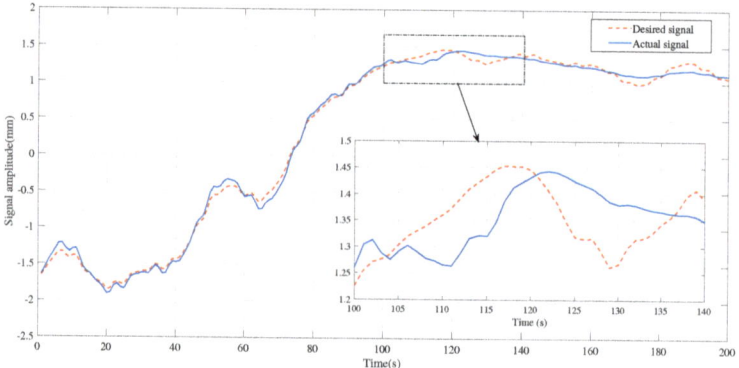

Figure 7. Expected values and actual values when the operator was operating.

For the BLS-based tremor filter, the sparse regularization parameter C was set to 2^{-30}, while the reduction parameter s for the enhanced node was set to 0.8. The broad learning system consists of $N11 = 10$ feature nodes, $N2 = 80$ feature node windows, and $N33 = 200$ enhanced nodes for each window. The number of added feature nodes was $m1 = 10$, the number of enhancement nodes related to the incremental feature nodes per increment step was $m2 = 20$, and the number of enhancement nodes in each incremental learning was $m3 = 50$. The activation function $\xi(\cdot) = tanh(\cdot)$ was used for mapping the feature nodes to the enhanced nodes. For the SVM-based filter, a method based on epsilon support vector regression was used with a loss function parameter p set to 0.4, which indicated the penalty degree for the input data. Moreover, the radial basis function (RBF) was selected as the kernel function of the network.

6. Tremor Forecast Results

In this section, the comparison simulation experiments w.r.t. the broad learning system and support vector machine were achieved under the simulated physiological tremor signal. As shown in Figure 8, the simulated manipulator was built in our MATLAB platform, and the motion trajectory and the joint angle of the manipulator with tremors and without tremors are given. In millimeters, we can see that the joint angle and the motion trajectory with tremors deviated from the desired trajectory.

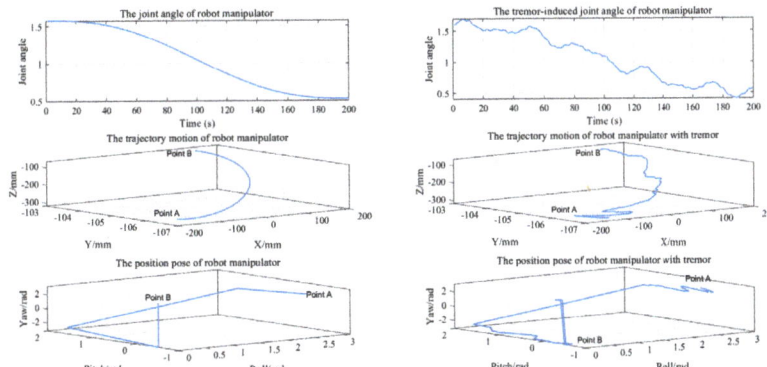

Figure 8. Joint angle and motion trajectory of robot manipulator with tremors and without tremors.

Figure 9a shows the prediction and estimation ability for the broad learning system and the support vector machine, where four curves are shown in the figure, which are the broad learning system, the incremental broad learning system, the support vector machine, and the actual tremor-induced offset. Since the ability of the broad learning system to predict the tremor-induced offset trajectory reached a saturation state, there was still no obvious effect improvement after the reinforcement of incremental learning, which indicates that the ability of the broad learning system to learn time sequence signals is limited to some extent. To compare the performance of different algorithms, we can observe the error curve shown in Figure 9b. As we can see, the SVM error fluctuated, whereas the BLS error decreased gradually over time. In summary, the proposed BLS algorithm outperformed the SVM. This conclusion is supported by the evaluation metrics in Table 2, which indicate that a good regression model should have a high determination coefficient R^2. In Figure 10, we can observe the recovery trajectory of a tele-operation system under the influence of various filters. By comparing the recovery trajectories across multiple filters, we can gain insights into the relative effectiveness and limitations of each filter in achieving the desired outcome. As shown in Figure 11, the position and velocity control were achieved by applying a sliding mode controller. First, the position and velocity references were input into the controller, which generated a control signal that was then applied to the system. The sliding mode controller ensures that the system tracks the desired position trajectory while also regulating the velocity. The resulting system behavior is shown in the position and velocity plots.

Overall, the teleoperation robot system applied the sliding mode controller to obtain a good performance, while our proposed filter was efficient in canceling tremors.

Table 2. Canceling tremor results of different filters.

Different Methods and Metrics	SSE	RMSE	R^2	Train Time
Broad learning system filter	0.0687	0.0026	80.06%	0.118
Incremental broad learning system filter	0.0587	0.0024	82.94%	0.122
Support vector machine filter	0.0918	0.0303	73.35%	0.278

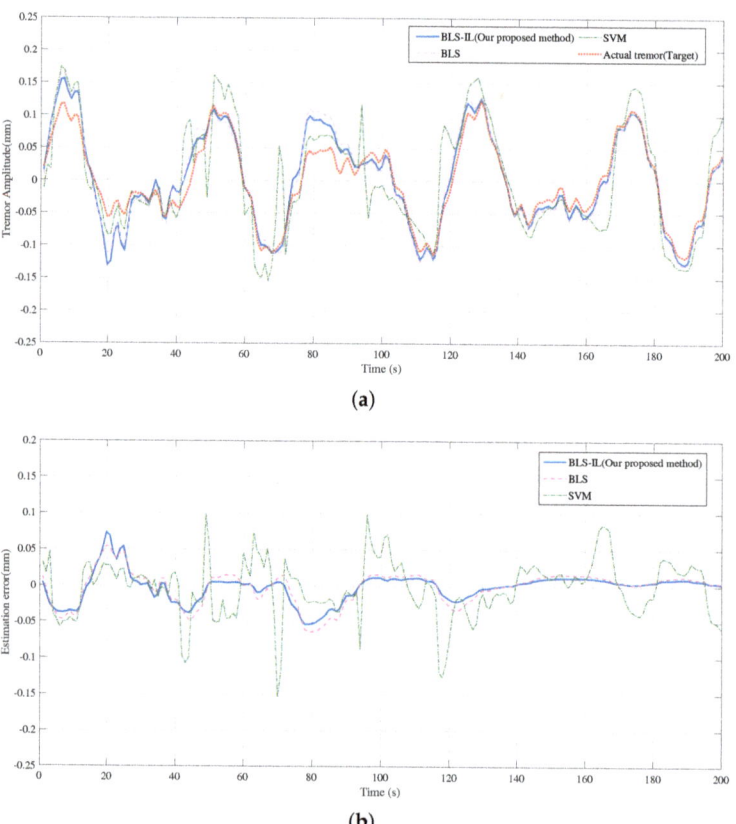

Figure 9. The results of canceling tremors based on different approaches. (**a**) In the case of tremors, prediction values based on different approaches. (**b**) In the case of tremors, the error between tremor values and prediction values based on different approaches.

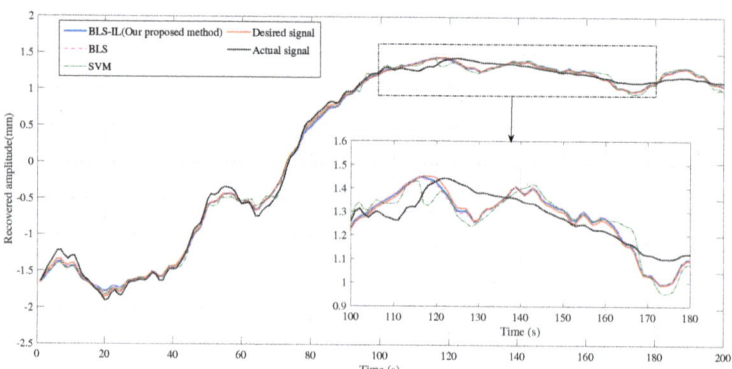

Figure 10. Tremor attenuation performance based on different approaches.

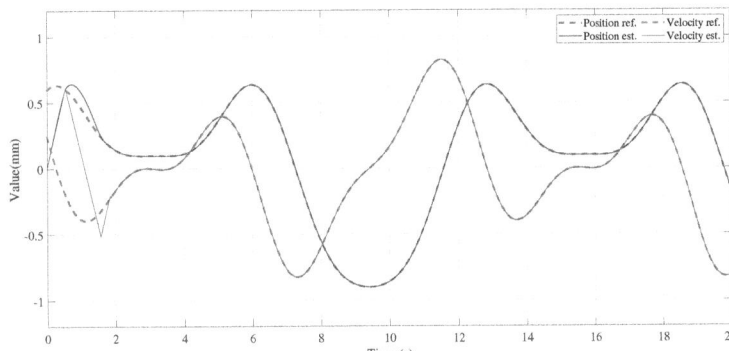

Figure 11. Position and velocity control resulting from applying sliding mode controller.

7. Conclusions

In this paper, we proposed a simple and efficient network model, the incremental broad learning system, as a tremor filter architecture for the current application issues of deep learning and machine learning in tele-operation. Unlike deep learning algorithms that have many hyperparameters, our proposed approach simplifies the learning process and avoids such complexities. Furthermore, the support vector machine often suffers from poor precision in regression tasks, and our novel architecture was designed to overcome this issue. We combined it with incremental learning algorithms to rapidly improve performance. Additionally, our proposed sliding mode controller provided greater stability and faster response performance when compared to traditional controllers. The simulation results and performance metrics demonstrated the effectiveness of our approach in attenuating tremors.

In future work, we will delve deeper into the feature extraction module of the broad learning system to improve its ability to eliminate physiological tremors to the best of our ability. Although our experimental results demonstrated the efficient elimination of physiological tremors by the broad-learning-system-based, we believe that there is still room for improvement.

Author Contributions: Conceptualization, G.L.; methodology, W.Y.; software, W.L.; validation, G.L. and H.Z.; formal analysis, W.L.; writing—original draft preparation, W.L.; writing—review and editing, Y.H. and Y.Z. All authors have read and agreed to the published version of the manuscript.

Funding: This research was funded by the Tertiary Education Scientific research project of the Guangzhou Municipal Education Bureau [No.202235364] and the Science and Technology Program of Guangzhou, China [No.202201010381], and the Special projects in key fields of colleges and universities in Guangdong Province [No.2021ZDZX1109], and the Research project of Guangzhou City Polytechnic [No.KYTD2023004], and the National Natural Science Foundation of China [No.6210021076].

Institutional Review Board Statement: Not applicable.

Data Availability Statement: Not applicable.

Conflicts of Interest: The authors declare no conflict of interest.

References

1. Xu, W.; Peng, J.; Liang, B.; Mu, Z. Hybrid modeling and analysis method for dynamic coupling of space robots. *IEEE Trans. Aerosp. Electron. Syst.* **2016**, *52*, 85–98. [CrossRef]
2. Chen, H.; Huang, P.; Liu, Z.; Ma, Z. Time delay prediction for space telerobot system with a modified sparse multivariate linear regression method. *Acta Astronaut.* **2020**, *166*, 330–341. . [CrossRef]
3. Ehrampoosh, A.; Shirinzadeh, B.; Pinskier, J.; Smith, J.; Moshinsky, R.; Zhong, Y. A Force-Feedback Methodology for Teleoperated Suturing Task in Robotic-Assisted Minimally Invasive Surgery . *Sensors* **2022**, *22*, 7829. [CrossRef] [PubMed]

4. Zhi, L.; Wu, Q.; Yun, Z.; Wang, Y.; Chen, C. Adaptive fuzzy wavelet neural network filter for hand tremor canceling in microsurgery. *Appl. Soft Comput.* **2011**, *11*, 5315–5329.
5. Liu, Z.; Mao, C.; Luo, J.; Zhang, Y.; Chen, C.P. A three-domain fuzzy wavelet network filter using fuzzy PSO for robotic assisted minimally invasive surgery. *Knowl. Based Syst.* **2014**, *66*, 13–27. [CrossRef]
6. Tatinati, S.; Veluvolu, K.C.; Ang, W.T. Multistep Prediction of Physiological Tremor Based on Machine Learning for Robotics Assisted Microsurgery. *IEEE Trans. Cybern.* **2015**, *45*, 328–339. [CrossRef]
7. Latt, W.T.; Veluvolu, K.C.; Ang, W.T. Drift-Free Position Estimation of Periodic or Quasi-Periodic Motion Using Inertial Sensors. *Sensors* **2011**, *11*, 5931–5951. [CrossRef]
8. Li, X.; Guo, S.; Shi, P.; Jin, X.; Kawanishi, M.; Suzuki, K. A Bimodal Detection-Based Tremor Suppression System for Vascular Interventional Surgery Robots. *IEEE Trans. Instrum. Meas.* **2022**, *71*, 1–12. [CrossRef]
9. Zheng, L.; Guo, S.; Zhang, L. Preliminarily Design and Evaluation of Tremor Reduction Based on Magnetorheological Damper for Catheter Minimally Invasive Surgery. In Proceedings of the 2019 IEEE International Conference on Mechatronics and Automation (ICMA), Tianjin, China, 4–7 August 2019 ; pp. 2229–2234. [CrossRef]
10. Guo, J.; Yang, S.; Guo, S. Study on the Tremor Elimination Strategy for the Vascular Interventional Surgical Robot. In Proceedings of the 2020 IEEE International Conference on Mechatronics and Automation (ICMA), Beijing, China, 2–5 August 2020 ; pp. 1547–1552. [CrossRef]
11. Mellone, S.; Palmerini, L.; Cappello, A.; Chiari, L. Hilbert–Huang-Based Tremor Removal to Assess Postural Properties from Accelerometers. *IEEE Trans. Biomed. Eng.* **2011**, *58*, 1752–1761. [CrossRef]
12. Riley, P.O.; Rosen, M.J. Evaluating manual control devices for those with tremor disability. *J. Rehabil. Res. Dev.* **1987**, *24*, 99.
13. Wei, T.A.; Khosla, P.K.; Riviere, C.N. Kalman filtering for real-time orientation tracking of handheld microsurgical instrument. In Proceedings of the 2004 IEEE/RSJ International Conference on Intelligent Robots and Systems (IROS), Sendai, Japan, 28 September–2 October 2004 .
14. Wang, Y.; Veluvolu, K.C. Time-frequency decomposition of band-limited signals with bmflc and kalman filter. In Proceedings of the 2012 7th IEEE Conference on Industrial Electronics and Applications (ICIEA), Singapore, 18–20 July 2012 .
15. Tatinati, S.; Veluvolu, K.C.; Hong, S.M.; Latt, W.T.; Ang, W.T. Physiological tremor estimation with autoregressive (ar) model and kalman filter for robotics applications. *IEEE Sensors J.* **2013**, *13*, 4977–4985. [CrossRef]
16. Veluvolu, K.C.; Tatinati, S.; Hong, S.M.; Ang, W.T. Multistep prediction of physiological tremor for surgical robotics applications. *IEEE Trans. Biomed. Eng.* **2013**, *60*, 3074–3082. [CrossRef]
17. Ghassab, V.K.; Mohammadi, A.; Ataszhar, S.F.; Patel, R.V. Dynamic estimation strategy for e-bmflc filters in analyzing pathological hand tremors. In Proceedings of the 2017 IEEE Global Conference on Signal and Information Processing (GlobalSIP), Montreal, QC, Canada, 14–16 November 2017 .
18. Yang, C.; Luo, J.; Pan, Y.; Liu, Z.; Su, C.Y. Personalized variable gain control with tremor attenuation for robot teleoperation. *IEEE Trans. Syst. Man Cybern. Syst.* **2018**, *48*, 1759–1770. [CrossRef]
19. Liu, Z.; Luo, J.; Wang, L.; Zhang, Y.; Philip, Chen, C.L.; Chen, X. A time-sequence-based fuzzy support vector machine adaptive filter for tremor cancelling for microsurgery. *Int. J. Syst. Sci.* **2015**, *46*, 1131–1146. [CrossRef]
20. Du, F.; Zhang, J.; Ji, N.; Shi, G.; Zhang, C. An effective hierarchical extreme learning machine based multimodal fusion framework. *Neurocomputing* **2018**, *322*, 141–150. [CrossRef]
21. Yue, B.; Wang, S.; Liang, X.; Jiao, L. An external learning assisted self-examples learning for image super-resolution. *Neurocomputing* **2018**, *312*, 107–119. [CrossRef]
22. Salaken, S.M.; Khosravi, A.; Nguyen, T.; Nahavandi, S. Extreme learning machine based transfer learning algorithms: A survey. *Neurocomputing* **2017**, *267*, 516–524. [CrossRef]
23. Liu, W.; Lai, G.; Liu, A. Tremor Attenuation For Robot Teleoperation By A Broad Learning System-Based Approach. In Proceedings of the 2021 China Automation Congress (CAC), Beijing, China, 22–24 October 2021 ; pp. 7627–7632. [CrossRef]
24. Taylor, R.H.; Lavealle, S.; Burdea, G.C.; Mosges, R. *Computer-Integrated Surgery: Technology and Clinical Applications*, 1st ed.; MIT Press: Cambridge, MA, USA, 1995.
25. Yang, C.; Wang, X.; Li, Z.; Li, Y.; Su, C.Y. Teleoperation Control Based on Combination of Wave Variable and Neural Networks. *IEEE Trans. Syst. Man Cybern. Syst.* **2017**, *47*, 2125–2136. [CrossRef]
26. Al-Mouhamed, M.A.; Nazeeruddin, M.; Merah, N. Design and Instrumentation of Force Feedback in Telerobotics. *IEEE Trans. Instrum. Meas.* **2009**, *58*, 1949–1957.
27. Ju, Z.; Yang, C.; Li, Z.; Cheng, L.; Ma, H. Teleoperation of humanoid baxter robot using haptic feedback. In Proceedings of the 2014 International Conference on Multisensor Fusion and Information Integration for Intelligent Systems (MFI), Beijing, China, 28–29 September 2014 ; pp. 1–6.
28. Kang, B.W.; Kang, S.G. The sliding mode controller with an additional proportional controller for fast tracking control of second order systems. In Proceedings of the 2012 IEEE International Conference on Systems, Man, and Cybernetics (SMC), Seoul, Republic of Korea, 14–17 October 2012 ; pp. 2383–2388.
29. Tokat, S.; Eksin, I.; Güzelkaya, M. New approaches for on-line tuning of the linear sliding surface slope in sliding mode controllers. *Turk. J. Electr. Eng. Comput. Sci.* **2003**, *11*, 45–54.
30. Yang, Q.; Liang, K.; Su, T.; Geng, K.; Pan, M. Broad learning extreme learning machine for forecasting and eliminating tremors in teleoperation. *Appl. Soft Comput.* **2021**, *112*, 107863. [CrossRef]

31. Chen, C.L.P.; Liu, Z.; Feng, S. Universal Approximation Capability of Broad Learning System and Its Structural Variations. *IEEE Trans. Neural Netw. Learn. Syst.* **2019**, *30*, 1191–1204. [CrossRef] [PubMed]
32. Chen, C.; Liu, Z. Broad Learning System: An Effective and Efficient Incremental Learning System Without the Need for Deep Architecture. *IEEE Trans. Neural Netw. Learn. Syst.* **2018**, *29*, 10–24. [CrossRef] [PubMed]
33. Xu, Z.; Chang, X.; Xu, F.; Zhang, H. L-1/2 regularization: A thresholding representation theory and a fast solver. *IEEE Trans. Neural Netw. Learn. Syst.* **2012**, *23*, 1013–1027. [PubMed]
34. Yang, W.; Gao, Y.; Shi, Y.; Cao, L. Mrm-lasso: A sparse multiview feature selection method via low-rank analysis. *IEEE Trans. Neural Netw. Learn. Syst.* **2015**, *26*, 2801–2815. [CrossRef] [PubMed]
35. Tibshirani, R. Regression Shrinkage and Selection via the Lasso. *J. R. Stat. Soc. Ser. B* **1996**, *58*, 267–288. [CrossRef]

Disclaimer/Publisher's Note: The statements, opinions and data contained in all publications are solely those of the individual author(s) and contributor(s) and not of MDPI and/or the editor(s). MDPI and/or the editor(s) disclaim responsibility for any injury to people or property resulting from any ideas, methods, instructions or products referred to in the content.

Article

Adaptive Orbital Rendezvous Control of Multiple Satellites Based on Pulsar Positioning Strategy

Qiang Chen, Yong Zhao and Lixia Yan *

School of Automation Science and Electrical Engineering, Beihang University (BUAA), Beijing 100191, China; qiangchen@buaa.edu.cn (Q.C.); zhaoyong1996@buaa.edu.cn (Y.Z.)
* Correspondence: yanlixia@buaa.edu.cn

Abstract: This paper addresses the orbital rendezvous control for multiple uncertain satellites. Against the background of a pulsar-based positioning approach, a geometric trick is applied to determine the position of satellites. A discontinuous estimation algorithm using neighboring communications is proposed to estimate the target's position and velocity in the Earth's Centered Inertial Frame for achieving distributed rendezvous control. The variables generated by the dynamic estimation are viewed as virtual reference trajectories for each satellite in the group, followed by a novel saturation-like adaptive control law with the assumption that the masses of satellites are unknown and time-varying. The rendezvous errors are proven to be convergent to zero asymptotically. Numerical simulations considering the measurement fluctuations validate the effectiveness of the proposed control law.

Keywords: pulsar navigation; nonlinear observer; relative satellite dynamics; adaptive control

1. Introduction

In recent years, the cooperative rendezvous of orbital satellites has become a popular research topic among the academic community [1]. Advanced positioning and rendezvous control techniques are building a firm base for the potential applications, such as orbital maintenance, orbital refueling, and orbital assembly [2], via steering multiple orbital satellites to achieve rendezvous at a certain target. Particularly, research on the pulsar-based cooperative rendezvous control is of practical significance as the pulsar sources feature high-precision yet stable timing properties for determining the position coordinates of orbital vehicles [3,4].

The pulsar-based positioning technology allows the orbital vehicles to locate themselves by comparing the received signals from pulsar sources with a database of known pulsars and locations [5]. It is the next-generation navigation technology for orbiting or interplanetary spacecrafts [6] and an alternative calibrated source for GNSS (Global Navigation Satellite System) [7]. Compared with the positioning technique, the control technology plays a more important role as the mission success can only be achieved with a robust control design that provides orbital vehicles with robust properties towards the external disturbance, measurement noise, and modeling uncertainty [8–13]. In Ref. [8], the extended Kalman filter is applied to denoising the virtual noisy pulsar signal within Poisson distribution and the state observing criteria of a linearized pulsar model via using pulsar data is also proposed. With the help of orbit information and the long-term observation of a single pulsar, navigation algorithms developed based on the adaptive divided different Kalman filter under scenarios of 1–3 orbital satellites are reported in [9]. For improving the reliability, robust control design features the capacity of overcoming the problem of either parameter variation or external disturbances [10,11,14]. Considering the limitation of fuel storage of orbital spacecraft described by Clohessy–Wiltshire equations of motion, the robust L_1 control strategy shown in [10] has achieved better fuel efficiency during orbit transferring even with parameter variations. One other fuel optimization

Citation: Chen, Q.; Zhao, Y.; Yan, L. Adaptive Orbital Rendezvous Control of Multiple Satellites Based on Pulsar Positioning Strategy. *Entropy* **2022**, *24*, 575. https://doi.org/10.3390/e24050575

Academic Editor: Katalin M. Hangos

Received: 25 January 2022
Accepted: 14 April 2022
Published: 19 April 2022

Publisher's Note: MDPI stays neutral with regard to jurisdictional claims in published maps and institutional affiliations.

Copyright: © 2022 by the authors. Licensee MDPI, Basel, Switzerland. This article is an open access article distributed under the terms and conditions of the Creative Commons Attribution (CC BY) license (https://creativecommons.org/licenses/by/4.0/).

strategy based on the saw-like control updating algorithm can be found in [12]. As the command from the ground control station to the orbital satellites features an obvious delay phenomenon, advanced control schemes such as [15] can be applicable to fill this gap. Sometimes the trade-off problem between impulsive efficiency and task complexity should even be predetermined before launching the spacecraft into the space [13]. Though the results reported in [8–15] have solved various control problems, only solutions for a single satellite are developed without considering cooperative control of multiple orbital satellites.

As the complexity increases in space missions [16], the cooperative missions of multiple orbital vehicles have received lots of attention [17–23]. The formation control of numerous spacecrafts is studied in [17] under the J2 perturbation caused by the oblateness of the Earth. Robotic arms and wheeled mobile robots are generally used for ground algorithm investigations [18,19]. Two ground 6-DOF robotic arms are used for performing lidar-based rendezvous and docking control design for orbital satellites [18], in which the real-time target pose estimation and tracking algorithms are applied. A wheeled satellite simulator testbed with experimental docking discussions is reported in [19], illustrating the basic idea of maneuvering two spacecraft to achieve rendezvous in the final phase. In addition, applying atmosphere lift and drag force for satellites in low Earth orbits is proved to be an alternative to complete rendezvous and docking missions [20]. It is worth noting that the control schemes proposed in [17–20] convert the orbital formation into a local tracking problem and only achieve the formation control of two satellites. This, however, is not applicable for more general tasks such as orbital assembly that requires multiple orbital vehicles to rendezvous at the target orbit simultaneously [21–23]. In Ref. [21], the authors propose an adaptive fuzzy control law that ensures the spacecrafts' rendezvous errors to be ultimately bounded by small numbers. In Ref. [22], the Lyapunov barrier functions are applied to solve the inner-agent collision avoidance problem of multiple orbital spacecrafts for safe maneuver. Considering the constrained field of view of vision-based devices, the feasible path generating algorithm reported in [23] ensures that the target spacecraft maintains the view of the camera during the rendezvous process. However, the cooperative algorithms developed for multiple satellites in [21–23] only achieve small attraction regions of the closed-loop system, i.e., the rendezvous errors must be initialized as tens of meters for maintaining the effectiveness of the control law or the model simplifications. In practice, the orbital vehicles launched for cooperative missions are orbiting in different orbits, which means that the initial distance between them might be hundreds of kilometers, which cannot be dealt with by the control laws reported in [21–23]. Additionally, the cooperative control schemes reported in [17–23] contain the assumption that all orbital vehicles can know the information of the reference target, which is not a necessity for achieving cooperative missions.

Motivated by the discussions above, this paper makes a further endeavor to solve the adaptive rendezvous control problem of multiple uncertain orbital satellites with the background of the pulsar-based positioning method. First, we introduce a direct geometric trick to determine orbital coordinates via assuming simultaneous observations of three pulsars for each satellite. Second, we consider the scenario that only part of the satellites in the group have access to the states of the orbital reference vehicle and propose a discontinuously distributed estimation algorithm to estimate the reference orbital vehicle. Specifically, the estimation errors are globally exponentially convergent to zero. Third, we design a novel saturation-like adaptive control strategy via viewing the estimation algorithm as a virtual reference trajectory for each networked satellite. It is proven in the Lyapunov sense that the rendezvous errors of all networked satellites converge to zero asymptotically.

The main contribution of this paper is presenting the design and analysis of a novel adaptive rendezvous strategy for multiple uncertain orbital satellites with an initial pulsar-based positioning method. In comparison with control schemes developed for the single orbital satellites in [8–13] or the rendezvous of two spacecrafts in [17–20], the cooperative control algorithm in this paper allows numerous satellites to achieve rendezvous at the reference target. The adaptive capacity towards unknown and time-varying masses provided by the presented control scheme also features more practical significance than that only considers constant mass in [23]. Additionally, the proposed rendezvous algorithm still works even if the initial rendezvous errors are initialized in hundreds of kilometers, largely expanding the efficient range of a region of attraction in comparison with the control laws reported in [21–23].

The remainder is organized as follows. Section 2 presents some useful preliminaries for control formulation. In particular, the pulsar-based position scheme is introduced. Section 3 elaborates on the designs of the estimation algorithm and adaptive control law. Numerical simulations are included in Section 4. Section 5 concludes the work briefly.

2. Preliminaries

2.1. Notations and Definitions

In this paper, $\mathbb{R}$ denotes the real number set, $\|\cdot\|$ is the Euclidean norm, $|\cdot|$ represents the absolute value of a scalar, $\text{diag}\{\cdot\}$ forms a diagonal matrix, I_n is an n-dimensional identity matrix within a vector, 1_n is an n-dimensional identity row vector, 0_n denotes an n-dimensional zero row vector. For a square matrix, $\lambda(\cdot)$, $\lambda_m(\cdot)$ and $\lambda_M(\cdot)$ represent the eigenvalue, the smallest and largest eigenvalue, respectively. The time t is sometimes omitted without making confusion.

2.2. Pulsar-Based Positioning

Position measurement for orbital satellites is essential for the onboard control systems to maintain stability. It highly relies on the Telemetry, Tracking, and Command (TT&C) stations built on the ground or other orbital GNSS satellites, which requires continuous yet enormous investment. In comparison with that, pulsars are natural high-precision timing sources and are suitable for determining the position coordinates of satellites in both orbital and interplanetary missions. Henceforth, using pulsars as positioning sources is of practical significance for rendezvous control.

In what follows, an initial pulsar-based positioning method is presented.

Assumption 1. *The solar system barycenter coincides with the center of mass of the Sun.*

Assumption 2. *The axial precession of the Earth is neglected and the Earth orbit is a pure Keplerian orbit without being affected by other celestial bodies except the Sun.*

With Assumptions 1 and 2, the geometric relationship among pulsars, Sun, Earth, and satellites can be shown in Figure 1. The Sun-Centered-Inertial (SCI) frame $O_s - X_s Y_s Z_s$ that Earth is rotating around is viewed as an inertial frame, where O_s is the center of mass of the Sun. The X_s-axis coincides with the direction from the vernal equinox to the O_s, the Z_s-axis is perpendicular to the Earth's orbit plane, and the Y_s-axis completes the right-hand law according to $Z \times X$. For the Earth-Centered-Inertial (ECI) plane $O_E - X_E Y_E Z_E$ that is centered by the Earth, the X_E-axis features the direction to the vernal equinox, the Z_E-axis points the rotating axis of the Earth upwards, and the Y_E satisfies the right-hand law of $Z_E \times X_E$. In $O_s - X_s Y_s Z_s$ frame, the Earth orbit around the Sun satisfies the following Keplerian dynamics [24],

$$\dot{\vec{r}}_{s,e} = \vec{v}_{s,e}, \quad \dot{\vec{v}}_{s,e} = -\frac{\mu_s}{r_{s,e}^3}\vec{r}_{s,e}, \tag{1}$$

where $\vec{r}_{s,e}$ and $\vec{v}_{s,e}$ denote the position and velocity of the Earth, respectively.

Let α_j and λ_j denote the right ascension and declination in frame $O_s - X_s Y_s Z_s$ of the pulsar j, respectively. The radiation signal of the pulsar j has the following direction [9]:

$$\vec{q}_j = [\cos\alpha_j \cos\lambda_j, \cos\alpha_j \sin\lambda_j, \sin\alpha_j]. \tag{2}$$

Suppose that there are n satellites and the coordinate of i-th satellite is given by row vector $\vec{r}_{s,i} = [x_{s,i}, y_{s,i}, z_{s,i}]$. Generally, the projection of the $\vec{r}_{s,i}$ on the vector $\vec{q}_j$ is supposed to be measurable [5,24], i.e., $\vec{s}_{i,j} = (\vec{r}_{s,i} \cdot \vec{q}_j)\vec{q}_j$ is known, where $r_{s,i} = \|\vec{r}_{s,i}\| = c(t_{st} - t_{ssb})$ with c being the light speed, t_{st} the time instant that the satellite receives the pulsar signal, and t_{ssb} the time of solar system barycenter. Henceforth, the plane perpendicular to the vector $\vec{s}_{i,j}$ and passing the i-th satellite can be given by:

$$\vec{s}_{i,j} \cdot (\vec{r}_{s,i} - \vec{s}_{i,j}) = 0. \tag{3}$$

Actually, the position vector of the satellite $\vec{r}_{s,i}$ in the $O_s - X_s Y_s Z_s$ frame can be uniquely determined via solving the linear equation set in the form of Equation (3) if there are more than three pulsars in the detecting area of the satellite. This is because three non-parallel two-dimensional planes would uniquely determine one single point in the three-dimensional surface [25]. For this sake, we make the following assumption:

Assumption 3. *For each satellite, at least three pulsars' signals can be observed.*

Remark 1. *Note that the state-of-the-art instruments designed for observing pulsar sources, such as those launched by NICER and XPNAV missions, can only perform observation on a single pulsar during certain time windows. Assumption 3 is built upon future technology wherein the onboard equipment allows the orbital satellites to observe multiple pulsar signals simultaneously, which would be achieved by improving sensor sensitivity and minimization packaging.*

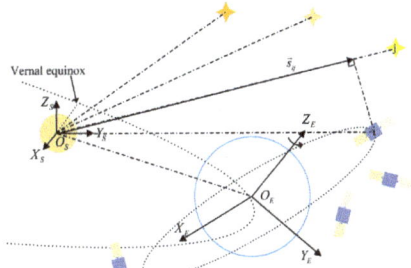

Figure 1. Geometric relationships among pulsars, the Sun, the Earth and the satellites.

2.3. Relative Satellite Dynamics in ECI Frame

Considering n orbital satellites under ECI coordinates, let row vectors $\vec{r}_i = [x_{e,i}, y_{e,i}, z_{e,i}]$ and $\vec{v}_{e,i} = \dot{\vec{r}}_{e,i} = [\dot{x}_{e,i}, \dot{y}_{e,i}, \dot{z}_{e,i}]$ denote the ECI position and velocity of the i-th satellite, respectively. The satellites can generate arbitrary thrust force to control their positions and the position control is decoupled from attitude loop. Via ignoring the J2 perturbation caused by Earth oblateness, the unperturbed orbital satellites can be described by [24]:

$$\dot{\vec{r}}_{e,i} = \vec{v}_{e,i}, \dot{\vec{v}}_{e,i} = -\frac{\mu_e}{r_{e,i}^3}\vec{r}_{e,i} + \frac{\tau_i}{m_i(t)}, i \in \mathcal{N} \triangleq \{1,2,\ldots,n\}, \tag{4}$$

where μ_e is the gravitational constant of the Earth, $m_i(t)$ denotes the mass of the i-the satellite, and $\tau_i \in \mathbb{R}^3$ represents the control input. We suppose that $\dot{m}_i = \alpha\|\tau_i\|$, where $\alpha < 0$ is a negative constant related to specific impulse coefficient. Via observing Equations (1), (2) and (4), the position of the i-th satellite satisfies

$$\vec{r}_{e,i} = \vec{r}_{s,i} - \vec{r}_{s,e}, \tag{5}$$

which reveals the basic principle of orbital satellite positioning by measuring pulsar signals. We assume that the satellites can obtain Earth's position vector in the SCI frame. In reality, the satellite's total mass is unknown and decreasing due to generating certain thrust forces and torques via burning the fuel. For this sake, the mass $m_i(t)$ and its ratio of change $\dot{m}_i(t)$ are supposed to satisfy:

Assumption 4. For all $i \in \mathcal{N}$, the mass $m_i(t)$ is an unknown variable and satisfies

$$m_i(t) \geq \underline{m}_i, \tag{6}$$

where $\underline{m}_i$ is a known positive constant, and the ratio of change of mass $\dot{m}_i(t)$ is a known non-positive bounded variable and $\ddot{m}_i(t)$ is bounded.

Besides n satellites, we suppose that there is one single free-flying reference orbital vehicle indexed by 0 and described by:

$$\dot{\vec{r}}_{e,0} = \vec{v}_{e,0}, \dot{\vec{v}}_{e,0} = -\frac{\mu_e}{r_{e,0}^3}\vec{r}_{e,0}. \tag{7}$$

Actually, the state vector of the satellite described by Equation (7) can be uniquely determined by its orbital elements that include the eccentricity, semimajor axis, inclination, longitude of the ascending node, the argument of perigee, and the true anomaly, as well as other equivalent expressions [24]. For endowing the control formulation with the distributed feature, we assume:

Assumption 5. There is at least one satellite in $\mathcal{N}$ has access to the position and velocity of the 0-th satellite described by Equation (7).

Construct a local vertical and local horizontal (LVLH) coordinate frame $O_L - x_L y_L z_L$ according to position vector $\vec{r}_{e,ir} = [x_{e,ir}, y_{e,ir}, z_{i,r}]$ and velocity vector $\vec{v}_{e,ir} = [\dot{x}_{e,ir}, \dot{y}_{e,ir}, \dot{z}_{i,r}]$, where the local z_L axis points towards the Earth center, the local y_L axis perpendicular downwards to the orbital plane and the local x_L axis completes the right hand law from y_L to z_L. Without losing generality, the unit axis vectors can be given as $\vec{x}_L = \vec{y}_L \times \vec{z}_L, \vec{y}_L = -\frac{\vec{r}_{e,ir} \times \vec{v}_{e,ir}}{\|\vec{r}_{e,ir} \times \vec{v}_{e,ir}\|}, \vec{z}_L = -\frac{\vec{r}_{e,ir}}{\|\vec{r}_{e,ir}\|}$. The angular velocity of the LVLH frame $O_L - x_L y_L z_L$ with respect to the ECI frame is:

$$\vec{\omega}_{ir} = \frac{\vec{r}_{e,ir} \times \vec{v}_{e,ir}}{r_{e,ir}^2}, \tag{8}$$

where $r_{e,ir} = \|\vec{r}_{e,ir}\|$. Define the relative position between the i-th satellite and $\vec{r}_{e,ir}$ in $O_L - x_L y_L z_L$ frame by $\vec{r}_{ir} = (\vec{r}_{e,i} - \vec{r}_{e,ir}) \cdot [\vec{x}_L; \vec{y}_L; \vec{z}_L]^T$, whose derivative is [24]:

$$\vec{v}_{ir} \triangleq \dot{\vec{r}}_{ir} = \dot{\vec{r}}_{e,i} - \dot{\vec{r}}_{e,ir} - \vec{\omega}_{ir} \times \vec{r}_{ir}. \tag{9}$$

Then, the associated relative acceleration can be given by:

$$\dot{\vec{v}}_{ir} = -\frac{\mu_e}{r_{e,i}^3}\vec{r}_{e,i} + \frac{\tau_i}{m_i} + \frac{\mu_e}{r_{e,ir}^3}\vec{r}_{e,ir} - \dot{\vec{\omega}}_{ir} \times \vec{r}_{ir} - \vec{\omega}_{ir} \times (\vec{\omega}_{ir} \times \vec{r}_{ir}) - 2\vec{\omega}_{ir} \times \vec{v}_{ir}, \tag{10}$$

where $\dot{\vec{\omega}}_{ir} = -2\frac{\vec{v}_{e,ir}\vec{r}_{e,ir}^T}{r_{e,ir}^2}\vec{\omega}_{ir}$ denotes the angular acceleration.

2.4. Graph Theory

A graph $\mathcal{G} = \{\mathcal{N}, \mathcal{E}, \mathcal{A}\}$ can be used to describe the interaction among satellites, where $\mathcal{E} \subseteq \mathcal{N} \times \mathcal{N}$ is the edge set and $\mathcal{A}$ denotes the adjacency matrix [26]. An edge $\{(i,j) : i \neq j\} \in \mathcal{E}$ denotes that the j-th satellite can send information to the i-th satellite via wireless devices. The adjacency matrix is defined by $\mathcal{A} = \{a_{ij}\} \in \mathbb{R}^{n \times n}$, where $a_{ij} = 1$ if $(i,j) \in \mathcal{E}$, otherwise $a_{ij} = 0$. Self connection is not allowed, i.e., $a_{ii} = 0, \forall i \in \mathcal{N}$. For an undirected graph, $a_{ij} = 1 \Leftrightarrow a_{ji} = 1$ holds, denoting that satellite i and satellite j can exchange information with each other. The path from satellite i to satellite j denotes an edge sequence $\{(i, j_1), (j_2, j_3), \ldots, (j_*, j)\}$. The graph $\mathcal{G}$ is called connected if there is a path between any two distinct nodes. The in-degree matrix is given by:

$\mathcal{D} = \mathrm{diag}\{[l_{11}, l_{22}, \ldots, l_{nn}]\}, l_{ii} = \sum_{j=1}^{n} a_{ij}, \forall i, j \in \mathcal{N}$, and the Laplacian matrix is: $\mathcal{L} = \mathcal{D} - \mathcal{A}$. As reported, the matrix $\mathcal{L}$ is semi-positive definite and has only one zero eigenvalue and $n-1$ positive eigenvalues provided that the graph $\mathcal{G}$ is undirected and connected [26]. Define $a_{i0} = 1$ if the there is a valid information flow from the spacecraft to the i-th satellite, otherwise $a_{i,0} = 0$. Define:

$$\mathcal{H} = \mathcal{L} + \mathcal{B}, \tag{11}$$

where $\mathcal{B} = \mathrm{diag}\{[a_{10}, \ldots, a_{n0}]\}$.

Assumption 6. *The graph $\mathcal{G}$ is undirected and connected, and $\mathcal{B} \neq 0_{n \times n}$.*

Lemma 1 ([26]). *Under Assumption 6, the matrix $\mathcal{H}$ is positive definite.*

2.5. Problem Formulation

Until now, the control objective can be formally stated as: given Assumptions 1–6, find a control law $\tau_i, \forall i \in \mathcal{N}$ so that the networked satellites described by Equation (4) achieve rendezvous at the orbit trajectory generated by Equation (7), i.e.,

$$\lim_{t \to \infty} \vec{r}_{e,i} - \vec{r}_{e,0} = 0_3. \tag{12}$$

3. Control Design and Stability Analysis

3.1. Estimation Algorithm

Under Assumption 5, the states of the reference orbital vehicle are not available to all satellites, which might obstruct achieving the control objective Equation (12). To overcome this problem, we plan to construct an algorithm to estimate the $[\vec{r}_{e,0}, \vec{v}_{e,0}]$ via the neighboring communications among satellites in the ECI frame.

Define $\vec{\eta}_{e,ir} = [x_{e,ir}, y_{e,ir}, z_{e,ir}], \vec{\varphi}_{e,ir} = [\dot{x}_{e,ir}, \dot{y}_{e,ir}, \dot{z}_{e,ir}]$ and $\vec{\rho}_{e,ir} = [\ddot{x}_{e,ir}, \ddot{y}_{e,ir}, \ddot{z}_{e,ir}]$ as the estimated position, velocity and acceleration of the reference orbital vehicle for the i-th satellite. The updating algorithm for $\vec{\eta}_{e,ir}, \vec{\varphi}_{e,ir}, \vec{\rho}_{e,ir}$ is designed as:

$$\begin{cases} \dot{\vec{\eta}}_{e,ir} = \vec{\varphi}_{e,ir} \\ \dot{\vec{\varphi}}_{e,ir} = \vec{\rho}_{e,ir} \\ \dot{\vec{\rho}}_{e,ir} = -g_1 \vec{\varphi}_{e,ir} - g_2 \vec{\rho}_{e,ir} - g_3 s_{e,ir} - g_4 \mathrm{sign} s_{e,ir} \end{cases}, \tag{13}$$

where $g_1 > 0, g_2 > 0, g_3 > 0, g_4 > \sup_{t \geq 0} \left\| \vec{r}^{(3)}_{e,0} + g_1 \ddot{\vec{r}}_{e,0} + g_2 \dot{\vec{r}}_{e,0} \right\|$, $\mathrm{sign}(\cdot)$ denotes sign function and

$$s_{e,ir} = \sum_{j=1}^{n} a_{ij}(\vec{\rho}_{e,ir} - \vec{\rho}_{e,jr}) + a_{i0}(\vec{\rho}_{e,ir} - \ddot{\vec{r}}_{e,0}) + g_1 \sum_{j=1}^{n} a_{ij}(\vec{\varphi}_{e,ir} - \vec{\varphi}_{e,jr}) + g_1 a_{i0}(\vec{\varphi}_{e,ir} - \dot{\vec{r}}_{e,0})$$
$$+ g_2 \sum_{j=1}^{n} a_{ij}(\vec{\eta}_{e,ir} - \vec{\eta}_{e,jr}) + g_2 a_{i0}(\vec{\eta}_{e,ir} - \vec{r}_{e,0}). \tag{14}$$

Lemma 2. *Given Assumption 6, the gains satisfying $g_1 > 0, g_2 > 0, g_3 > 0$, $g_4 > \sup_{t \geq 0} \left\| \vec{r}^{(3)}_{e,0} + g_1 \ddot{\vec{r}}_{e,0} + g_2 \dot{\vec{r}}_{e,0} \right\|$ and any initial states, the estimation algorithm given by Equations (13) and (14) ensures that:*

$$\lim_{t \to \infty} \vec{\eta}_{e,ir} = \vec{r}_{e,0}, \lim_{t \to \infty} \vec{\varphi}_{e,ir} = \dot{\vec{r}}_{e,0}, \lim_{t \to \infty} \vec{\rho}_{e,ir} = \ddot{\vec{r}}_{e,0}. \tag{15}$$

Proof. Define $3n$-dimensional stacked row vectors,

$$\eta_{e,r} = [\vec{\eta}_{e,1r}, \ldots, \vec{\eta}_{e,nr}]^T, \varphi_{e,r} = [\vec{\varphi}_{e,1r}, \ldots, \vec{\varphi}_{e,nr}]^T,$$
$$\rho_{e,r} = [\vec{\rho}_{e,1r}, \ldots, \vec{\rho}_{e,nr}]^T, s_{e,r} = [s_{e,1r}, \ldots, s_{e,nr}]^T. \quad (16)$$

It then, by Equations (13) and (14) follows that:

$$\begin{cases} \dot{\eta}_{e,r} = \varphi_{e,r} \\ \dot{\varphi}_{e,r} = \rho_{e,r} \\ \dot{\rho}_{e,r} = -g_1 \varphi_{e,r} - g_2 \rho_{e,r} - g_3 s_{e,r} - g_4 P_e s_{e,r} \end{cases}, \quad (17)$$

where

$$P_e = \text{diag}\{\text{signs}_{e,1r}, \ldots, \text{signs}_{e,nr}\} \in \mathbb{R}^{3n \times 3n}. \quad (18)$$

Define estimation error

$$\varepsilon_{e,r} = \vec{\eta}_{e,r} - \mathbf{1}_n \otimes \vec{r}_{e,0}, \quad (19)$$

where $\otimes$ denotes the Kronecker product. Within Equation (19), the $s_{e,r}$ in Equation (16) can be arranged as:

$$s_{e,r} = \mathcal{H} \otimes I_3(\ddot{\varepsilon}_{e,r} + g_2 \dot{\varepsilon}_{e,r} + g_1 \varepsilon_{e,r}), \quad (20)$$

with time-derivative along the trajectory of Equation (14) being given by:

$$\dot{s}_{e,r} = \mathcal{H} \otimes I_3(\dddot{\vec{r}}_{e,r} + g_2 \ddot{\varepsilon}_{e,r} + g_1 \dot{\varepsilon}_{e,r})$$
$$= \mathcal{H} \otimes I_3 \left[-g_3 s_{e,r} - g_4 P_e s_{e,r} - \mathbf{1}_n \otimes (\dddot{\vec{r}}_{e,0} + g_2 \ddot{\vec{r}}_{e,0} + g_1 \dot{\vec{r}}_{e,0}) \right]. \quad (21)$$

Given Assumption 6 and Lemma 1, we know that the matrix $\mathcal{H}$ is positive definite. Choose a Lyapunov candidate as:

$$V_1 = 0.5 s_{e,r}^T (\mathcal{H} \otimes I_3)^{-1} s_{e,r}, \quad (22)$$

which, by the relationship $\lambda(\mathcal{H} \otimes I_3) = \lambda(\mathcal{H})$, leads to:

$$\frac{1}{2\lambda_M(\mathcal{H})} \|s_{e,r}\|^2 \leq V_1 \leq \frac{1}{2\lambda_m(\mathcal{H})} \|s_{e,r}\|^2. \quad (23)$$

Then, calculate the time-derivative of V_1 as:

$$\dot{V}_1 = -g_3 s_{e,r}^T s_{e,r} - g_4 s_{e,r}^T P_e s_{e,r} - s_{e,r}^T \mathbf{1}_n \otimes (\dddot{\vec{r}}_{e,0} + g_2 \ddot{\vec{r}}_{e,0} + g_1 \dot{\vec{r}}_{e,0})$$
$$= -g_3 s_{e,r}^T s_{e,r} - \sum_{i=1}^n \left[g_4 s_{e,ir} \text{signs}_{e,ir}^T + s_{e,ir} (\dddot{\vec{r}}_{e,0} + g_2 \ddot{\vec{r}}_{e,0} + g_1 \dot{\vec{r}}_{e,0})^T \right] \quad (24)$$

Let $\vartheta \triangleq \sup_{t \geq 0} \{\|\dddot{\vec{r}}_{e,0} + g_2 \ddot{\vec{r}}_{e,0} + g_1 \dot{\vec{r}}_{e,0}\|\}$. The Equation (24) satisfies:

$$\dot{V}_1 \leq -g_3 \|s_{e,r}\|^2 - \sum_{i=1}^n |s_{e,ir}|(g_4 - \vartheta)$$
$$\leq -g_3 \|s_{e,r}\|^2. \quad (25)$$

Therefore, $s_{e,r}$ would decay to zero with exponential decaying rate $-0.5 g_3$ [27]. We then transform Equation (20) into

$$\ddot{\varepsilon}_{e,r} + g_2 \dot{\varepsilon}_{e,r} + g_1 \varepsilon_{e,r} = (\mathcal{H} \otimes I_3)^{-1} s_{e,r}. \quad (26)$$

The vectors $\varepsilon_{e,r}, \dot{\varepsilon}_{e,r}$ and $\ddot{\varepsilon}_{e,r}$ will converge to zero exponentially due to $s_{e,r}$ decays to zero exponentially and the matrix $\mathcal{H} \otimes I_3$ is positive definite [28]. By definition Equation (19), the claims in the lemma follows. This completes the proof. □

Via the information exchange with connected satellites, all satellites achieve the estimate of $[\vec{r}_{e,0}, \vec{v}_{e,0}]$. One should notice that the generated velocity $\vec{\varphi}_{e,ir}$ and acceleration $\vec{\rho}_{e,ir}$ are bounded because the associated components of the orbital reference vehicle are bounded, and the exponential convergence to zero of the estimation errors. The generated signals $\vec{\eta}_{e,ir}, \vec{\varphi}_{e,ir}$ and $\vec{\rho}_{e,ir}$ stated in Lemma 2 will be viewed as a reference signal for the i-th satellite for $i \in \mathcal{N}$. As can be seen, the efficacy of the proposed estimation algorithm relies only on partial access to $[\vec{r}_{e,0}, \vec{v}_{e,0}]$ and undirected connected communications, which gains a more robust property of the group as a whole when compared with the centralized ones [17–23] that require all satellites in the group to know the information of the reference target.

3.2. Feedback Control

In this subsection, we will take the modeling uncertainty into account and design a control law for each satellite $i \in \mathcal{N}$ so as to track the reference signal generated by Equations (13) and (14).

Define tracking errors by:

$$\omega_{e,i} = [\vec{r}_{e,i} - \vec{\eta}_{e,ir}] \cdot \left[\frac{\vec{\eta}_{e,ir} \times \vec{\varphi}_{e,ir}}{\|\vec{\eta}_{e,ir} \times \vec{\varphi}_{e,ir}\|} \times \frac{\vec{\eta}_{e,ir}}{\|\vec{\eta}_{e,ir}\|} \quad -\frac{\vec{\eta}_{e,ir} \times \vec{\varphi}_{e,ir}}{\|\vec{\eta}_{e,ir} \times \vec{\varphi}_{e,ir}\|} \quad -\frac{\vec{\eta}_{e,ir}}{\|\vec{\eta}_{e,ir}\|} \right]^T \tag{27}$$

Following the derivations Equations (8)–(10), the first- and second-order derivatives of Equation (27) can be calculated as:

$$\dot{\omega}_{e,i} = \dot{\vec{r}}_{e,i} - \varphi_{e,ir} - \zeta_{e,ir} \times \omega_{e,i},$$

$$\ddot{\omega}_{e,i} = -\frac{\mu_e}{r_{e,i}^3}\vec{r}_{e,i} + \frac{\tau_i}{m_i(t)} + \frac{\mu_e}{\|\eta_{e,ir}\|^3}\eta_{e,ir} - \dot{\zeta}_{e,ir} \times \omega_{e,i} - \zeta_{e,ir} \times (\zeta_{e,ir} \times \omega_{e,i}) - 2\zeta_{e,ir} \times \dot{\omega}_{e,i}, \tag{28}$$

where $\zeta_{e,ir} = \frac{\eta_{e,ir} \times \varphi_{e,ir}}{\|\eta_{e,ir}\|}$ and $\dot{\zeta}_{e,ir} = -2\frac{\varphi_{e,ir}\eta_{e,ir}^T}{r_{e,ir}^2}\zeta$ represent the angular velocity and angular acceleration of the frame formed by the reference signal $[\vec{\eta}_{e,ir}, \vec{\varphi}_{e,ir}]$.

Define an intermediate variable:

$$\delta_{e,i} = -\frac{\mu_e}{r_{e,i}^3}\vec{r}_{e,i} + \frac{\mu_e}{\|\eta_{e,ir}\|^3}\eta_{e,ir} - \dot{\zeta}_{e,ir} \times \omega_{e,i} - \zeta_{e,ir} \times (\zeta_{e,ir} \times \omega_{e,i})$$
$$- 2\zeta_{e,ir} \times \dot{\omega}_{e,i} + k_3 \tanh(\dot{\omega}_{e,i} + k_2\omega_{e,i}) + k_1 \tanh(\dot{\omega}_{e,i}), \tag{29}$$

where $k_1, k_2, k_3 > 0$. We then propose the following control law with a parameter updating algorithm,

$$\begin{cases} \tau_i = -\hat{m}_i(t)\delta_{e,i} \\ \dot{\hat{m}}_i = \dot{m}_i(t) + k_4[k_3 \tanh(\dot{\omega}_{e,i} + k_2\omega_{e,i}) + k_1 \tanh \dot{\omega}_{e,i} + k_2\dot{\omega}_{e,i}]\delta_{e,i}^T \end{cases} \tag{30}$$

where $k_4 > 0$ and $\hat{m}_i(t)$ denotes the estimated value for m_i. The estimated value $\hat{m}_i(t)$ is updated via the integration $\hat{m}_i(t) = \hat{m}_i(0) + \int_0^t \dot{\hat{m}}_i(\chi) d\chi$.

Theorem 1. *If the control gains are selected such that $k_1 > 0, k_2 > 0, k_3 > 0, k_4 > 0$, the adaptive control law shown in Equation (30) guarantees that the tracking error $\omega_{e,i}$ converges to zero globally asymptotically. In addition, the control objective Equation (12) is achieved.*

Proof. Substituting Equation (29) into $\ddot{\omega}_{e,i}$ in Equation (28) results in:

$$\ddot{\omega}_{e,i} = -k_3 \tanh(\dot{\omega}_{e,i} + k_2\omega_{e,i}) - k_1 \tanh(\dot{\omega}_{e,i}) + \frac{\tau_i}{m_i(t)} + \delta_{e,i}. \tag{31}$$

Define estimation error as $\tilde{m}_i(t) = m_i(t) - \hat{m}_i(t)$ and taking the control law Equation (30) into Equation (31), we obtain:

$$\ddot{\omega}_{e,i} = -k_3 \tanh(\dot{\omega}_{e,i} + k_2 \omega_{e,i}) - k_1 \tanh(\dot{\omega}_{e,i}) + \frac{\tilde{m}_i}{m_i(t)} \delta_{e,i}. \tag{32}$$

Choose a positive definite function $V_2 = V_2(\omega_i, \dot{\omega}_i, \tilde{m}_i, t)$ as follows:

$$V_2 = k_3 m_i \ln \cosh(\dot{\omega}_{e,i} + k_2 \omega_{e,i}) + k_1 m_i \ln \cosh \dot{\omega}_{e,i} + 0.5 k_2 m_i \dot{\omega}_{e,i} \dot{\omega}_{e,i}^T + \frac{\tilde{m}_i^2}{2k_4}. \tag{33}$$

The time-derivative of Equation (33) along the solution trajectory of Equation (32) can then be calculated as:

$$\dot{V}_2 = \dot{m}_i \left[k_3 \ln \cosh(\dot{\omega}_{e,i} + k_2 \omega_{e,i}) + k_1 \ln \cosh \dot{\omega}_{e,i} + 0.5 k_2 \dot{\omega}_{e,i} \dot{\omega}_{e,i}^T \right]$$
$$+ k_3 m_i \tanh(\dot{\omega}_{e,i} + k_2 \omega_{e,i})(\ddot{\omega}_{e,i} + k_2 \dot{\omega}_{e,i})^T + k_1 m_i \tanh \dot{\omega}_{e,i} \ddot{\omega}_{e,i}^T + k_2 m_i \dot{\omega}_{e,i} \ddot{\omega}_{e,i}^T \tag{34}$$
$$+ \frac{\tilde{m}_i}{k_4}(\dot{\tilde{m}}_i - \dot{\hat{m}}_i).$$

As $\dot{m}_i \leq 0, \forall t \geq 0$, the first row of Equation (34) is non-positive and hence the $\dot{V}_2$ satisfies:

$$\dot{V}_2 \leq k_3 m_i \tanh(\dot{\omega}_{e,i} + k_2 \omega_{e,i})(\ddot{\omega}_{e,i} + k_2 \dot{\omega}_{e,i})^T + k_1 m_i \tanh \dot{\omega}_{e,i} \ddot{\omega}_{e,i}^T + k_2 m_i \dot{\omega}_{e,i} \ddot{\omega}_{e,i}^T$$
$$+ \frac{\tilde{m}_i}{k_4}(\dot{\tilde{m}}_i - \dot{\hat{m}}_i). \tag{35}$$

Substitute $\ddot{\omega}_{e,i}$ of Equation (32) into Equation (35),

$$\dot{V}_2 \leq m_i(t) \left\{ -k_3^2 \tanh(\dot{\omega}_{e,i} + k_2 \omega_{e,i}) [\tanh(\dot{\omega}_{e,i} + k_2 \omega_{e,i})]^T - k_1^2 \tanh \dot{\omega}_{e,i} \tanh (\dot{\omega}_{e,i})^T \right\}$$
$$+ m_i(t) \left\{ -k_1 k_2 \dot{\omega}_{e,i} \tanh (\dot{\omega}_{e,i})^T - 2k_3 k_1 \tanh(\dot{\omega}_{e,i} + k_2 \omega_{e,i}) \tanh (\dot{\omega}_{e,i})^T \right\} \tag{36}$$
$$+ [k_3 \tanh(\dot{\omega}_{e,i} + k_2 \omega_{e,i}) + k_1 \tanh \dot{\omega}_{e,i} + k_2 \dot{\omega}_{e,i}] \tilde{m}_i \delta_{e,i}^T + \frac{\tilde{m}_i}{k_4}(\dot{\tilde{m}}_i - \dot{\hat{m}}_i).$$

Taking the updating algorithm for $\hat{m}_i$ in Equation (30) into Equation (36) and performing some direct calculations, we obtain:

$$\dot{V}_2 \leq -m_i(t) \|k_3 \tanh(\dot{\omega}_{e,i} + k_2 \omega_{e,i}) + k_1 \tanh \dot{\omega}_{e,i}\|^2 - k_1 k_2 m_i(t) \dot{\omega}_{e,i} \tanh (\dot{\omega}_{e,i})^T$$
$$\leq -\underline{m}_i \|k_3 \tanh(\dot{\omega}_{e,i} + k_2 \omega_{e,i}) + k_1 \tanh \dot{\omega}_{e,i}\|^2 - k_1 k_2 \underline{m}_i \dot{\omega}_{e,i} \tanh (\dot{\omega}_{e,i})^T \tag{37}$$
$$\leq 0.$$

It is obvious that $\dot{V}_2$ is semi-negative definite, which means that $\omega_{e,i}, \dot{\omega}_{e,i}$ and $\tilde{m}_i$ are bounded. With direct calculation, one would find out that $\delta_{e,i}$ is bounded. Henceforth, $\ddot{\omega}_{e,i}$ is bounded. All these bounded variables demonstrate that:

$$\ddot{V}_2 = \dot{m}_i(t) \left[k_3 \ln \cosh(\dot{\omega}_{e,i} + k_2 \omega_{e,i}) + k_1 \ln \cosh \dot{\omega}_{e,i} + 0.5 k_2 \dot{\omega}_{e,i} \dot{\omega}_{e,i}^T \right]$$
$$- \dot{m}_i(t) \|k_3 \tanh(\dot{\omega}_{e,i} + k_2 \omega_{e,i}) + k_1 \tanh \dot{\omega}_{e,i}\|^2 - k_1 k_2 \dot{m}_i(t) \dot{\omega}_{e,i} \tanh (\dot{\omega}_{e,i})^T$$
$$- 2 m_i(t) [k_3 \tanh(\dot{\omega}_{e,i} + k_2 \omega_{e,i}) + k_1 \tanh \dot{\omega}_{e,i}] \times \left[\frac{k_3 \ddot{\omega}_{e,i} + k_3 k_2 \dot{\omega}_{e,i}}{1 + \|\dot{\omega}_{e,i} + k_2 \omega_{e,i}\|^2} + \frac{k_1 \ddot{\omega}_{e,i}}{1 + \|\dot{\omega}_{e,i}\|^2} \right]^T \tag{38}$$
$$- k_1 k_2 m_i(t) \ddot{\omega}_{e,i} \tanh (\dot{\omega}_{e,i})^T - k_1 k_2 m_i(t) \frac{\dot{\omega}_{e,i} \ddot{\omega}_{e,i}^T}{1 + \|\dot{\omega}_{e,i}\|^2}$$

is bounded. The boundedness of $\dot{V}_2$ implies that $\dot{V}_2$ is uniformly continuous, which, according to Barbalat's lemma [27], proves that $\omega_{e,i}$ and $\dot{\omega}_{e,i}$ converge to zero globally asymptotically. Rewrite Equation (27) as follows:

$$\vec{r}_{e,i} - \vec{r}_{e,0} = \omega_{e,i} + (\vec{\eta}_{e,ir} - \vec{r}_{e,0}). \tag{39}$$

Because $\lim_{t \to \infty} \omega_{e,i} = 0$ and $\lim_{t \to \infty} (\vec{\eta}_{e,ir} - \vec{r}_{e,0}) = 0$ hold simultaneously, we use a simple theory of a cascade system [28] and obtain that $\lim_{t \to \infty} (\vec{r}_{e,i} - \vec{r}_{e,0}) = 0$. This completes the proof. □

The above analysis shows that the satellites in $\mathcal{N}$ would achieve rendezvous at the reference orbital vehicle. One might note that the safe orbit height is not taken into account, i.e., the satellite must be orbiting above a certain height with respect to the Earth's surface. We here propose an initial solution for this issue without presenting the detailed proof. Choose two positive numbers h_1 and h_2 satisfying $h_2 > h_1 > 0$, define a continuous and derivable function $f(x) : (h_1, +\infty) \mapsto \mathbb{R}_{\geq 0}$ as:

$$f(x) = \begin{cases} -\alpha_1 \ln \alpha_2 (x - h_1) + \alpha_3 & , \forall h_1 < x \leq 0.5(h_1 + h_2); \\ \alpha_4 (x - h_2)^2 & , \forall 0.5(h_1 + h_2) < x \leq h_2; \\ 0 & , \forall h_2 < x, \end{cases} \tag{40}$$

where the constants are chosen as $\alpha_1 = 0.5\alpha_4 (h_2 - h_1)^2, \alpha_2 = \dfrac{1}{2(h_2 - h_1)}, \alpha_3 = 0.5\alpha_1, \alpha_4 > 0$. It is direct to verify that $f(x)$ is derivable and continuous for all $x \in (h_1, +\infty)$, so we omit the proof here. To ensure the orbital satellites orbiting above the h_1 and $\|\vec{r}_{e,i}\| > h_1, \forall t \geq 0$, we borrow the basic idea from an artificial potential field approach [29] and modify the control law Equation (30) into:

$$\begin{cases} \tau_i = -\hat{m}_i \delta_{e,i} + f(\|\vec{r}_{e,i}\|) \dfrac{\vec{r}_{e,i}}{\|\vec{r}_{e,i}\|} \tau_{i,c} \\ \dot{\hat{m}}_i = \dot{m}_i(t) + k_4 [k_3 \tanh(\dot{\omega}_{e,i} + k_2 \omega_{e,i}) + k_1 \tanh \dot{\omega}_{e,i} + k_2 \dot{\omega}_{e,i}] \delta_{e,i}^T \end{cases} \tag{41}$$

with $\tau_{i,c}$ being a constant. The introduced term $f(\|\vec{r}_{e,i}\|) \dfrac{\vec{r}_{e,i}}{\|\vec{r}_{e,i}\|} \tau_{i,c}$ would produce a reversing force in the direction $\vec{r}_{e,i}$, pointing outwards from the Earth's center, so that the $\|\vec{r}_{e,i}\| > h_1, \forall t \geq 0$. In addition, as can be seen from Equations (40) and (41), the term becomes efficient only when $\|\vec{r}_{e,i}\| \in (h_1, h_2)$. This means that the modified control law Equation (41) reduces to Equation (30) if $\|\vec{r}_{e,i}\| \geq h_2$.

4. Numerical Simulation

This section presents the numerical simulation to validate the proposed control design. Two parts are involved. First, we discuss how to propagate the projection vector $\vec{s}_{i,j}$ of the satellite position and calculate the satellite position for feedback control. Second, we validate the leader estimation algorithm and feed the disturbed state signals into the control loop to verify the effectiveness of the adaptive control algorithm.

4.1. Pulsar-Based Positioning and Simulation Setup

For simulation use, we select three pulsars as positioning sources. The pulsar positions are listed in Tables 1 and 2. The galactic coordinates shown in Table 1 are transformed into SCI coordinates in Table 2 via the online coordinate calculator provided by https://ned.ipac.caltech.edu/coordinate_calculator, accessed on 11 November 2021.

Table 1. Pulsar Position in Galactic Frame [30].

j	Name	Longitude	Latitude
1	PSR B0531+21	184.56°	−5.78°
2	PSR B1821− 24	7.8°	−5.58°
3	PSR B1937+21	57.51°	−0.29°

Table 2. Pulsars Information in SCI Frame.

j	Name	Right Ascension	Declination
1	PSR B0531+21	84.102438550°	−1.29446370°
2	PSR B1821− 24	275.56845533°	−1.54719148°
3	PSR B1937+21	301.97479547°	42.29726047°

Within the pulsar position in Table 2, we use Equation (2) to calculate the fixed vectors formed by pulsar radiations in the SCI frame as follows:

$$\vec{q}_1 = [0.1027, -0.0023, -0.0226],$$
$$\vec{q}_2 = [0.0970, -0.0026, -0.0270], \qquad (42)$$
$$\vec{q}_3 = [0.3917, 0.3564, 0.6730].$$

As discussed before, the projection of satellite position vector $\vec{r}_{s,i}, i \in \mathcal{N}$ in the SCI frame on $\vec{q}_j, j = \{1, 2, 3\}$ is measurable. Given three pulsars, i.e., $j = 3$, we define the measured row vectors for the i-th satellite by $\bar{s}_{i,1}, \bar{s}_{i,2}, \bar{s}_{i,3}$, which can be propagated by the following equations:

$$\bar{s}_{i,j} = [(\vec{r}_{e,i} + \vec{r}_{s,e}) \cdot \vec{q}_j]\vec{q}_j, j = \{1, 2, 3\}, \qquad (43)$$

where $\vec{r}_{e,i}$ and $\vec{r}_{s,e}$ are solved by the associated Keplerian equations of motion. Then we use the two-dimensional plane given in Equation (3) and obtain the following linear equations:

$$\begin{bmatrix} \bar{s}_{i,1} \\ \bar{s}_{i,2} \\ \bar{s}_{i,3} \end{bmatrix} \vec{r}_{s,i}^T = [\|\bar{s}_{i,1}\|^2, \|\bar{s}_{i,2}\|^2, \|\bar{s}_{i,3}\|^2]^T. \qquad (44)$$

Equation (44) suggests that the satellite position in the SCI frame using pulsar measurements can be solved by:

$$\vec{r}_{s,i} = \left[\begin{bmatrix} \bar{s}_{i,1} \\ \bar{s}_{i,2} \\ \bar{s}_{i,3} \end{bmatrix}^{-1} [\|\bar{s}_{i,1}\|^2, \|\bar{s}_{i,2}\|^2, \|\bar{s}_{i,3}\|^2]^T \right]^T. \qquad (45)$$

To investigate the robust property of the proposed algorithm regarding the disturbance of position and velocity measurement, we inject the unknown time-varying disturbance into the state variables for feedback control. The overall process can be summarized as the figure shown below (Figure 2), in which the satellite ECI coordinate II is fed into the control loop.

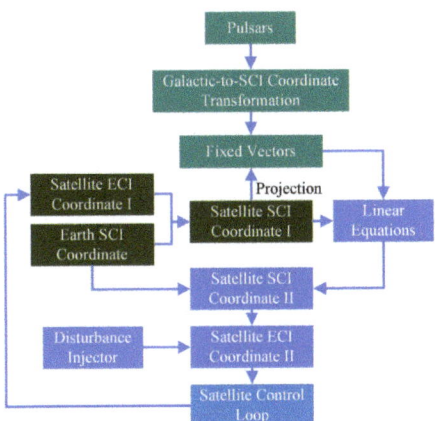

Figure 2. Pulsar-based Positioning Flowchart For Simulations.

4.2. Estimate of $\vec{r}_{e,0}$ and $\vec{v}_{e,0}$.

For validating the proposed algorithm, we suppose that four orbital satellites transporting supplies to the international space station (ISS) are indexed by 0, which accords with the rendezvous scenario. The initial orbital elements of ISS and satellites are listed in Table 3.

Table 3. Initial Orbital Elements.

i	A	ECC	INCL	RAAN	AANP	APP
0	6792 km	0.005426	51.6438°	38.8886°	23.0560°	63°
1	6881 km	0.006340	50.3210°	40.0100°	20.2022°	60°
2	6922 km	0.005924	60.5380°	39.4500°	25.1991°	64°
3	7238 km	0.007020	52.6225°	43.1526°	28.5234°	58°
4	7055 km	0.009070	40.4819°	42.5128°	23.4040°	55°

Where the definitions for letters in Table 3 are listed below, A: semi-major axis, ECC: orbit eccentricity, INCL: Inclination, RAAN: right ascension of ascending node, AANP: angle between ascending node and periapsis, APP: angle between periapsis and the satellite.

According to the initial orbital elements in Table 3, we calculate $\vec{r}_{e,i}(0)$ and $\vec{v}_{e,i}(0), \forall i \in \mathcal{N}$ as follows:

$$\vec{r}_{e,1}(0) = [\ 1880.5, 1055.8, 5201.9]\ \text{km},\ \vec{v}_{e,1}(0) = [-6.3076, -4.1772, 1.0317]\ \text{km/s},$$
$$\vec{r}_{e,2}(0) = [-2082.9, 2683.1, 6010.5]\ \text{km},\ \vec{v}_{e,2}(0) = [-5.9198, -4.7778, 0.1278]\ \text{km/s},$$
$$\vec{r}_{e,3}(0) = [-2669.4, 3486.7, 5719.6]\ \text{km},\ \vec{v}_{e,3}(0) = [-5.6280, -4.8637, 0.3940]\ \text{km/s},$$
$$\vec{r}_{e,4}(0) = [-2493.6, 4807.7, 4463.1]\ \text{km},\ \vec{v}_{e,4}(0) = [-6.2565, -4.1120, 1.0216]\ \text{km/s}.$$

The initial orbits of the satellites are depicted in Figure 3. As can be seen, the satellite $i, \forall i \in \mathcal{N}$ and the ISS are orbiting the Earth with independent orbits.

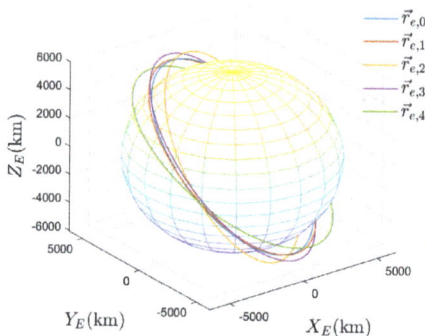

Figure 3. The initial orbits of satellites in the ECI coordinate frame.

For achieving distributed estimation about $[\vec{r}_{e,0}, \vec{v}_{r,0}]$, the satellites in $\mathcal{N}$ are supposed to exchange information via the communication topology shown in Figure 4.

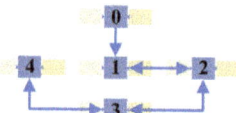

Figure 4. The communication topology.

The coefficients are chosen as $g_1 = 1, g_2 = 0, 1, g_3 = 1, g_{4x} = g_{4y} = g_{4z} = 1.2$. Set initial values by $\vec{\eta}_{e,ir}(0) = \vec{r}_{e,i}(0)$, $\vec{\varphi}_{e,ir}(0) = \vec{v}_{e,i}(0)$ and $\vec{\rho}_{e,ir}(0) = 0_3$. The simulation time length is the same as the next subsection. However, we only plot the error trajectory in the time interval $t \in [0, 300]$ for showing the convergence clearly. The simulation results are depicted in Figure 5. It can be found out that all estimation errors converge to zero asymptotically. Within $t \geq 50$ s, the signals $\vec{\eta}_{e,1r}, \forall i \in \mathcal{N} = \{1,2,3,4\}$ feature stable shape without any chattering phenomenon occurring.

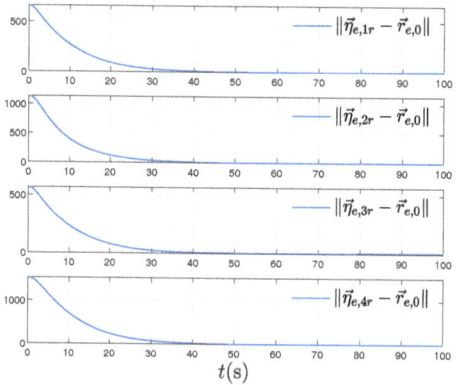

Figure 5. The estimation errors.

4.3. Control Validation

For achieving cooperative tasks, we plan to steer all satellites to the orbit $\vec{r}_{e,0}$ via the proposed control law and the pulsar-based positioning method. Within the initial condition shown in Table 3, we set the control coefficients by $k_1 = 0.2, k_2 = 0.4, k_3 = 0.2, k_4 = 0.01$.

The real satellite masses are set as $m_1 = 450$ kg, $m_2 = 423$ kg, $m_3 = 467$ kg and $m_4 = 433$ kg. These masses are unknown for each satellite in $\mathcal{N}$. The ratio of change of mass is chosen as $\alpha_i = -0.1, \forall i \in \{1,2,3,4\}$. The initial estimated mass values are set as $\hat{m}_i(0) = (1+6\%)m_i$. Set the Sun gravitational constant as $\mu_s = 1.32712439935 \times 10^{11}$ kg^3/s^2 and the Earth gravitational constant as $\mu_e = 3.986005 \times 10^5$ kg^3/s^2. The semi-major axis of Earth rotating round the Sun is chosen as $a_{es} = 1.496 \times 10^8$ km within eccentricity given by $ecc_e = 0.0167$. The initial anomaly of the Earth is set as zero. The Earth's radius is set as $R_e = 6371$ (km). Accordingly, we set $h_1 = R_e + 300$ (km) $= 6671$(km), $h_2 = R_e + 350$ (km) $= 6721$ (km), $\alpha_1 = 125, \alpha_2 = 0.04, \alpha_3 = 62.5, \alpha_4 = 0.1$ and $\tau_{i,c} = 100$. This setting means that we want all satellites are always orbiting 300 km above the Earth's surface. In addition, the position and velocity measurements used for feedback control are supposed to suffer from disturbances $\vec{\Delta}_{e,r,i}$ and $\vec{\Delta}_{e,v,i}$ respectively, defined as follows:

$\vec{\Delta}_{e,r,1} = 50[\sin 0.01t, \cos 0.05t, \sin 0.02t]$ (m), $\vec{\Delta}_{e,r,2} = 50[\sin 0.03t, \sin 0.04t, \cos 0.02t]$ (m),

$\vec{\Delta}_{e,r,3} = 50[\cos 0.02t, \sin 0.01t, \cos 0.03t]$ (m), $\vec{\Delta}_{e,r,4} = 50[\sin 0.015t, \cos 0.02t, \sin 0.012t]$ (m),

$\vec{\Delta}_{e,v,1} = 2[\sin 0.023t, \cos 0.03t, \sin 0.012t]$ (m/s), $\vec{\Delta}_{e,v,2} = 2[\sin 0.016t, \sin 0.035t, \cos 0.022t]$ (m/s),

$\vec{\Delta}_{e,v,3} = 2[\cos 0.026t, \sin 0.013t, \cos 0.03t]$ (m/s), $\vec{\Delta}_{e,v,4} = 2[\sin 0.018t, \cos 0.02t, \sin 0.032t]$ (m/s).

More precisely, the position measurement errors are fluctuated within ± 75.7 m, while the fluctuation for velocity measurement is ± 3.4 m/s. We set simulation time as 5000 s and start the control update after $t \geq 100$ s according to the convergence of the estimation errors stated previously. The simulation results are depicted in Figures 6–9.

As depicted in Figure 6a, all satellites in $\mathcal{N}$ move to the orbit of $\vec{r}_{e,0}$, achieving a rendezvous control scenario from initial orbits. In Figure 6a, the Earth is rotating around the Sun while the satellites are rotating with the Earth. The collective helix curves demonstrate that the control algorithm is effective even though the controller Equation (30) is developed in ECI coordinate frame. Figure 6b further supports this point of view and demonstrates that the position rendezvous errors converge into small regions enclosing the zero.

Due to position and velocity measurement disturbances during the transient phase, the rendezvous errors are not steered converging to zero. We plot the final relative position and velocity with respect to $[\vec{r}_{e,0}, \vec{v}_{e,0}]$ in Figure 7. The simulation resulting in the steady-state of $t \geq 3000$ s shows that the final relative position is maintained to be less than 90 m and the relative velocity to be less than 3 m/s. More precisely, the final relative position and velocity mean values are calculated and summarized in Table 4. As can be seen, the relative position and velocity in the steady-state phase are maintained at a reasonable level. The robust property of the proposed control algorithm towards measurement disturbances is validated. For such small ranges of final relative position and velocity difference, high-precision onboard measurement units can then be used to guide the satellites to accomplish tasks such as docking and circumventing.

Table 4. The mean values of final relative states for $t \geq 3000$ s.

i	1	2	3	4
mean$\{\|\vec{r}_{e,i} - \vec{r}_{e,0}\|\}$	60.45 m	59.65 m	61.64 m	62.03 m
mean$\{\|\vec{v}_{e,i} - \vec{v}_{e,0}\|\}$	1.83 m/s	1.84 m/s	1.36 m/s	1.01 m/s

By observing Figure 8, we find that the satellites are always over the safe height h_1, i.e., they would not collide with the Earth's surface and guarantee $\|\vec{r}_i\| > h_1, \forall i \in \mathcal{N}, t \geq 0$.

The change of masses due to fuel consumption is plotted in Figure 8. The mass losses are related to the initial rendezvous error. Via simple calculation, we obtain the initial rendezvous error for the first to the fourth satellites as 661.13 km, 1134.40 km, 569.04 km, and 1529.90 km, respectively. These initial errors explain the fuel consumption difference depicted in the figure. Additionally, the masses of satellites for $t \geq 3000$ s decrease slightly to overcome the measurement disturbances.

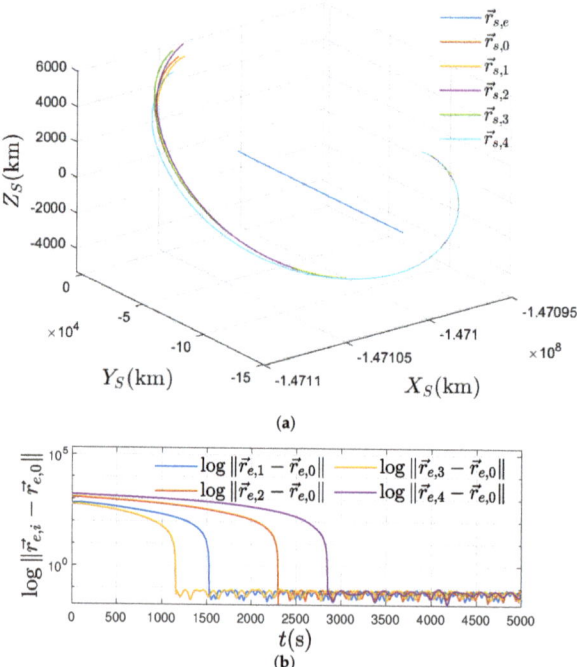

Figure 6. Simulation results of the rendezvous of multiple uncertain satellites. (**a**) The orbits of satellites in the SCI coordinate Frame; (**b**) The position rendezvous errors of semi-logarithmic scale.

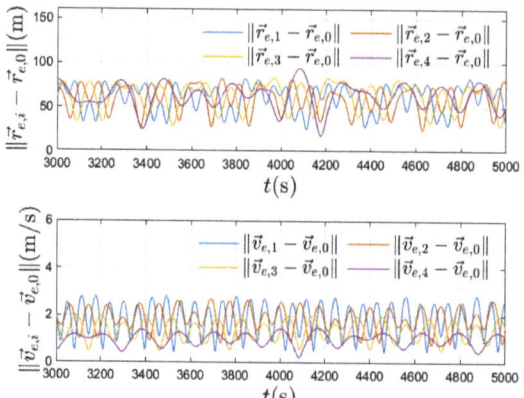

Figure 7. The relative position and velocity in steady state.

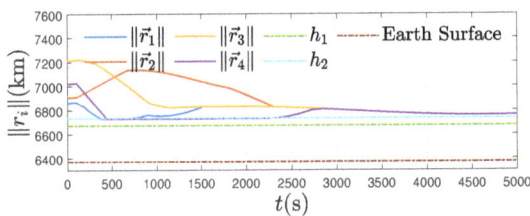

Figure 8. Orbital radius of satellites.

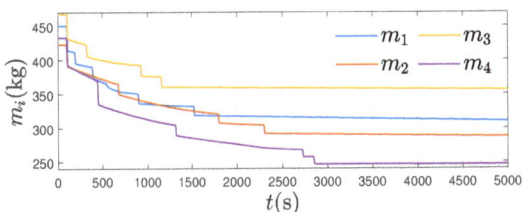

Figure 9. Masses of satellites.

5. Conclusions

This paper presents the design and analysis of an adaptive rendezvous control strategy for multiple orbital satellites. Nonlinear tools, including a discontinuous approach, the reduced-order method, and saturation-like adaptive control design, are applied to derive a novel control law that steers multiple uncertain orbital satellites to achieve rendezvous at a certain reference orbital vehicle. It is proven that the rendezvous errors are convergent to zero asymptotically. In the future, the authors will consider more problems for the pulsar-based rendezvous control, such as intermitted observation of pulsar sources, statistical measurements, and fuel optimizations.

Author Contributions: Conceptualization, Q.C.; Resources, Y.Z.; Software, L.Y.; Supervision, L.Y.; Writing—original draft, Y.Z.; Writing—review & editing, L.Y. All authors have read and agreed to the published version of the manuscript.

Funding: This work was supported by National Key R&D Program of China (No.2017YFB0503304).

Institutional Review Board Statement: Not applicable.

Informed Consent Statement: Not applicable.

Conflicts of Interest: The authors declare no conflict of interest.

References

1. Shirazi, A.; Ceberio, J.; Lozano, J.A. Spacecraft trajectory optimization: A review of models, objectives, approaches and solutions. *Prog. Aerosp. Sci.* **2018**, *102*, 76–98. [CrossRef]
2. Woffinden, D.C.; Geller, D.K. Navigating the road to autonomous orbital rendezvous. *J. Spacecr. Rocket.* **2007**, *44*, 898–909. [CrossRef]
3. Dal Moro, R. Satellite Localization for Gamma-Ray Pulsar Observations in Low Earth Orbit. Mater's Thesis, University Degli, Udine, Italy , 2020.
4. Shuai, P.; Chen, S.L.; Wu, Y.F.; Zhang, C.Q.; Li, M. Navigation principles using X-ray pulsars. *J. Astronaut.* **2007**, *28*, 1538–1543.
5. Zheng, W.; Wang, Y. *X-ray Pulsar-Based Navigation: Theory and Applications*; Springer Nature: Berlin/Heidelberg, Germany, 2020; Volume 5.
6. Buist, P.J.; Engelen, S.; Noroozi, A.; Sundaramoorthy, P.; Verhagen, S.; Verhoeven, C. Overview of pulsar navigation: Past, present and future trends. *Navigation* **2011**, *58*, 153–164. [CrossRef]
7. Qiu, W.; Yin, H.; Zhang, L.; Luo, X.; Yao, W.; Zhu, L.; Liu, Y. Pulsar-calibrated Timing Source for Synchronized Sampling. *IEEE Trans. Smart Grid* **2021**, early view. [CrossRef]

8. Chen, P.T.; Speyer, J.L.; Bayard, D.S.; Majid, W.A. Autonomous navigation using x-ray pulsars and multirate processing. *J. Guid. Control Dyn.* **2017**, *40*, 2237–2249. [CrossRef]
9. Hua, Z.; Rong, J.; Luping, X. Formation of a Satellite Navigation System Using X-Ray Pulsars. *Publ. Astron. Soc. Pac.* **2019**, *131*, 045002. [CrossRef]
10. Fujimoto, K.; Kodama, T.; Maruta, I. On Rendezvous Trajectory Design of Earth Orbiting Satellites with Robust L1-Optimal Control for Parameter Variations. *Trans. Jpn. Soc. Aeronaut. Space Sci. Aerosp. Technol. Jpn.* **2020**, *18*, 250–257.
11. Jia, Y. Alternative proofs for improved LMI representations for the analysis and the design of continuous-time systems with polytopic type uncertainty: A predictive approach. *IEEE Trans. Autom. Control* **2003**, *8*, 1413–1416.
12. Walther, M.; Traub, C.; Herdrich, G.; Fasoulas, S. Improved success rates of rendezvous maneuvers using aerodynamic forces. *CEAS Space J.* **2020**, *12*, 463–480. [CrossRef]
13. Brentari, M.; Urbina, S.; Arzelier, D.; Louembet, C.; Zaccarian, L. A hybrid control framework for impulsive control of satellite rendezvous. *IEEE Trans. Control Syst. Technol.* **2018**, *27*, 1537–1551. [CrossRef]
14. Jia, Y. Robust control with decoupling performance for steering and traction of 4WS vehicles under velocity-varying motion. *IEEE Trans. Control Syst. Technol.* **2000**, *8*, 554–569.
15. Jia, Y. General solution to diagonal model matching control of multiple-output-delay systems and its applications in adaptive scheme. *Prog. Nat. Sci.* **2009**, *19*, 79–90. [CrossRef]
16. Kristiansen, R.; Nicklasson, P.J. Spacecraft formation flying: A review and new results on state feedback control. *Acta Astronaut.* **2009**, *65*, 1537–1552. [CrossRef]
17. Kuiack, B.J. Spacecraft Formation Guidance and Control on J2-Perturbed Eccentric Orbits. Mater's Thesis, Carleton University, Ottawa, ON, Canada, 2018.
18. Jasiobedzki, P.; Se, S.; Pan, T.; Umasuthan, M.; Greenspan, M. Autonomous satellite rendezvous and docking using LIDAR and model based vision. *Spaceborne Sens. II* **2005**, *5798*, 54–65.
19. Cookson, J.J. Experimental investigation of spacecraft rendezvous and docking by development of a 3 degree of freedom satellite simulator testbed. Master's Thesis, York University, Toronto, ON, Canada, 2020.
20. Horsley, M.; Nikolaev, S.; Pertica, A. Small satellite rendezvous using differential lift and drag. *J. Guid. Control Dyn. Control* **2013**, *36*, 445–453. [CrossRef]
21. Sun, L.; He, W.; Sun, C. Adaptive fuzzy relative pose control of spacecraft during rendezvous and proximity maneuvers. *IEEE Trans. Fuzzy Syst.* **2018**, *26*, 3440–3451. [CrossRef]
22. Wang, W.; Li, C.; Guo, Y. Relative position coordinated control for spacecraft formation flying with obstacle/collision avoidance. *Nonlinear Dyn.* **2021**, *104*, 1329–1342. [CrossRef]
23. Zhao, X.; Zhang, S. Adaptive saturated control for spacecraft rendezvous and docking under motion constraints. *Aerosp. Sci. Technol.* **2021**, *114*, 106739. [CrossRef]
24. Curtis, H. *Orbital Mechanics for Engineering Students*; Butterworth-Heinemann: Oxford, UK, 2013.
25. Hartshorne, R. *Algebraic Geometry*; Springer Science & Business Media: Berlin/Heidelberg, Germany, 2013.
26. Wei, R.; Beard, R.W.; Atkins, E.M. A survey of consensus problems in multi-agent coordination. In Proceedings of the 2005 American Control Conference, Portland, OR, USA, 8–10 June 2005; Volume 3, pp. 1859–1864.
27. Khalil, H.K. *Nonlinear Systems*; Prentice Hall: Upper Saddle River, NJ, USA, 2002.
28. Sundarapandian, V. Global asymptotic stability of nonlinear cascade systems. *Appl. Math. Lett.* **2002**, *15*, 275–277. [CrossRef]
29. Park, M.G.; Jeon, J.H.; Lee, M.C. Obstacle avoidance for mobile robots using artificial potential field approach with simulated annealing. *IEEE Int. Symp. Ind. Electron. Proc.* **2001**, *3*, 1530–1535.
30. Sheikh, S.I.; Pines, D.J.; Ray, P.S.; Wood, K.S.; Lovellette, M.N.; Wolff, M.T. Spacecraft navigation using X-ray pulsars. *J. Guid. Control Dyn.* **2006**, *29*, 49–63. [CrossRef]

MDPI
St. Alban-Anlage 66
4052 Basel
Switzerland
www.mdpi.com

Entropy Editorial Office
E-mail: entropy@mdpi.com
www.mdpi.com/journal/entropy

Disclaimer/Publisher's Note: The statements, opinions and data contained in all publications are solely those of the individual author(s) and contributor(s) and not of MDPI and/or the editor(s). MDPI and/or the editor(s) disclaim responsibility for any injury to people or property resulting from any ideas, methods, instructions or products referred to in the content.

www.ingramcontent.com/pod-product-compliance
Lightning Source LLC
LaVergne TN
LVHW070742100526
838202LV00013B/1287

9 783725 801503